LOW-POWER DESIGN TECHNIQUES AND CAD TOOLS FOR ANALOG AND RF INTEGRATED CIRCUITS

Low-Power Design Techniques and CAD Tools for Analog and RF Integrated Circuits

Edited by

Piet Wambacq
IMEC

Georges Gielen
Katholieke Universiteit Leuven

and

John Gerrits
CSEM System Engineering

Series Editors

Rene van Leuken
Alexander de Graaf
Reinder Nouta
TU Delft/DIMES, Delft, The Netherlands

KLUWER ACADEMIC PUBLISHERS
BOSTON / DORDRECHT / LONDON

A C.I.P. Catalogue record for this book is available from the Library of Congress.

ISBN 0-7923-7432-0

Published by Kluwer Academic Publishers,
P.O. Box 17, 3300 AA Dordrecht, The Netherlands.

Sold and distributed in North, Central and South America
by Kluwer Academic Publishers,
101 Philip Drive, Norwell, MA 02061, U.S.A.

In all other countries, sold and distributed
by Kluwer Academic Publishers,
P.O. Box 322, 3300 AH Dordrecht, The Netherlands.

Printed on acid-free paper

Printed in the Netherlands.

Contents

List of Figures

List of Tables

Contributing Authors

Piet Wambacq was born in Asse, Belgium in 1963. He obtained the M.Sc and Ph.D degrees from the Katholieke Universiteit Leuven, Belgium, in 1986 and 1996, respectively. Since the end of 1996 he is working at IMEC in Leuven, Belgium, where he is a principal scientist in the field of design and methodologies for mixed-signal and RF applications. He has authored or co–authored more than 50 papers in edited books, international journals and conference proceedings and he is the author of one book on nonlinear distortion analysis of analog ICs.

Georges G.E. Gielen received the MSc and PhD degrees in Electrical Engineering from the Katholieke Universiteit Leuven, Belgium, in 1986 and 1990, respectively. After being a visiting lecturer at the department of Electrical Engineering and Computer Science of the University of California, Berkeley, U.S.A., he became a Faculty member at the ESAT-MICAS laboratory of the Department of Electrical Engineering of the Katholieke Universiteit Leuven, where he is currently a full-time professor. His research interests are in the design of analog and mixed-signal integrated circuits, and especially in analog and mixed-signal CAD tools and design automation (modeling, simulation and symbolic analysis, analog synthesis, analog layout generation, analog and mixed-signal testing). He has authored or coauthored two books and more than 150 papers in edited books, international journals and conference proceedings. He regularly is a member of the Program Committees of international conferences and a member of the editorial board of international journals. He was the 1997 Laureate of the Belgian Royal Academy of Sciences, Literature and Arts, in the category of engineering sciences.

John F.M. Gerrits was born in Leiden, the Netherlands, on September 16, 1963. He received the M.Sc. degree in electrical engineering in 1987 from Delft University of Technology, the Netherlands. From 1988 to 1991 he was with Philips T&M, Enschede, the Netherlands, where he designed various high-speed bipolar integrated circuits for digital oscilloscope applications. In 1991 he joined the CSEM Centre Suisse d'Electronique et de Microtechnique S.A., Neuchâtel, Switzerland. He has been involved in the design of various low-voltage, low-power wireless IC products for the VHF and UHF bands. His current activities focus on RF system design and RF measurement methodology. He holds 3 US patents.

Jaijeet Roychowdhury received a B.Tech degree from IIT Kanpur in 1987, and M.S. and Ph.D degrees from the University of California at Berkeley in 1989 and 1993, all in electrical engineering. From 1993 to 1995, he was with the CAD Laboratory of AT&T Bell Labs in Allentown, PA. From 1995 to 2000, he was with the Research Division of Lucent's Bell Labs in Murray Hill, NJ. He is currently with Celight, Inc.. He received a Distinguished Paper Award at ICCAD 1991, and Best Paper Awards at DAC 1997, ASP-DAC 1997 and ASP-DAC 1999. His research interests include design and analysis of wireless and optical communication systems.

Gerd Vandersteen was born in Belgium in 1968. He received his degree of Electrical Engineering in 1991 from the Free University of Brussels (VUB), Brussels, Belgium. He received his Ph.D. degree in Electrical Engineering in 1997 from the Free University of Brussels (VUB) on the topic of Identification of Nonlinear Systems in an Errors-In-Variables framework. He is presently working as a principle scientist in the mixed signal design group at IMEC/DESICS (Leuven). His main interest are in the field of modelling, measurement and simulation of nonlinear microwave devices.

Piet Vanassche was born in Poperinge, Belgium, in 1974. He received the degree of M.Sc. in Electronics in 1997 from the Katholieke Universiteit Leuven, Belgium. The subject of his the thesis was research on high level synthesis and optimization of continuous-time filters. Currently, he is a research assistant at the ESAT-MICAS laboratories of the Katholieke Universiteit Leuven, where he's working towards a Ph.D. degree on high-level exploration and trade-off analysis for mixed signal telecommunication systems. For this work he obtained a fellowship from the Flemish Institute for Science and Technology (IWT).

Erik Lauwers was born in Leuven, Belgium, in 1973. He received the M.Sc. degree in electrical engineering in 1997 from the Katholieke Universiteit Leuven, Belgium. Since 1997, he has been working toward the Ph.D. degree on analog power estimators as a research assistant at the ESAT-MICAS laboratories of the Katholieke Universiteit Leuven. His research interests are mainly in analog power estimators and analog continuous-time filters.

Bart De Smedt was born 1972 in Hasselt, Belgium. He received the M.Sc. degree in electrical engineering in 1995 from the Katholieke Universiteit Leuven, Belgium. He is currently working towards the Ph.D. degree in the field of analog design automation. In 1994 he spent a summer internship at the DESY-EMBL Laboratory in Hamburg, Germany, working on a data acquisition module. He is now a research assistant at the ESAT-MICAS Laboratory at the Katholieke Universiteit Leuven. His research interests are in modeling and simulation techniques for front-end telecommunication applications.

Wim Verhaegen was born in Lier, Belgium, in 1973. He received the degree of M.Sc. in Electrical Engineering in 1996 from the Katholieke Universiteit Leuven, Belgium. The subject of his MS thesis was the automated test generation for analog integrated circuits. Currently, he is a research assistant at the CAD group of the MICAS laboratories of the Katholieke Universiteit Leuven. He is working towards a Ph.D. degree on symbolic analysis of and test generation for analog integrated circuits.

Lars Hedrich was born in Hannover, Germany, on February 19, 1966. He graduated (Dipl.-Ing.) in electrical engineering at the University of Hannover in 1992. Since 1992 he is with the Institute of Microelectronic Circuits and Systems at the Department of Electrical Engineering of the University of Hannover, where he received his Ph.D. degree in electrical engineering for his research on approaches to formal verification of analog circuits in 1997. His research interests include several areas of analog design automation: symbolic analysis of linear and nonlinear circuits, behavioral modeling, circuit synthesis and formal verification.

Catherine Dehollain received the degree of electrical engineer in 1982, and the Ph.D. degree in 1995, both from the Swiss Federal Institute of Technology (EPFL), Lausanne. From 1982 to 1984, she was a Research Assistant at the Electronics Laboratories (LEG) of EPFL. In 1984, she joined the Motorola European Center for Research and Development as a Design Engineer in telecommunications. In 1990,she returned to EPFL as a Senior Assistant at the Chaire des Circuits et Systèmes (CIRC) where she worked on broadband impedance matching. Since 1995, she has been responsible at the LEG for radio frequency activities. In particular, she is the Project Manager of the Esprit project SUPREGE and she is Partner of the IST project SODERA. Since 1998, she has been a Lecturer at EPFL in the area of RF circuits and techniques. She is the author or co-author of 2 scientific books and of 25 scientific publications.

Andrea Baschirotto graduated, summa cum laude, in 1989 and received the Ph.D. degree in 1994. In 1994, he joined University of Pavia as Assistant Professor. In 1998, he joined the University of Lecce as Associate Professor. Since 1989 he collaborated with STMicroelectronics for the design of ASIC. Since 1991 he is associated with I.N.F.N. for the design and realization of read-out channel for High-Energy Physics (HEP) Experiments and space experiments. He collaborated with Medico for the design of a low-power front-end for implantable devices. His main research interests are in the design of mixed analog/digital integrated circuits in particular for low-power and/or high-speed signal processing. He authored and co-authored more than 40 papers in International Journals, more than 50 presentations at International Conferences, 2 book chapters, 10 patents, and co-authored more than 120 papers within research collaborations on HEP Experiments. He is an associate editor for IEEE Transactions on Circuits and Systems - part II, and he is an IEEE senior member.

Francesco Svelto was born in Milano, Italy, in 1966. He received the electrical engineering laurea degree and the Ph.D. degree in electronics and computer science from Pavia University, Italy, in 1991 and 1995 respectively. His Ph.D. thesis was focused on low-noise design for instrumentation. During 1997-2000 he has been Assistant Professor at Bergamo University, Italy. In 2000 he joined Pavia University as an Associate Professor. Since 2000 he is Member of the Technical Program Committee of the IEEE Custom Integrated Circuits Conference. His current research interests are in the field of high frequency integrated circuits for telecommunications. He has authored or co-authored more than 40 papers published in international journals or conference proceedings.

Foreword

This book is the second in a series on novel low power design architectures, methods and design practices. It results from of a large European project started in 1997, whose goal is to promote the further development and the faster and wider industrial use of advanced design methods for reducing the power consumption of electronic systems.

Low power design became crucial with the wide spread of portable information and communication terminals, where a small battery has to last for a long period. High performance electronics, in addition, suffers from a permanent increase of the dissipated power per square millimeter of silicon, due to the increasing clock-rates, which causes cooling and reliability problems or otherwise limits the performance.

The European Union's Information Technologies Programme 'Esprit' did therefore launch a 'Pilot action for Low Power Design', which eventually grew to 19 R&D projects and one coordination project, with an overall budget of 14 million EURO. It is meanwhile known as European Low Power Initiative for Electronic System Design (ESD-LPD) and will be completed in the year 2002. It involves to develop or demonstrate new design methods for power reduction, while the coordination project takes care that the methods, experiences and results are properly documented and publicised.

The initiative addresses low power design at various levels. This includes system and algorithmic level, instruction set processor level, custom processor level, RT-level, gate level, circuit level and layout level. It covers data dominated and control dominated as well as asynchronous architectures. 10 projects deal mainly with digital, 7 with analog and mixed-signal, and 2 with software related aspects. The principal application areas are communication, medical equipment and e-commerce devices.

The following list describes the objectives of the 20 projects. It is sorted by decreasing funding budget.

CRAFT CMOS Radio Frequency Circuit Design for Wireless Application

- Advanced CMOS RF circuit design including blocks such as LNA, down converter mixers & phase shifters, oscillator and frequency synthesiser, integrated filters delta sigma conversion, power amplifier
- Development of novel models for active and passive devices as well as fine-tuning and validation based on first silicon fabricates
- Analysis and specification of sophisticated architectures to meet in particular low power single chip implementation

PAPRICA Power and Part Count Reduction Innovative Communication Architecture

- Feasibility assessment of DQIF, through physical design and characterisation of the core blocks
- Low-power RF design techniques in standard CMOS digital process
- RF design tools and framework; PAPRICA Design Kit.
- Demonstration of a practical implementation of a specific application

MELOPAS Methodology for Low Power Asic design

- To develop a methodology to evaluate the power consumption of a complex ASIC early on in the design flow
- To develop a hardware/software co-simulation tool
- To quickly achieve a drastic reduction on the power consumption of electronic equipment

TARDIS Technical Coordination and Dissemination

- To organise the communication between design experiments and to exploit their potential synergy
- To guide the capturing of methods and experiences gained in the design experiments
- To organise and promote the wider dissemination and use of the gathered design know-how and experience

LUCS Low Power Ultrasound Chip Set.

- Design methodology on low power ADC, memory and circuit design
- Prototype demonstration of a handheld medical ultrasound scanner

ALPINS Analog Low Power Design for Communications Systems

- Low-voltage voice band smoothing filters and analog-to-digital and digital-to-analog converters for an analog front-end circuit of a DECT system
- High linear transconductor-capacitor (gm-C) filter for GSM Analog Interface Circuit operating at supply voltages as low as 2.5V
- Formal verification tools, which will be implemented in the industrial partners design environment. These tools support the complete design process from system level down to transistor level

SALOMON System-level analog-digital trade-off analysis for low power

- A general top-down design flow for mixed-signal telecom ASICs
- High-level models of analog and digital blocks and power estimators for these blocks
- A prototype implementation of the design flow with particular software tools to demonstrate the general design flow

DESCALE Design Experiment on a Smart Card Application for Low Energy

- The application of highly innovative handshake technology
- Aiming at some 3 to 5 times less power and some 10 times smaller peak currents compared to synchronously operated solutions

SUPREGE A low power SUPerREGEnerative transceiver for wireless data transmission at short distances

- Design trade-offs and optimisation of the micro power receiver/transmitter as a function of various parameters (power consumption, area, bandwidth, sensitivity, etc)
- Modulation / demodulation and interface with data transmission systems
- Realisation of the integrated micro power receiver / transmitter based on the super-regeneration principle

PREST Power REduction for System Technologies

- Survey of contemporary Low Power Design techniques and commercial power analysis software tools
- Investigation of architectural and algorithmic design techniques with a power consumption comparison
- Investigation of Asynchronous design techniques and Arithmetic styles
- Set-up and assessment of a low power design flow
- Fabrication and characterisation of a Viterbi demonstrator to assess the most promising power reduction techniques

DABLP Low Power Exploration for Mapping DAB Applications to Multi-Processors

- A DAB channel decoder architecture with reduced power consumption
- Refined and extended ATOMIUM methodology and supporting tools

COSAFE Low Power Hardware-Software Co-Design for Safety-Critical Applications

- The development of strategies for power efficient assignment of safety critical mechanisms to hardware or software
- The design and implementation of a low-power, safety-critical ASIP, which realises the control unit of a portable infusion, pump system

AMIED Asynchronous Low-Power Methodology and Implementation of an Encryption/Decryption System

- Implementation of the IDEA encryption/decryption method with drastically reduced power consumption
- Advanced low power design flow with emphasis on algorithm and architecture optimisations
- Industrial demonstration of the asynchronous design methodology based on commercial tools

LPGD A Low-Power Design Methodology/Flow and its Application to the Implementation of a DCS1800-GSM/DECT Modulator/Demodulator

- To complete the development of a top-down, low power design methodology/flow for DSP applications
- To demonstrate the methods at the example of an integrated GFSK/GMSK Modulator-Demodulator (MODEM) for DCS1800-GSM/DECT applications

SOFLOPO Low Power Software Development for Embedded Applications

- Develop techniques and guidelines for mapping a specific algorithm code onto appropriate instruction subsets
- Integrate these techniques into software for the power-conscious ARM-RISC and DSP code optimisation

I-MODE Low Power RF to Base band Interface for Multi-Mode Portable Phone

- To raise the level of integration in a DECT/DCS1800 transceiver, by implementing the necessary analog base band low-pass filters and data converters in CMOS technology using low power techniques

COOL-LOGOS Power Reduction through the Use of Local don't Care Conditions and Global Gate Resizing Techniques: An Experimental Evaluation.

- To apply the developed low power design techniques to the existing 24-bit DSP, which is already fabricated
- To assess the merit of the new techniques using experimental silicon through comparisons of the projected power reduction (in simulation) and actually measured reduction of new DSP; assessment of the commercial impact

LOVO Low Output VOltage DC/DC converters for low power applications

- Development of technical solutions for the power supplies of advanced low power systems, comprising the following topics
- New methods for synchronous rectification for very low output voltage power converters

PCBIT Low Power ISDN Interface for Portable PC's

- Design of a PC-Card board that implements the PCBIT interface
- Integrate levels 1 and 2 of the communication protocol in a single ASIC
- Incorporate power management techniques in the ASIC design:
  - system level: shutdown of idle modules in the circuit
  - gate level: precomputation, gated-clock FSMs

COLOPODS Design of a Cochlear Hearing Aid Low-Power DSP System

- Selection of a future oriented low-power technology enabling future power reduction through integration of analog modules
- Design of a speech processor IC yielding a power reduction of 90% compared to the 3.3 Volt implementation

The low power design projects have achieved the following results:

- Projects, who have designed a prototype chip, can demonstrate a power reduction of 10 to 30 percent.
- New low power design libraries have been developed.
- New proven low power RF architectures are now available.
- New smaller and lighter mobile equipment is developed.

Instead of running a number of Esprit projects at the same time independently of each other, during this pilot action the projects have collaborated strongly. This is achieved mostly by the novelty of this action, which is the presence and role of the coordinator: DIMES - the Delft Institute of Microelectronics and Submicron-technology, located in Delft, the Netherlands (http://www.dimes.tudelft.nl). The task of the coordinator is to co-ordinate, facilitate, and organize:

- The information exchange between projects.
- The systematic documentation of methods and experiences.
- The publication and the wider dissemination to the public.

The most important achievements, credited to the presence of the coordinator are:

- New personnel contacts have been made, and as a consequence the resulting synergy between partners resulted in better and faster developments.
- The organization of low power design workshops, special sessions at conferences, and a low power design web site, http://www.esdlpd.dimes.tudelft.nl. At this site all public reports of the projects can be found and all kind of information about the initiative itself.
- The used design methodology, design methods and/or design experience are disclosed, are well documented and available.

  Based on the work of the projects, in cooperation with the projects, the publication of a low power design book series is planned. Written by members of the projects this series of books on low power design will disseminate novel design methodologies and design experiences, which were obtained during the runtime of the European Low Power Initiative for Electronic System Design, to the general public.

In conclusion, the major contribution of this project cluster is that, except the already mentioned technical achievements, the introduction of novel knowledge on low power design methods into the mainstream development processes is accelerated.

We would like to thank all project partners from all the different companies and organizations who make the Low Power Initiative a success.

Rene van Leuken, Reinder Nouta, Alexander de Graaf, Delft, April 2001

INTRODUCTION

In the world of electronics, high-performance systems often need to be designed with a low power consumption. Complex, versatile and flexible electronic systems are nowadays most often realized with digital circuits. For this type of circuits systematic design approaches can be followed, leading to superb high-performance, low-power solutions in a short time. To decrease cost, size and power consumption of such systems, the use of micro-electronics has proven to be essential.

However, digital circuits, which nowadays constitute the heart of many electronic systems, need interface electronics for communication with the outside world. This interface is realized with analog electronics. An example of such interface that is representative for this book is a front-end of a receiver for digital telecommunications. This front-end circuit, which consists of several analog functional blocks — some of these blocks can even be digitally controlled — has to bring down a high-frequency signal that carries some information, to a frequency that is low enough such that this signal can be digitized and then further processed in the digital domain. At the same time, this front-end needs to amplify the possibly weak signal of interest while at the same time unwanted signals must be rejected. Although the analog functionality of this front-end is in essence very simple compared to some of the complicated tasks of the digital part of a system, the design of the analog part is often more time-consuming than the digital part. To shorten the time-consuming process, good methodologies are essential.

This edited book addresses design methodologies for low-power analog electronics. Compared to digital circuitry, a systematic design approach for analog circuits is less straightforward. The reasons for this will be explained in the first chapter of this book. Nevertheless, the analog design community has been trying for many years to systematize many aspects of analog circuit design. This has led to many useful insights in complicated circuit aspects and methodologies, which can be reused, and to the development of interesting analog CAD tools. This book unifies some of these insights with CAD tools and design stories of analog or mixed-signal integrated circuits.

The combination of a selection of relevant designs with methodologies and CAD is a unique aspect of this book that is seldom found in other books. However, the combination of these two worlds is very valuable. For example, good design practice is best supported by reliable methodologies and a positive attitude versus CAD tools.

P. Wambacq et al. (eds.),
Low-Power Design Techniques and CAD Tools for Analog and RF Integrated Circuits, xxxi.

On the other hand, the best CAD programs are written by people who understand design issues, such that the programs solve relevant problems.

Almost all designs that are addressed in this book, either as the main subject of the chapter or as illustration for a method or a simulation approach, are related to the analog circuitry in transceivers for digital telecommunications. This is not a surprise, since in the last few years the number of micro-electronics realizations in GSM, DECT, ADSL, Bluetooth, WLAN, GPS, ... has grown enormously. Several of these applications were the driver behind the RF and analog related projects that belong to the European Low-Power Initiative for Electronic System Design. This book contains material from these projects, more specifically from the projects ALPINS, CRAFT, SALOMON, and SUPREGE. The contents of Chapter 7 has been generated in the European (ESPRIT) project AMADEUS. Further, material generated outside the European Low-Power Initiative, such as the contents of Chapter 2, has already been presented on a workshop organized within the European Low-Power Initiative.

The involvement of European industry in these projects guarantees the relevance of the work described in this book. Therefore, we believe that this book will be of interest to industrial people. Designers will not only be interested by design-related issues, but they will also find their way in the tutorial aspects of the chapters that are more related to CAD.

Although this book does not cover many important insights and circuit types, we believe that this book solves some crucial problems that are hot topics in the era where telecommunications and micro-electronics have found each other. Therefore, this book will also inspire academic people.

Most of the material was already known by the editors from the workshops and meetings that have been organized in the frame of the European Low-Power Initiative. In the past four years the people from DIMES, Delft, who is the coordinator of this initiative, succeeded to render the partners of the Low-Power Initiative projects enthusiastic to disseminate interesting project results to an extent that exceeded the normal standards of distribution of project results. We believe that an intensive dissemination of relevant project results to the complete design community is one of the most rewarding activities that can be done within sponsored projects! This book is one of the proofs of this wide spread of information.

Although the compilation of a book together with other dissemination activities within the cluster of projects of the European Low-Power Initiative, have taken much of our time, it was a valuable and pleasant experience that has given rise to many contacts and to technical discussions, that in turn could yield new insights. We hope that in a similar way the reader will find new insights when reading this book.

The different authors of this book together with the editors would like to thank DIMES and everyone who contributed to this book and to the success of the RF and analog related projects in the European Low-Power Initiative.

The author of Chapter 2 would like to thank Alper Demir, Peter Feldmann, David Long, Bob Melville, Onuttom Narayan and Joel Phillips for collaborations and discussions on RF simulation. The author of Chapter 10 would like to thank R. Castello for his contribution. This chapter is mainly based on the paper: A. Baschirotto, R. Castello, " 1V Switched-capacitor filters ", Workshop on Advances in Analog Circuit Design (AACD '98) Copenhagen, 28-30 April 1998 - pp. 3.1-3.13.

Piet Wambacq, Georges Gielen, John Gerrits, April 2001

1 MOTIVATION, CONTEXT AND OBJECTIVES

Georges Gielen

ESAT-MICAS,
Katholieke Universiteiten Leuven, Belgium
E-mail: Georges.Gielen@esat.kuleuven.ac.be

Piet Wambacq

IMEC, Heverlee, Belgium
E-mail: wambacq@imec.be

John F.M. Gerrits

Centre Suisse d'Electronique et de Microtechnique S.A., Neuchâtel, Switzerland
E-mail: john.gerrits@csem.ch

Driven by cost-constrained applications such as telecommunications, computing and consumer/multimedia and facilitated by the continuing miniaturization in the CMOS ULSI technology, the micro-electronics IC market is characterized by an ever increasing level of integration complexity. Today complete systems, that previously occupied one or more boards, are integrated on a few chips or even on one single multi-million transistor chip – a so called System-on-Chip (SoC). Examples are single-chip cameras

P. Wambacq et al. (eds.),
Low-Power Design Techniques and CAD Tools for Analog and RF Integrated Circuits, 1–7.

or new generations of integrated telecommunication systems that include analog, digital and eventually radio-frequency (RF) sections on one chip. Although most functions in such integrated systems are implemented with digital or digital signal processing (DSP) circuitry, the analog circuits needed at the interface between the electronic system and the continuous-valued outside world are also being integrated on the same die for reasons of cost and performance. Modern System-on-Chip designs are therefore increasingly mixed-signal designs, and this will even be more prevalent if we move towards the intelligent homes, the mobile road/air offices and the wireless workplaces of the future.

Unfortunately, just as deep-submicron CMOS technologies have complicated the design of digital functions with issues such as design complexity management and predictable timing closure, likewise these technologies complicate the analog subsystems on SoC designs. Since analog circuits exploit (rather than abstract away) the low-level physics of the fabrication process, they remain difficult and costly to design, validate, reuse. In addition, many of the current application markets are characterized by shortening product life cycles and tightening time-to-market constraints, necessitating the use of systematic design methodologies and supporting computer-aided design (CAD) tools to manage the design complexity in the available design time. In addition, the high levels of integration (moving towards 100 million transistors per chip clocked at ever higher frequencies) as well as the increasing need for mobile and therefore battery-operated devices necessitates the reduction of the power consumption of these chips. This impacts the design at all levels, from system architecture down to circuit level, and requires the development of appropriate low-power design tools and design methodologies.

In the digital domain, CAD tools are fairly well developed and commercially available today, ranging from HDL-based high-level and logic synthesis to standard-cell based place & route. Timing and power information is being incorporated in the synthesis tools to allow a secured timing closure. Research interest is now moving in the direction of system synthesis where an object-oriented system-level specification is translated into a hardware-software co-architecture with high-level specifications for both the hardware, the software and the interfaces. In addition, reuse and platform-based design methodologies are being developed to further reduce the design effort for complex systems. Of course, the level of automation is far from the push-button stage, but the developments are keeping up reasonably well with the chip complexity offered by the technology.

The story is quite different on the analog side. There are not yet any robust commercial CAD tools to support or automate analog circuit design apart from circuit simulators (in most cases some flavor of the ubiquitous SPICE simulator) and layout editing environments and their accompanying tools (e.g., some limited optimization capabilities around the simulator, or layout verification tools). Some of the main reasons for this lack of automation are that analog design in general is perceived as less systematic and more heuristic and knowledge-intensive in nature than digital design, and that it has not yet been possible for analog designers to establish a higher level of abstraction that shields all the device-level and process-level details from the higher-level design. Analog IC design is a complex endeavor, requiring specialized knowledge and circuit design skills acquired through many years of experience. The variety of circuit schematics and the number of conflicting requirements and corresponding diversity of device sizes is also much larger. In addition, analog circuits are more sensitive to nonidealities and all kinds of higher-order effects and parasitic disturbances (crosstalk, substrate noise, supply noise, etc.). These differences from digital design also explain why analog CAD tools cannot simply adapt the digital algorithms, but why specific analog solutions need to be developed that are targeted to the analog design paradigm and complexity.

The practical result is that due to the lack of adequate and mature commercially available analog CAD tools, analog designs today are still largely being hand-crafted with only a SPICE-like simulation shell and an interactive layout environment as supporting facilities, resulting in a long and error-prone design cycle. Therefore, although analog circuits typically occupy only a small fraction of the total area of mixed-signal ICs, their design is often the bottleneck in mixed-signal systems, both in design time and effort as well as test cost, and they are often responsible for design errors and expensive reruns. In the emerging era of integrated systems-on-chip, this practice of hand-crafted, one-transistor-at-a-time analog design is increasingly at odds with the need for more analog design productivity, practical circuit synthesis and reuse, and reliable verification at all levels of the mixed-signal hierarchy. To keep pace with the digital side in these mixed-signal integrated systems and to fully exploit the potential offered by the present deep-submicron CMOS ULSI technologies, boosting analog design productivity is a major concern in the industry today. The design time and cost for analog circuits from specification to successful silicon has to be reduced drastically. The risk for design errors impeding first-pass functional (and possibly also parametricly correct) chips has to be eliminated. In addition, analog CAD tools can also help to increase the quality of the resulting designs. Before starting detailed circuit implementation, more higher-level explorations and optimizations should be performed at the system architectural level, preferably across the analog-digital boundary, since decisions at those levels have a much larger impact on key overall system parameters such as power consumption and chip area. Likewise, designs at lower levels should be "automated" where possible. Also, the continuous pressure of technology updates and process migrations is a large burden on analog designers. CAD tools could take over large part of the technology retargeting effort, and could make analog design easier to port or migrate to new technologies. This need for analog CAD tools beyond simulation has also clearly been identified in the SIA and MEDEA EDA roadmaps, where for example analog synthesis is predicted to take off somewhere beyond the year 2001.

Despite the dearth of really commercial analog CAD tools, analog CAD and design automation has been a field of profound academic and industrial research activity over the past fifteen years, resulting in a slow but steady progress. Some of the aspects of the analog CAD field are fairly mature today, some are ready for commercialization – in recent years a bunch of start-ups with initial offerings have entered the marketplace building upon these academic results – while other aspects are still in the process of exploration and development. The simulation area has been particularly well developed since the advent of the SPICE simulator, which has led to the development of many simulators including timing simulators in the digital field and the newer generation of mixed-signal and multi-level commercial simulators. Standardized analog and mixed-signal hardware description languages like VHDL-AMS and VERILOG-AMS provide here the link between the analog, the digital and the system domains, as needed in designing future mixed analog-digital integrated systems. Also, analog circuit and layout synthesis has shown extremely promising results at the research level in recent years, and commercial solutions based on these results have just started to appear on the marketplace.

The purpose of this book is to provide an overview of very recent research results that have been achieved as part of the Low-Power Initiative of the European Union, in the field of analog, RF and mixed-signal design methodologies and CAD tools. It is a representative sampling of the current state of the art in this area, with special focus on low-power design methodologies and tools for analog and RF circuits and architectures. This volume complements other similar volumes that focus on RF circuit and architectural design, as well as on digital low-power design techniques. The book consists of eleven contributions that we will briefly introduce next.

Simulation is one of the key techniques to analyze and verify the performance of any electronic system that must implement a desired functionality under specified performance requirements. This is true at the circuit level but also at the architectural level. Before any trade-offs can be explored between different alternative design solutions, the performance of these alternatives must be checked. Before any layout can be sent for tape-out, the performance of the circuit has to be verified extensively. For RF applications this however requires special simulation techniques, since the standard SPICE algorithms are too time-consuming. *Chapter 2* therefore gives an overview of analysis and simulation techniques for RF subsystems, some of which have only been developed in very recent years. The chapter includes periodic steady-state analysis methods such as harmonic balance and shooting methods, multi-time analysis methods and special techniques for the analysis of noise (mixing noise, phase noise) in RF designs.

In order to reduce the overall power consumption of an electronic system, power-reducing decisions and techniques must be used at all levels of the design hierarchy. However, decisions taken at the architectural level have a much larger impact than decisions taken at the circuit level. Therefore, a system architectural exploration environment is needed that allows to explore different architectural alternatives, check their performance and compare their power and/or area consumption. Current methodologies and corresponding tools suffer from drawbacks such as lack of accuracy, long simulation times, etc. *Chapter 3* presents a new methodology for the efficient simulation, at the architectural level, of mixed-signal front-ends of digital telecom transceivers. The efficient execution is obtained using a local multirate, multicarrier signal representation together with a dataflow simulation scheme that dynamically switches to the most efficient signal processing technique available. The methodology has been implemented in the program FAST (Front-end Architecture Simulator for digital Telecom applications). Simulation examples show both excellent runtimes and a high accuracy for realistic front-end architectures.

The importance of telecommunication systems justifies research targeted to develop dedicated simulation algorithms that increase simulation speed by incorporating properties of telecom systems and their signals into the algorithm. *Chapter 4* presents such an approach based upon a complex damped exponential signal model that incorporates the typical properties of digitally modulated telecom signals, like their many different time constants, in a natural way. This allows to construct simple signal models (containing only a few base functions) that are valid over a long time interval. This results into a significant increase of the simulation time step, speeding up the simulations. This complex damped exponential signal model is combined with a runtime Volterra series expansion into an algorithm that is particularly well suited for the simulation of weakly nonlinear (telecom) systems. The algorithm also allows to compute wanted and unwanted signals separately. This, together with the natural relationship between the exponential signal model and frequency content, greatly facilitates analysis of the results by the designer, which is also an important asset provided by CAD tools.

Another element that is needed to perform system-level architectural explorations and power minimization, is the availability of power estimation models that allow to assess the power consumption of an analog block, given only its performance specifications, not knowing the detailed circuit implementation. In high-level system design such power (and area) estimators provide a criterion, minimal total power consumption for the overall architecture, that has to be optimized when comparing different architectural alternatives. *Chapter 5* introduces this topic and presents different approaches for the construction of analog power estimator functions: bottom-up and top-down. This is illustrated with examples of practical power estimators, covering both analog-to-digital converters and continuous-time filters. The level at which topology-specific information has is used varies in the two cases, but inversely affects the genericity of the resulting estimator.

One of the key blocks in a wireless telecom system is the frequency synthesizer. Increasing the design productivity for this block is therefore one of the key challenges for analog CAD tools. *Chapter 6* presents models and analysis techniques for the systematic design and verification of frequency synthesizers used in telecommunication applications. The presented models are tuned towards the evaluation of the trade–off between the loop settling time and the phase noise performance of the synthesizer. Both models for top–down design and bottom–up verification are presented. The latter concentrates on the analysis of phase noise and spurious tones in the spectrum of the synthesizer. The validity of the models is illustrated using a practical 1.8 GHz CMOS frequency synthesizer.

Analog and RF circuit design tends to rely very much on knowledge and experience built up over many years. Improving and speeding up the understanding of analog designers in the behavior of an analog or RF circuit is therefore a key component to increasing analog design productivity. Distortion and intermodulation are important performance specifications in many systems, including wireless communications, that are however extremely difficult to understand by analog designers. Distortion and intermodulation are either unwanted, as is the case in linear building blocks like opamps or filters, or they are explicitly wanted to obtain a signal shifted in frequency, as is the case with mixers. In both cases, distortion and intermodulation need to be assessed accurately, which requires time–consuming simulations using the classical numerical approaches. In addition, these numerical simulations do not provide any insight in the actual nonlinear behavior of the circuit. *Chapter 7* therefore presents an alternative approach based on symbolic analysis techniques. An algorithm is presented that generates symbolic formulas for the distortion and intermodulation behavior of weakly nonlinear analog and RF circuits as a function of the frequencies involved and of the small-signal parameters of the devices in the circuit. A multitude of simplification techniques is needed to get interpretable yet accurate symbolic results. The method is illustrated for several practical circuits.

Verifying the performance of an analog circuit is typically done using numerical simulations. These however only provide a sampling of the performance space of the circuit, and are never a 100% guarantee that the circuit will perform correctly under all circumstances. In the digital world formal verification is therefore used as alternative to simulation. It has some important advantages over traditional validation methods like circuit simulation, in that it gives a strong mathematical correctness proof of the entire circuit behavior. Currently, this technique is evolving to a widely used verification method, indicated by a growing number of commercial vendors. For analog circuits no comparable techniques are known yet. However, the same problems driving the development of such tools in the digital world can be found for analog circuits. *Chapter 8* therefore presents algorithms for the formal verification of analog circuits that compare two system descriptions on different levels of abstraction. They prove/disprove that the systems have functionally similar input-output behavior. Additionally, in case of nonlinear dynamic circuits, an algorithm for generating transient stimuli is outlined to help the designer finding the design flaws with well known transient simulations. Some examples show the feasibility of the approaches.

Large power savings can be obtained by choosing the proper system architecture for the application at hand. Different architectures allow to trade off different specifications for the building blocks, which then results in a different power consumption for the overall architecture. Indeed, for many applications the required performance is not equal. A signal transmitted by a garage door opener does not need to have the same spectral purity as the signal produced by a portable phone. However, the user still expects his door to open within a reasonable time. Research in alternative architectures for different applications is therefore key. *Chapter 9* presents the super-regenerative radio receiver architecture discovered in the 1920s by Armstrong. This architecture was recently rediscovered and made more robust and performing by implementing it in an integrated circuit. This is a solution that can be used in applications that require low power, however at the expense of lower receiver selectivity.

Filtering is one of the most important signal processing tasks in analog transceivers. It selects the wanted signal from all received signals. Electrical filters started as discrete-element passive filters using inductors and capacitors. When filters have to be tunable, discrete-element solutions lack flexibility. This problem can be solved in a number of ways. For instance in a superheterodyne radio receiver, frequency selectivity is realized at one or two fixed intermediate frequency values. This allows the use of fixed-frequency discrete filters. Another solution is to electronically tune the filter elements and realize the inductors electronically. Unlike their passive counterparts, electronic inductors are not noise-free due to the noise originating from the active circuits. One step further is to design filters using only capacitors and opamps.

A resistor can be emulated by a switched capacitor. *Chapter 10* addresses low-voltage switched-capacitor filters that operate on supply voltages as low as the supply voltage of the digital electronics.

Another key component in wireless receivers is the low-noise amplifier (LNA). Building a fully-integrated LNA at RF frequencies is not trivial. On-chip inductors with reasonable quality factor have recently become available. They can be used to implement narrowband resonant loads and matching networks. Noise and power matching at the LNA input is crucial for obtaining a low noise figure. *Chapter 11* gives an introduction into the main LNA design issues, with particular emphasis on CMOS implementations using today's active and passive on-chip components. The inductively degenerated CMOS LNA proves to be best suited to achieve low noise under matching conditions. Power consumption can be lowered by current reuse. This is illustrated with a practical fully-differential 900-MHz circuit that has a 2 dB noise figure, 16 dB gain with 8 mA current consumption from a 2 V supply.

Finally, it is the task of an oscillator to generate the carrier signal needed in telecom applications. Combining in an intelligent way a passive resonator and an active circuit yields an oscillator circuit. *Chapter 12* shows that theoretically the power consumption of an oscillator circuit is determined by the quality factor of the resonator and the required phase noise performance. In practice, it appears that it is not always straightforward to approach this fundamental limit. Resonator element values cannot be chosen randomly, but span a limited solution space. Resonator voltage and currents are not necessarily compatible with the active circuit's voltage and current headroom. By using two-port resonators and tapping techniques, a good compromise is usually available. Two design examples illustrate oscillator design in practice.

We wish the reader much pleasure in exploring the different chapters in this book, and in adopting the presented techniques in his/her daily practice to reduce the power consumption of analog circuits in future electronic systems, making way for more and more mobile applications that don't need recharging every evening.

2 ANALYSIS AND SIMULATION OF RF SUBSYSTEMS

Jaijeet Roychowdhury

Celight, Inc., USA

E-mail: jaijeet@yahoo.com

Simulation/verification is a crucial step in the design of modern RF systems, which often stretch the capabilities of established tools like SPICE. In this chapter, several specialized methods for common RF design verification tasks are described. The steady-state calculation methods harmonic balance (HB) and shooting are reviewed, and fast techniques for handling large circuits are described. Multitime techniques for analyzing circuits with disparate rates of signal variation are described. Techniques for creating accurate macromodels of RF circuits automatically from SPICE-like descriptions are outlined. Finally, concepts and techniques in mixing and phase noise analysis are reviewed.

2.1 STEADY-STATE ANALYSIS

It is often important in RF design to find the periodic steady state of a circuit driven by one or more periodic inputs. For example, a power amplifier driven to saturation by a large single-tone input is operating in a periodic steady state. A variant is the quasiperiodic steady state, *i.e.*, when the circuit is driven by more than one signal tone; for example, an amplifier driven by two closely spaced sinusoidal tones at 1 GHz and 990 MHz. Such excitations are closer approximations to real-life signals than pure tones, and are useful for estimating intermodulation distortion.

The workhorse of analog verification, SPICE (and its derivatives), can of course be applied to find the (quasi)periodic steady state of a circuit, simply by performing a timestepping integration of the circuit's differential equations ("transient analysis") long enough for the transients to subside and the circuit's response to become (quasi)periodic. This approach has several disadvantages, however. In typical RF circuits, the transients take thousands of periods to die out, hence the procedure can be very inefficient. Further, harmonics are typically orders of magnitude smaller than the fundamental, hence long transient simulations are not well suited for their accu-

P. Wambacq et al. (eds.),
Low-Power Design Techniques and CAD Tools for Analog and RF Integrated Circuits, 9–42.

rate capture, because numerical errors from timestepping integration can mask them. These issues are exacerbated in the presence of quasiperiodic excitations, because simulations need to be much longer – *e.g.*, for excitations of 1 GHz and 990 MHz, the system needs to be simulated for thousands of multiples of the common period, 1/10 MHz, yet the simulation timesteps must be much smaller than 1ns, the period of the fastest tone. For these reasons, more efficient and accurate specialized techniques have been developed. We will focus on two different methods with complementary properties, harmonic balance (HB) and shooting.

For concreteness, we will use the following differential algebraic equation (DAE) form for describing a circuit or system:

$$\dot{q}(x) + f(x) = b(t). \tag{2.1}$$

The nonlinear functions $q(\cdot)$ and $f(\cdot)$ represent the charge/flux and resistive parts of the circuit, respectively, while $b(t)$ is a forcing term representing external inputs from independent current and voltage sources.

2.1.1 Harmonic Balance and Shooting

In the *Harmonic Balance (HB)* method (*e.g.*, [1, 2, 3, 4, 5, 6, 7, 8, 9]), $x(t)$ and $b(t)$ of (2.1) are expanded in a Fourier series. The Fourier series can be one-tone for periodic excitations (*e.g.*, 1 GHz and its harmonics) or multitone in the case of quasiperiodic excitations (eg, 1 GHz and 990 MHz, their harmonics, and intermodulation mixes). The DAE is rewritten directly in terms of the Fourier coefficients of $x(t)$ (which are unknown) and of $b(t)$; the resulting system of nonlinear equations is larger by a factor of the number of harmonic/mix components used, but are algebraic (*i.e.*, there are no differential components). Hence they can be solved numerically e.g. using the Newton-Raphson method [10].

Example 2.1 *We illustrate the one-tone procedure with the following scalar DAE:*

$$\dot{x} + x - \epsilon x^2 - \cos(2\pi 1000t) = 0. \tag{2.2}$$

First, we expand all time variations in a Fourier series of three terms, namely the DC component, fundamental and second harmonic components:

$$x(t) = \sum_{i=-2}^{2} X_i e^{j2\pi i 10^3 t}. \tag{2.3}$$

In this equation X_i are the unknown Fourier coefficients of $x(t)$. For notational convenience, we express them as the vector $X = [X_2, \cdots, X_{-2}]^T$.

Similarly, $x^2(t)$ is also expanded in a Fourier series in t; the Fourier coefficients of this expansion are functions of the elements of X, which we denote by $F_i(X)$:

$$x^2(t) = \sum_{i=-2}^{2} \sum_{k=-2}^{2} X_i X_k e^{j2\pi(i+k)10^3 t} = \sum_{i=-2}^{2} F_i(X) e^{j2\pi i 10^3 t} + \textit{higher terms}. \tag{2.4}$$

In this case, where the nonlinearity is a simple quadratic, F_i can be obtained analytically; but in general, numerical techniques like those used in harmonic balance need to be employed for computing these functions. For convenience, we write $F(X) = [F_2(X), \cdots, F_{-2}(X)]^T$.

We also write the Fourier coefficients of the excitation $\cos(2\pi 1000t)$ *as a vector* $B = [0, \frac{1}{2}, 0, \frac{1}{2}, 0]^T$. *Finally, we write the differential term* $\dot{x}$ *also as a vector of Fourier coefficients. Because the differentiation operator is diagonal in the Fourier basis, this becomes simply* ΩX, *where* $\Omega = j2\pi 1000\, diag(2, 1, 0, -1\, -2)$ *is the diagonal frequency-domain differentiation matrix.*

Invoking the orthogonality of the Fourier basis, we now obtain the HB equations for our DAE:

$$H(X) \equiv \Omega X + X - \epsilon F(X) - B = 0. \tag{2.5}$$

This is a set of nonlinear algebraic equations in five unknowns, and can be solved by numerical techniques such as Newton-Raphson.

The above example illustrates that the size of the HB equations is larger than that of the underlying DAE, by a factor of the number of harmonic/mix components used for the analysis. In fact, the HB equations are not only larger in size than the DAE, but also considerably more difficult to solve using standard numerical techniques. The reason for this is the dense structure of the derivative, or Jacobian matrix, of the HB equations. If the size of the DAE is n, and a total of N harmonics and mix components are used for the HB analysis, the Jacobian matrix has a size of $Nn \times Nn$. Just the storage for the nonzero entries can become prohibitive for relatively moderate values of n and N; for example, $n = 1000$ (for a medium-sized circuit) and $N = 100$ (*e.g.*, for a two-tone problem with about 10 harmonics each) leads to 10 GB of storage for the matrix alone. Further, inverting the matrix, or solving linear systems with it, requires $O(N^3 n^3)$ operations, which is usually unfeasible for moderate to large-sized problems. Such linear solutions are typically required as steps in solving the HB equations, for example by the Newton-Raphson method. Despite this disadvantage, HB is a useful tool for small circuits and few harmonics, especially for microwave circuit design. Moreover, as we will see later, new algorithms have been developed for HB that make it much faster for larger problems.

Another technique for finding periodic solutions is the *shooting* method (*e.g.*, [11, 12, 13, 14]). Shooting works by finding an initial condition for the DAE that also satisfies the periodicity constraint. A guess is made for the initial condition, the system is simulated for one period of the excitation using timestepping DAE solution methods, and the error from periodicity used to update the initial condition guess, often using a Newton-Raphson scheme.

More precisely, shooting computes the *state transition function* $\Phi(t, x_0)$ of (2.1). This function represents the solution of the system at time t, given initial condition x_0 at time 0. Shooting finds an initial condition x^* that leads to the same state after one period T of the excitation $b(t)$; in other words, shooting solves the equation $H(x) \equiv \Phi(t, x) - x = 0$.

The shooting equation is typically solved numerically using the Newton-Raphson method, which requires evaluations of $H(x)$ and its derivative (or Jacobian) matrix.

Evaluation of $H(x)$ is straightforward using timestepping, *i.e.*, transient simulation, of (2.1). However, evaluating its Jacobian is more involved. The Jacobian matrix is of the same size n as the number of circuit equations, but it is dense, hence storage of its elements and linear solutions with it is prohibitive in cost for large problems. In this respect, shooting suffers from size limitations, similar to harmonic balance. In other respects, though, shooting has properties complementary to harmonic balance. The following list contrasts the main properties of HB and shooting:

- **Problem size:** The problem size is limited for both HB and shooting, due to the density of their Jacobian matrices. However, since the HB system is larger by a factor of the number of harmonics used, shooting can handle somewhat larger problems given the same resources. Roughly speaking, sizes of about 40 for HB and 400 for shooting represent practical limits.

- **Accuracy/Dynamic Range:** Because HB uses orthogonal Fourier bases to represent the waveform, it is capable of very high dynamic range – a good implementation can deliver 120 dB of overall numerical accuracy. Shooting, being based on timestepping solution of the DAE with timesteps of different sizes, is considerably poorer in this regard.

- **Handling of nonlinearities:** HB is not well suited for problems that contain strongly nonlinear elements. The main reason for this is that strong nonlinearities (*e.g.*, clipping elements) generate sharp waveforms that do not represent compactly in a Fourier series basis. Hence many harmonics/mix components need to be considered for an accurate simulation, which raises the overall problem size.

  Shooting, on the other hand, is well suited for strong nonlinearities. By approaching the problem as a series of initial value problems for which it uses timestepping DAE methods, shooting is able to handle the sharp waveform features caused by strong nonlinearities quite effectively.

- **Multitone problems:** A big attraction of HB is its ability to handle multitone or quasiperiodic problems as a straightforward extension of the one tone case, by using multitone Fourier bases to represent quasiperiodic signals. Shooting, on the other hand, is limited in this regard. Since it uses timestepping DAE solution, shooting requires an excessive number of timepoints when the waveforms involved have widely separated rates of change; hence it is not well suited for such problems.

2.1.2 Fast Methods

A disadvantage of both HB and shooting is their limitation to circuits of relatively small size. This was not a serious problem as long as microwave/RF circuits contained only a few nonlinear devices. Since the mid-90s, however, economic and technological developments have changed this situation. The market for cheap, portable wireless communication devices has expanded greatly, leading to increased competition and consequent cost pressures. This has spurred on-chip integration of RF communication circuits and the reduction of discrete (off-chip) components. On-chip design techniques favor the use of many integrated nonlinear transistors over even a few linear external components. Hence the need has arisen to apply HB and shooting to large circuits in practical times.

To address this issue, so-called *fast* algorithms have arisen to enable both HB and shooting to handle large circuits. The key property of these methods is that computation/memory usage grow approximately linearly with problem size. The enabling idea behind the improved speed is to express the dense Jacobian matrices as sums and products of simpler matrices that are either sparse, or have very regular structure so can be applied/inverted efficiently. Using these expansions for the Jacobian, special solution algorithms called *preconditioned iterative linear solvers* are applied to solve linear equations involving the Jacobian without forming it explicitly.

A detailed description of fast techniques is beyond the scope of this chapter. The interested reader is referred to [6, 8, 15, 9, 14] for further information. Here we outline the main ideas behind these methods in a simplified form using Example 2.1 for illustration, and summarize their main properties.

From Example 2.1, the Jacobian matrix of the HB system is

$$J = \frac{\partial H(X)}{\partial X} = \Omega + I - \epsilon \frac{\partial F(X)}{\partial X}. \tag{2.6}$$

Now, $F(X)$ in this case represents the vector of Fourier coefficients of the nonlinear term $f(x) = x^2$. One way in which these can be computed numerically is to 1) use the inverse Fast Fourier Transform (FFT) to convert the Fourier coefficients X to samples of the time-domain waveform $x(t)$, then 2) evaluate the nonlinear function $f(x) = x^2(t)$ at each of these samples in the time domain, and finally, 3) use the FFT to reconvert the time-domain samples of $f(x)$ back to the frequency domain, to obtain $F(X)$. The derivative of these operations can be expressed as

$$\frac{\partial F(X)}{\partial X} = DGD^*, \tag{2.7}$$

where D is a block-diagonal matrix with each block equal to the Discrete Fourier Transform matrix, D^* is its inverse, and G is a diagonal matrix with entries $f'()$ evaluated at the time-domain samples of $x(t)$. Hence the overall Jacobian matrix can be represented as:

$$J = \Omega + I - \epsilon DGD^*. \tag{2.8}$$

Observe that each of the matrices in this expansion is either sparse or consists of DFT matrices. Hence multiplication of J with a vector is efficient, since the sparse matrices can be applied in approximately linear time, and the DFT matrix and its inverse can be applied in $N \log N$ time using the FFT, where N is the number of harmonics. It is this key property, that multiplications of J with a vector can be performed in almost-linear computation despite its dense structure, that enables the use of preconditioned iterative linear methods for this problem.

Preconditioned iterative linear methods (*e.g.*, [16, 17, 18]) are a set of numerical techniques for solving linear systems of the form $Jc = d$. Modern iterative solvers like QMR [17] and GMRES [16] use Krylov-subspace techniques for superior performance. The key feature of these solvers is that the only way in which J is used is in matrix-vector products Jz. This contrasts with traditional methods for linear solution, which use Gaussian elimination or variants like LU factorizations directly on elements of J. Due to this property of iterative linear solvers, it is not necessary to even form J explicitly in order to solve linear systems with it, so long as a means is available for computing matrix-vector products with it.

As we have seen above, products with the HB Jacobian can be conveniently computed in almost-linear time without having to build the matrix explicitly. Hence preconditioned linear iterative techniques are well-suited to solving the linear systems that arise when solving the nonlinear HB equations using the Newton-Raphson method. If the iterative linear methods use only a few matrix-vector products with the Jacobian to compute the linear system's solution, and the Newton-Raphson is well-behaved, the overall cost of solving the HB equations remains almost-linear in problem size.

An important issue with preconditioned iterative linear solvers, especially those based on Krylov subspace methods, is that they require a good *preconditioner* to converge reliably in a few iterations. Convergence of the iterative linear method is accelerated by applying a preconditioner $\tilde{J}$, replacing the original system $Jc = d$ with the *preconditioned* system $\tilde{J}^{-1}Jc = \tilde{J}^{-1}d$, which has the same solution. For robust and efficient convergence, the preconditioner matrix $\tilde{J}$ should be, in some sense, a good approximation of J, and also "easy" to invert, usually with a direct method such as LU factorization. Finding good preconditioners that work well for a wide variety of circuits is a challenging task, especially when the nonlinearities become strong.

The ideas behind the fast techniques outlined above are applicable not just to HB but also to shooting [14]. Jacobian matrices from shooting can be decomposed as products and sums of the sparse circuit Jacobians matrices. Preconditioned linear iterative techniques can then applied to invert the Jacobian efficiently.

As an example of the application of the fast methods, consider the HB simulation of an RFIC quadrature modulator reported in [9]. The circuit, of about 9500 devices, was simulated by fast HB with a three-tone excitation, with a baseband signal at 80 kHz, and local oscillators at 178 MHz and 1.62 GHz. The size of the circuit's DAE was $n = 4800$; the three tones, their harmonics and mixes totalled $N = 4320$ components. Simulating a circuit with these specifications is completely unfeasible using traditional HB techniques.

Using fast HB, the simulation required only 350MB of memory, and took 5 days of computation on an SGI 150 MHz R4400 machine. The results of the the simulation are shown in Figure 2.1.

2.2 MULTITIME ANALYSIS

In the previous section, we noted that HB and shooting have complementary strengths and weaknesses, stemming from their use of Fourier and time-domain bases, respectively. While HB is best suited for multitone problems that are only mildly nonlinear, shooting is best for single tone problems that can be strongly nonlinear. Neither method is suitable for circuits that have both multitone signals (*i.e.*, widely-

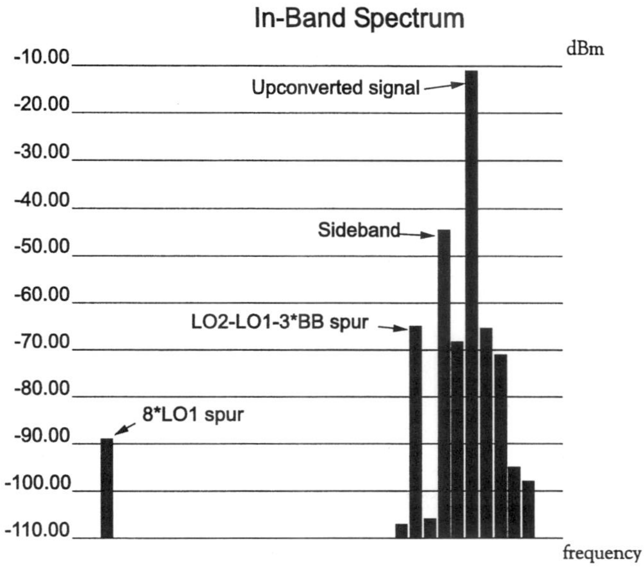

Figure 2.1 Quadrature modulator spectrum.

separated time scales of variation) *and* strongly nonlinear components. With greater RF integration, tools that can analyze precisely this combination of circuit characteristics effectively are required. In this section, we review a family of techniques, based on partial differential equations (PDEs) using multiple artificial time scales [19, 7, 20, 21, 22, 23], that hold this promise.

Consider a simple two-tone signal (see Figure 2.2) signal given by

$$y(t) = \sin\left(\frac{2\pi}{T_1}t\right)\sin\left(\frac{2\pi}{T_2}t\right), \qquad T_1 = 0.02\text{s}, \quad T_2 = 1\text{s}. \tag{2.9}$$

The two tones are at frequencies $f_1 = 1/T_1 = 50\,\text{Hz}$ and $f_2 = 1/T_2 = 1\,\text{Hz}$, *i.e.*, there are fifty fast-varying cycles of period $T_1 = 0.02\text{s}$ modulated by a slowly-varying sinusoid of period $T_2 = 1\text{s}$. If each fast cycle is sampled at n points, the total number of timesteps needed for one period of the slow modulation is $n\frac{T_2}{T_1}$. To generate Figure 2.2, fifteen points were used per cycle, hence the total number of samples was 750. This number can be much larger in applications where the rates are more widely separated, *e.g.*, separation factors of 1000 or more are common in electronic circuits. Now consider a multivariate representation of $y(t)$ using two artificial time scales, as follows: for the 'fast-varying' parts of $y(t)$, t is replaced by a new variable t_1; for the 'slowly-varying' parts, by t_2. The resulting function of two variables is denoted by

$$\hat{y}(t_1, t_2) = \sin\left(\frac{2\pi}{T_1}t_1\right)\sin\left(\frac{2\pi}{T_2}t_2\right). \tag{2.10}$$

The plot of $\hat{y}(t_1, t_2)$ on the rectangle $0 \leq t_1 \leq T_1$, $0 \leq t_2 \leq T_2$ is shown in Figure 2.3. Observe that $\hat{y}(t_1, t_2)$ does not have many undulations, unlike $y(t)$ in Fig-

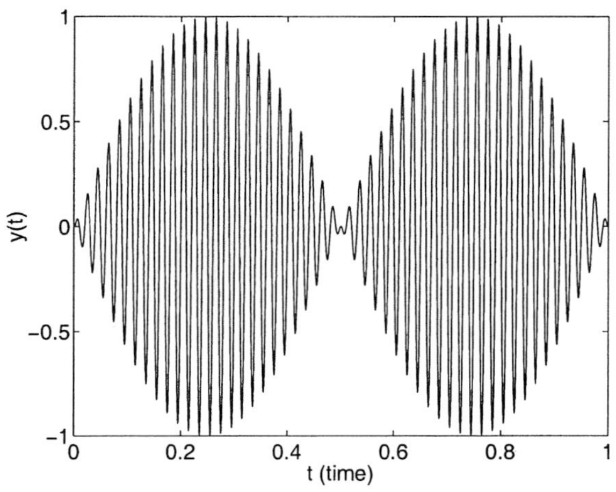

Figure 2.2 Example twotone quasiperiodic signal $y(t)$.

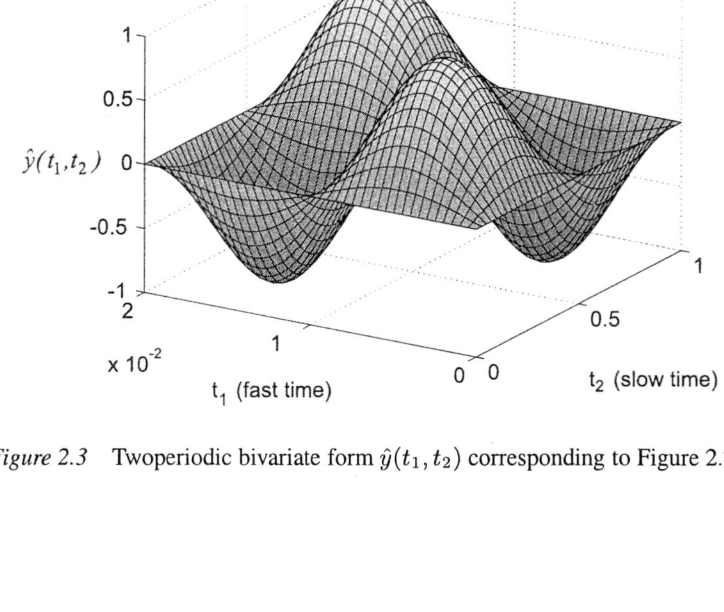

Figure 2.3 Twoperiodic bivariate form $\hat{y}(t_1, t_2)$ corresponding to Figure 2.2.

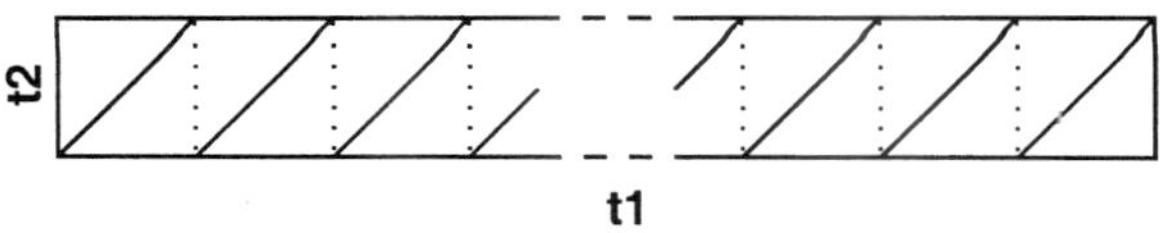

Figure 2.4 Path in the t_1-t_2 plane.

ure 2.2. *Hence it can be represented by relatively few points, which, moreover, do not depend on the relative values of* T_1 *and* T_2. Figure 2.3 was plotted with 225 samples on a uniform 15×15 grid – three times fewer than for Figure 2.2. This saving increases with increasing separation of the periods T_1 and T_2.

Further, note that $\hat{y}(t_1, t_2)$ is periodic with respect to both t_1 and t_2, *i.e.*, $\hat{y}(t_1 + T_1, t_2 + T_2) = \hat{y}(t_1, t_2)$. This makes it easy to recover $y(t)$ from $\hat{y}(t_1, t_2)$, simply by setting $t_1 = t_2 = t$, and using the fact that $\hat{y}$ is bi-periodic. It is easy, from direct inspection of the three-dimensional plot of $\hat{y}(t_1, t_2)$, to visualize what $y(t)$ looks like. As t increases from 0, the path given by $\{t_i = t \bmod T_i\}$ traces the sawtooth path shown in Figure 2.4. By noting how $\hat{y}$ changes as this path is traced in the t_1-t_2 plane, $y(t)$ can be traced. *When the time-scales are widely separated, therefore, inspection of the bivariate waveform directly provides information about the slow and fast variations of* $y(t)$ *more naturally and conveniently than* $y(t)$ *itself.*

We observe that the bivariate form can require far fewer points to represent numerically than the original quasiperiodic signal, yet it contains all the information needed to recover the original signal completely. This observation is the basis of the partial differential formulation to be introduced shortly. the waveforms in a circuit are represented in their bivariate forms (or multivariate forms if there are more than two time scales). The key to efficiency is to solve for these waveforms directly, without involving the numerically inefficient one-dimensional forms at any point. To do this, it is necessary to first describe the circuit's equations using the multivariate functions. If the circuit is described by the differential equations (2.1), then it can be shown that if $\hat{x}(t_1, t_2)$ and $\hat{b}(t_1, t_2)$ denote the bivariate forms of the circuit unknowns and excitations, then the following **M**ultitime **P**artial **D**ifferential **E**quation (MPDE) is the correct generalization of (2.1) to the bivariate case:

$$\frac{\partial q(\hat{x})}{\partial t_1} + \frac{\partial q(\hat{x})}{\partial t_2} + f(\hat{x}) = \hat{b}(t_1, t_2). \tag{2.11}$$

More precisely, if $\hat{b}$ is chosen to satisfy $b(t) = \hat{b}(t, t)$, and $\hat{x}$ satisfies (2.11), then it can be shown that $x(t) = \hat{x}(t, t)$ satisfies (2.1). Also, if (2.1) has a quasiperiodic solution, then (2.11) can be shown to have a corresponding bivariate solution.

By solving the MPDE numerically in the time domain, strong nonlinearities can be handled efficiently. Several numerical methods are possible, including discretization of the MPDE on a grid in the t_1-t_2 plane, or using a mixed time-frequency method in which the variation along one of the time scales is expressed in a short Fourier series. Quasiperiodic and envelope solutions can both be generated, by appropriate selection of boundary conditions for the MPDE. Sparse matrix and iterative linear methods are used to keep the numerical algorithms efficient even for large systems.

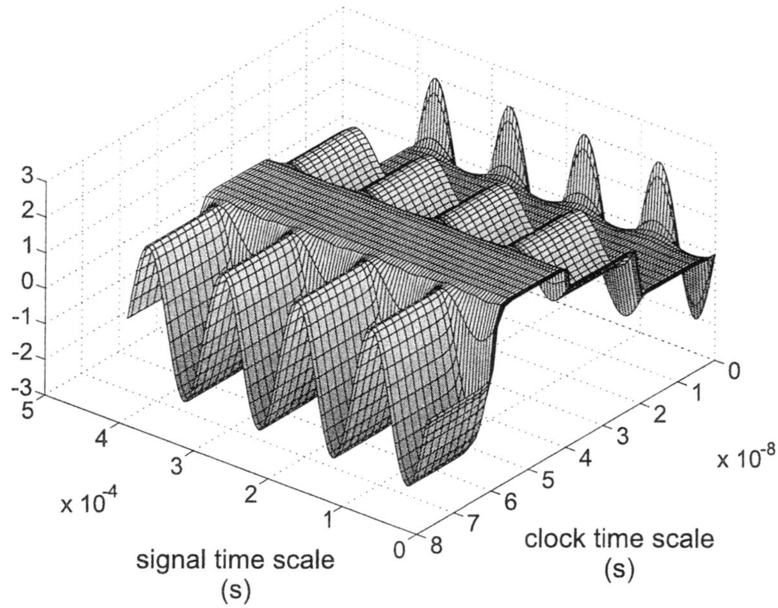

Figure 2.5 Multitime output waveform of an SC integrator.

As an example, Figure 2.5 depicts the output voltage of a switched-capacitor integrator block, obtained from a multitime simulation based on the above concepts. The cross-section parallel to the signal time scale represents the envelope of the signal riding on the switching variations. By moving these cross-sections to different points along the clock timescale, the signal envelope at different points of the clock waveform can be seen.

2.2.1 Autonomous Systems: the Warped MPDE

When the DAEs under consideration are oscillatory, frequency modulation (FM) can be generated. Unfortunately, FM cannot be represented compactly using multiple time scales as easily as the waveform in Figure 2.3. We illustrate the difficulty with an example. Consider the following prototype FM signal (see Figure 2.6)

$$x(t) = \cos\left(2\pi f_0 t + k\cos(2\pi f_2 t)\right), \qquad f_0 \gg f_2, \tag{2.12}$$

with instantaneous frequency

$$f(t) = f_0 - k f_2 \sin(2\pi f_2 t). \tag{2.13}$$

Following the same approach as for (2.9), a bivariate form can be defined to be

$$\hat{x}_1(t_1, t_2) = \cos\left(2\pi f_0 t_1 + k\cos(2\pi f_2 t_2)\right), \quad \text{with } x(t) = \hat{x}_1(t, t). \tag{2.14}$$

Note that $\hat{x}_1$ is periodic in t_1 and t_2, hence $x(t)$ is quasiperiodic with frequencies

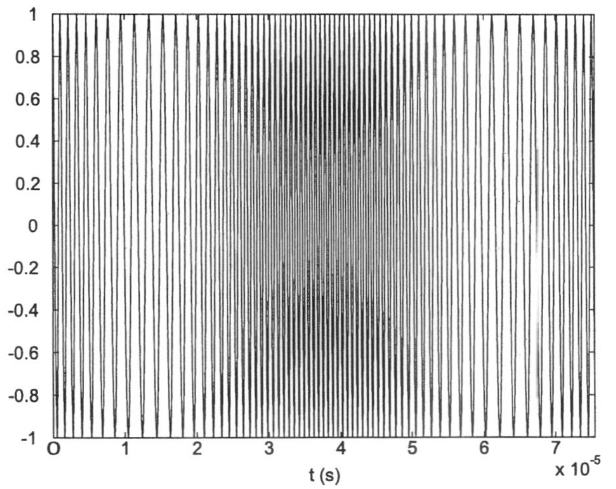

Figure 2.6 An FM signal $x(t) = \cos(2\pi f_0 t + k\cos(2\pi f_2 t))$, with $f_0 \gg f_2$, $f_0 = 1\,\text{MHz}$, $f_2 = 20\,\text{kHz}$, and modulation index $k = 8\pi$.

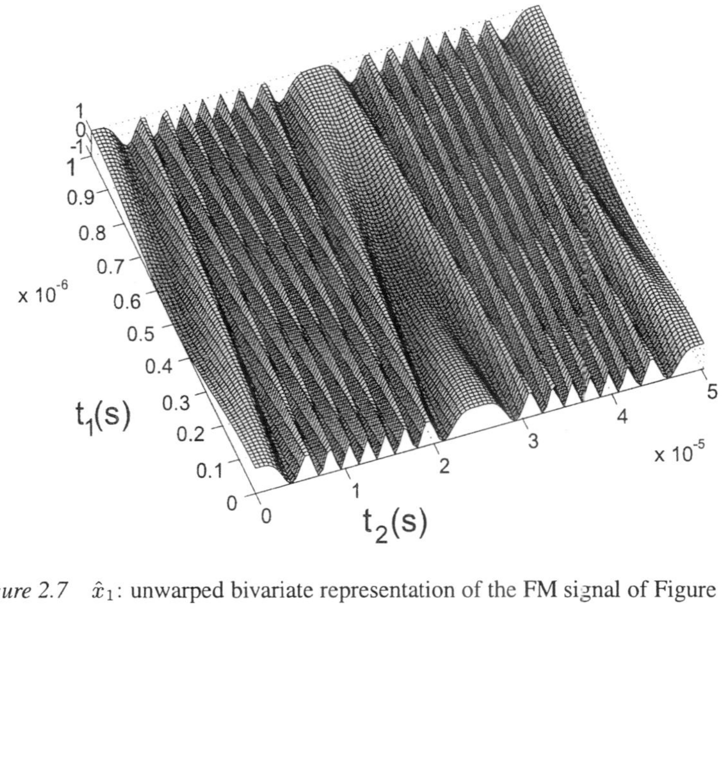

Figure 2.7 $\hat{x}_1$: unwarped bivariate representation of the FM signal of Figure 2.6.

f_0 and f_2. Unfortunately, $\hat{x}_1(t_1, t_2)$, illustrated in Figure 2.7, is not a simple surface with only a few undulations like Figure 2.3. When $k \gg 2\pi$, *i.e.*, $k \approx 2\pi m$ for some large integer m, then $\hat{x}_1(t_1, t_2)$ will undergo about m oscillations as a function of t_2 over one period T_2. In practice, k is often of the order of $f_0/f_2 \gg 2\pi$, hence this number of undulations can be very large. Therefore it becomes difficult to represent $\hat{x}_1$ efficiently by sampling on a two-dimensional grid. It turns out that resolving this problem requires the stretching, or warping, of one of the time scales. We illustrate this by returning to the prototype FM signal of (2.12). Consider the following new multivariate representation

$$\hat{x}_2(\tau_1, \tau_2) = \cos(2\pi\tau_1), \tag{2.15}$$

together with the warping function

$$\phi(\tau_2) = f_0\tau_2 + \frac{k}{2\pi}\cos(2\pi f_2 \tau_2). \tag{2.16}$$

We now retrieve our one-dimensional FM signal (*i.e.*, (2.12)) as

$$x(t) = \hat{x}_2(\phi(t), t). \tag{2.17}$$

Note that both $\hat{x}_2$ and ϕ, given in (2.15) and (2.16), can be easily represented with relatively few samples, unlike $\hat{x}_1$ in (2.14). What we have achieved with (2.16) is simply a stretching of the time axis differently at different times, to even out the period of the fast undulations in Figure 2.6. The extent of the stretching, or the derivative of $\phi(\tau_2)$, at a given point is simply the local frequency $\omega(\tau_2)$, which modifies the original MPDE to result in the **W**arped **M**ultirate **P**artial **D**ifferential **E**quation (WaMPDE) :

$$\omega(\tau_2)\frac{\partial q(\hat{x})}{\partial \tau_1} + \frac{\partial q(\hat{x})}{\partial \tau_2} + f(\hat{x}(\tau_1, \tau_2)) = b(\tau_2). \tag{2.18}$$

The usefulness of (2.18) lies in that specifying

$$x(t) = \hat{x}(\phi(t), t), \quad \phi(t) = \int_0^t \omega(\tau_2)\, d\tau_2 \tag{2.19}$$

results in $x(t)$ being a solution to the DAE given in (2.1). Furthermore, when (2.18) is solved numerically, the local frequency $\omega(\tau_2)$ is also obtained, which is desirable for applications such as VCOs and also difficult to obtain by any other means.

As an example, Figure 2.8 shows the changing local frequency in an LC tank VCO simulated with WaMPDE-based numerical techniques. The controlling input to the VCO was about 30 times slower than its nominal frequency. Figure 2.9 depicts the bivariate waveform of the voltage over the capacitor of the LC tank of the VCO. It is seen that the controlling voltage changes not only the local frequency, but also the amplitude and shape of the oscillator waveform.

The circuit was also simulated by traditional numerical ODE methods ("transient analysis") . The waveform from this simulation, together with the one-dimensional waveform obtained by applying (2.19) to Figure 2.9, are shown in Figure 2.10. Frequency modulation can be observed in the varying density of the undulations.

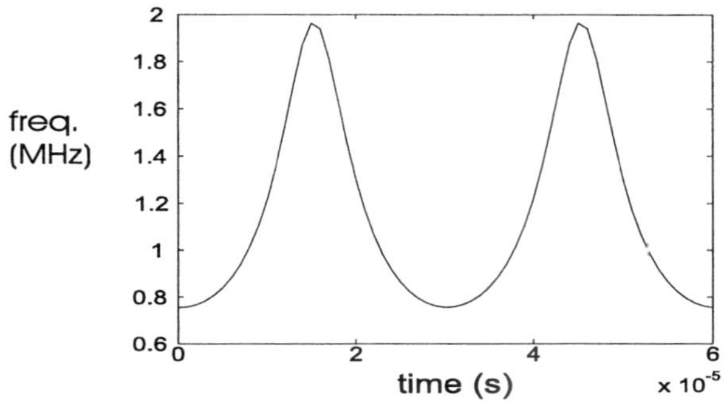

Figure 2.8 Frequency modulation of a VCO: oscillation frequency as a function of time.

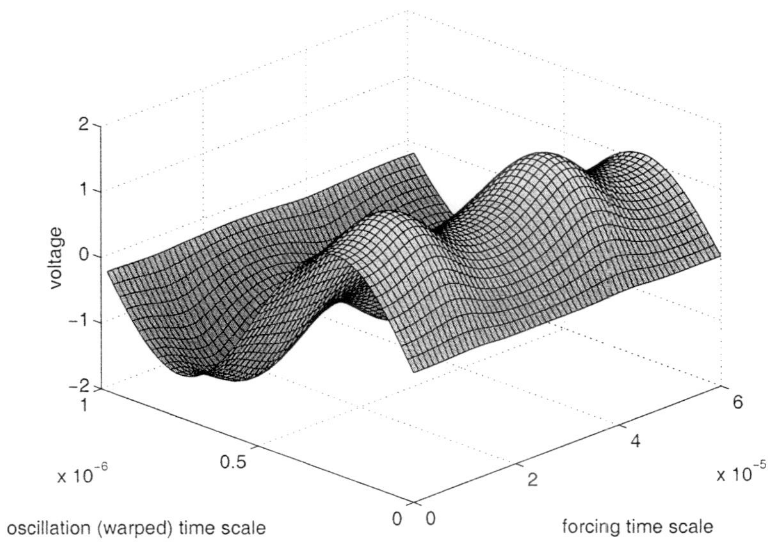

Figure 2.9 Bivariate representation of the capacitor voltage of an LC-tank VCO.

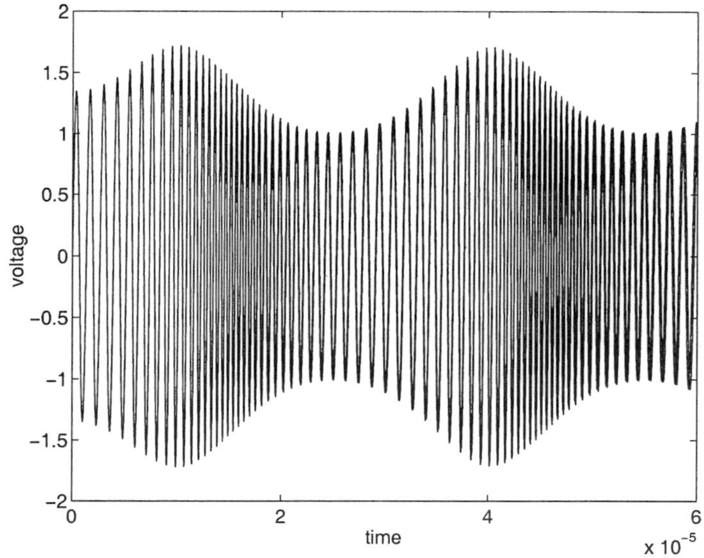

Figure 2.10 WaMPDE versus transient simulation for a VCO: almost no difference can be seen.

2.2.2 Macromodelling time-varying systems

Another useful application of multiple time scales is in macromodelling linear time-varying (LTV) systems [24, 25]. LTV approximations are adequate for many apparently nonlinear systems, like mixers and switched-capacitor filters, where the signal path is designed to be linear, even though other inputs (*e.g.*, local oscillators, clocks) cause "nonlinear" parametric changes to the system. LTV approximations of large systems with few inputs and outputs are particularly useful, because it is possible to automatically generate *macromodels* or *reduced-order models* of such systems. The macromodels are much smaller dynamical systems than the originals, but retain similar input-output behaviour to within a given accuracy. Such macromodels are useful in verifying systems hierarchically at different levels of abstraction, an important task in communication system design.

While mature techniques are available for the simpler task of reduced-order modelling of linear time-invariant (LTI) systems (*e.g.*, [26, 27, 28, 29, 30, 31, 32]), a difficulty in extending them to handle LTV systems has been the interference of the time-variations of the system and the input. By separating the two with artificial time variables, the MPDE provides an elegant solution to this problem.

The time-varying small-signal equations obtained by linearizing (2.1) around a steady-state solution are given by:

$$
\begin{aligned}
C(t)\dot{x}(t) + G(t)x(t) &= r\,u(t) \\
y(t) &= d^T x(t)
\end{aligned}
\tag{2.20}
$$

In (2.20), the input to the system is the scalar $u(t)$, while the output is $y(t)$. If the above equation is Laplace-transformed (following the LTI procedure), the system time variation in $C(t)$ and $G(t)$ interferes with the I/O time variation through a convolution. The LTV transfer function $H(t, s)$ is therefore hard to obtain; this is the difficulty alluded to earlier. The problem can be avoided by casting (2.20) as an MPDE:

$$\begin{aligned} C(t_1)\left[\frac{\partial \hat{x}}{\partial t_1}(t_1, t_2) + \frac{\partial \hat{x}}{\partial t_2}(t_1, t_2)\right] + G(t_1)\hat{x}(t_1, t_2) &= r\, u(t_2) \\ \hat{y}(t_1, t_2) = d^T \hat{x}(t_1, t_2), \qquad y(t) &= \hat{y}(t, t) \end{aligned} \tag{2.21}$$

Notice that the input and system time variables are now separated. By taking Laplace transforms in t_2 and eliminating $\hat{x}$, the time-varying transfer function $H(t_1, s)$ is obtained:

$$Y(t_1, s) = \underbrace{\left\{ d^T \left[C(t_1) \left\{ \frac{\partial}{\partial t_1} + s \right\} + G(t_1) \right]^{-1} [r] \right\}}_{H(t_1, s)} U(s) \tag{2.22}$$

Observe that $H(t_1, s)$ in (2.22) is periodic in t_1; hence, discretizing the t_1 axis, it can also be represented as *several* time-invariant transfer functions $H_i(s) = H(t_{1_i}, s)$. Or, a frequency-domain discretization using harmonics of the t_1-variation can be used. Once an equivalent system of LTI transfer functions has been obtained, existing reduced-order modelling techniques for LTI systems can be used to find a smaller system of equations, in the same form as (2.21), that have the same input-output relationship to within a given accuracy.

The reduced-order modelling technique (dubbed *Time-Varying Padé*, or TVP) was run on a RFIC I-channel mixer circuit of size about $n = 360$ nodes, excited by a local oscillator at 178 MHz [25]. A frequency-domain discretization of the t_1 axis in (2.21) was employed in the model reduction process. Figure 2.11 shows frequency plots of $H_1(s)$, the upconversion transfer function (the first harmonic *wrt* t_1 of $H(t_1, s)$). The points marked '+' were obtained by direct computation of the full system, while the lines were computed using the reduced models of size $q = 2$ and $q = 10$, respectively[1]. Even with $q = 2$, a size reduction of two orders of magnitude, the reduced model provides a good match up to the LO frequency. When the order of approximation is increased to 10, the reduced model is identical upto well beyond the LO frequency. The reduced models were more than three orders of magnitude faster to evaluate than the original system, hence they are useful for system-level verification.

The poles of the reduced models for $H_1(s)$, easily calculated on account of their small size, are shown in Table 2.1. These are useful in design because they constitute excellent approximations of the full system's poles, which are difficult to determine otherwise.

[1]The order q of the reduced model is the number of state variables in its differential equation description.

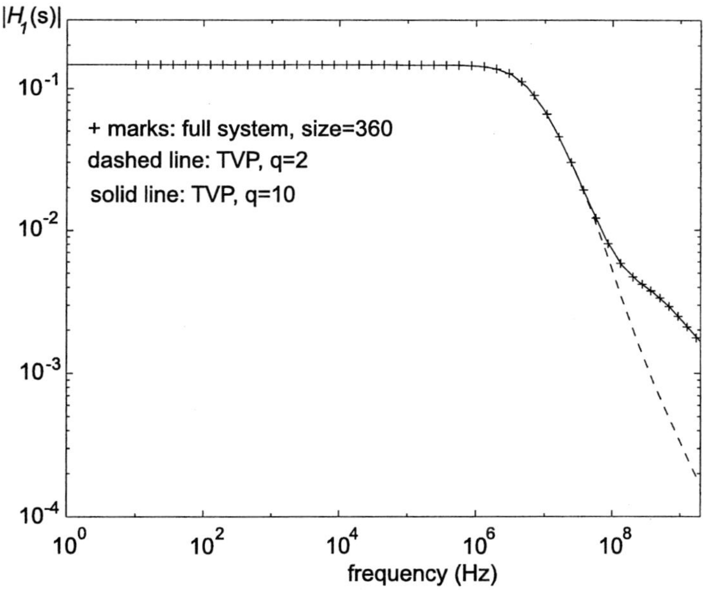

Figure 2.11 Upconversion transfer function $H_1(s)$ of an I-channel mixer : reduced system versus full system.

Table 2.1 Poles of $H_1(s)$ for the I-channel mixer.

TVP, $q = 2$	TVP, $q = 10$
-5.3951e+06	-5.3951e+06
-6.9196e+07 - j3.0085e+05	-9.4175e+06
	-1.5588e+07 - j2.5296e+07
	-1.5588e+07 + j2.5296e+07
	-6.2659e+08 - j1.6898e+06
	-1.0741e+09 - j2.2011e+09
	-1.0856e+09 + j2.3771e+09
	-7.5073e+07 - j1.4271e+04
	-5.0365e+07 + j1.8329e+02
	-5.2000e+07 + j7.8679e+05

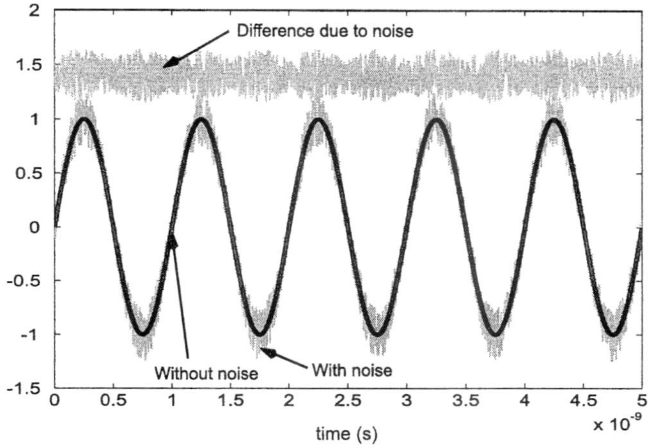

Figure 2.12 Time-domain view of mixing noise.

2.3 NOISE IN RF DESIGN

Predicting noise correctly in order to minimize its impact is central to RF design. Traditional circuit noise analysis is based on three assumptions: that noise sources and their effects are *small enough* not to change the operating point, that all noise sources are *stationary*, and that the small-signal linearization of the circuit is *time-invariant*. These assumptions break down when there are large signal variations, as is typical in RF circuits. Because of changing operating points, small-signal linearizations do not remain constant but become *time-varying*. In addition, noise sources that depend on operating point parameters (such as shot noise and flicker noise) also vary with time and no longer remain stationary. Finally, even though noise sources remain small, their impact upon circuit operation may or may not. In non-autonomous (driven) circuits, circuit effects of small noise remain small, allowing the use of linearized *mixing noise* analysis. In autonomous circuits (oscillators), however, noise creates frequency changes that lead to large deviations in waveforms over time – this phenomenon is called *phase noise*. Because of this, analysis based on linearization is not correct, and nonlinear analysis is required.

Figure 2.12 illustrates mixing noise. A periodic noiseless waveform in a circuit is shown as a function of time. The presence of small noise corrupts the waveform, as indicated. The extent of corruption at any given time remains small, as shown by the third trace, which depicts the difference between the noiseless and noisy waveforms. The noise power can, however, vary depending on the large signal swing, as indicated by the roughly periodic appearance of the difference trace – depicting *cyclostationary* noise, where the statistics of the noise are periodic.

Figure 2.13 illustrates oscillator phase noise. Note that the noisy waveform's frequency is now slightly different from the noise-free one's, leading to increasing deviations between the two with the progress of time. As a result, the difference between the two does not remain small, but reaches magnitudes of the order of the large signal

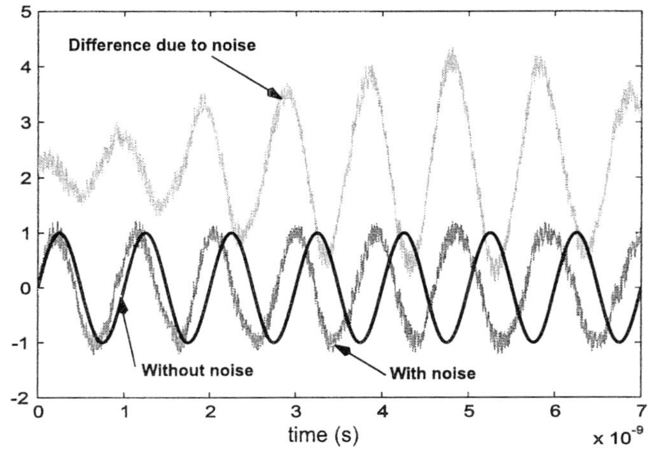

Figure 2.13 Time-domain view of phase noise.

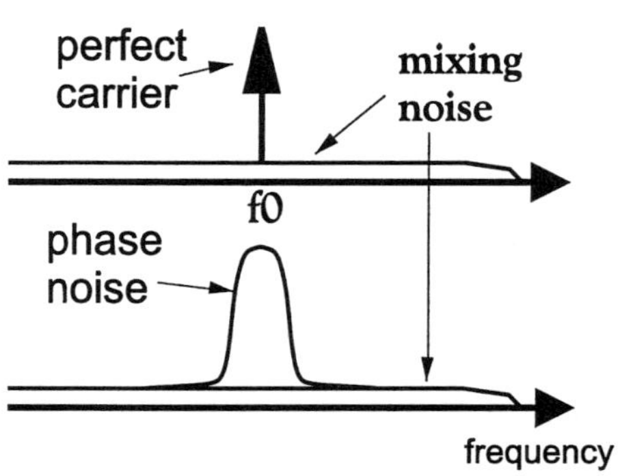

Figure 2.14 Frequency-domain view of mixing and phase noise.

itself. Small additive corruptions remain here also just as in the mixing noise case, but the main distinguishing characteristic of oscillator noise are the frequency deviations.

The difference between mixing and phase noise is also apparent in the frequency domain, shown in Figure 2.14. Noise-free periodic waveforms appear as the impulse in the upper graph. If this is corrupted by small mixing noise, the impulse is not modified, but a small, possibly broadband, noise floor appears. In the case of free-running oscillators, the impulse disappears in the presence of any noise, no matter how small. It is replaced by a continuous spectrum that peaks at the oscillation frequency, and retains the power of the noise-free signal. The width and shape of this *phase noise spectrum* (*i.e.*, the spread of power over neighbouring frequencies) is related to the amount and nature of noise in the circuit.

2.3.1 Mixing noise

Correct calculation of noise in nonlinear circuits with large signal swings (*e.g.*, mixers, gain-compressed amplifiers) requires a sufficiently powerful stochastic process model. In the following, we use cyclostationary time-domain processes (*e.g.*, [33, 34, 35, 36]), although a different but equivalent formulation, *i.e.*, that of correlated processes in the frequency domain (*e.g.*, [37, 4]), is often used. The statistics of cyclostationary processes (in particular the second-order statistics) are periodic or quasiperiodic, hence can be expressed in Fourier series. The coefficients of the Fourier series, termed *cyclostationary components*, capture the variations of noise power over time. The DC term of the Fourier series, or the *stationary* component, is typically the most relevant for design, since it captures the average noise power over a long time. It is important to realize, though, that calculating the correct value of the stationary component of noise over a circuit does require *all* the Fourier components to be properly accounted for. Basing calculations only on the stationary component at each node or branch current in the circuit will, in general, produce wrong results. This is analogous to computing the DC term of the product of two sinusodial waveforms by simply multiplying the DC terms of each.

We motivate the need for cyclostationary analysis with an example. The circuit of Figure 2.15 consists of a mixer, followed by a bandpass filter, followed by another mixer. This is a simplification of, e.g., the bias-dependent noise generation mechanism in semiconductor devices [38]. Both mixers multiply their inputs by a local oscillator of frequency f_0, i.e., by $\cos(2\pi f_0 t)$. The bandpass filter is centered around f_0 and has a bandwidth of $B \ll f_0$. The circuit is noiseless, but the input to the first mixer is stationary band-limited noise with two-sided bandwidth B.

A naïve attempt to determine the output noise power would consist of the following analysis, illustrated in Figure 2.15. The first mixer shifts the input noise spectrum by $\pm f_0$ and scales it by $1/4$. The resulting spectrum is multiplied by the squared magnitude of the filter's transfer function. Since this spectrum falls within the passband of the filter, it is not modified. Finally, the second mixer shifts the spectrum again by $\pm f_0$ and scales it by $1/4$, resulting in the spectrum with three components shown in the figure. The total noise power at the output, i.e., the area under the spectrum, is $1/4$ that at the input.

This common but simplistic analysis is inconsistent with the following alternative argument. Note that the bandpass filter, which does not modify the spectrum of its input, can be ignored. The input then passes through only the two successive mixers, resulting in the output noise voltage $o(t) = i(t)\cos^2(2\pi f_0 t)$. The output power is

$$o^2(t) = i^2(t)\left[\frac{3}{8} + \frac{\cos(2\pi\, 2 f_0 t) + \cos(2\pi\, 4 f_0 t)}{2}\right]. \tag{2.23}$$

The average output power consists of only the $3/8\, i^2(t)$ term, since the cosine terms time-average to zero. Hence the average output power is $3/8$ of the input power, 50% more than that predicted by the previous naïve analysis. This is, however, the correct result.

The contradiction between the arguments above underscores the need for cyclostationary analysis. The auto-correlation function of any cyclostationary process $z(t)$ (defined as $R_{zz}(t,\tau) = E\left[z(t)z(t+\tau)\right]$, $E[\cdot]$ denoting expectation) can be expanded in a Fourier series in t:

$$R_{zz}(t,\tau) = \sum_{i=-\infty}^{\infty} R_{z_i}(\tau)\, e^{j i 2\pi f_0 t}. \tag{2.24}$$

$R_{z_i}(\tau)$ are termed *harmonic autocorrelation functions*. The periodically time-varying power of $z(t)$ is its autocorrelation function evaluated at $\tau = 0$, i.e., $R_{zz}(t, 0)$. The quantities $R_{z_i}(0)$ represent the harmonic components of the periodically-varying power. The average power is simply the value of the DC or *stationary component*, $R_{z_0}(0)$[2]. The frequency-domain representation of the harmonic autocorrelations are termed *harmonic power spectral densities* (HPSDs) $S_{z_i}(f)$ of $z(t)$, defined as the Fourier transforms:

$$S_{z_i}(f) = \int_{-\infty}^{\infty} R_{z_i}(\tau)\, e^{-j\pi f \tau}\, d\tau. \tag{2.25}$$

Equations can be derived that relate the HPSDs at the inputs and outputs of various circuit blocks. By solving these equations, any HPSDs in the circuit can be determined.

Consider, for example, the circuit in Figure 2.15. The input and output HPSDs of a perfect cosine mixer with unit amplitude can be shown [39] to be related by:

$$\begin{aligned} S_{v_k}(f) &= \frac{S_{u_{k-2}}(f - f_0)}{4} + \frac{S_{u_k}(f - f_0) + S_{u_k}(f + f_0)}{4} \\ &\quad + \frac{S_{u_{k+2}}(f + f_0)}{4}, \end{aligned} \tag{2.26}$$

u and v denoting the input and output, respectively. The HPSD relation for a filter with transfer function $H(f)$ is given by [39]

$$S_{v_k}(f) = H(-f)H(f + k f_0)\, S_{u_k}(f). \tag{2.27}$$

[2] *Stationary* processes are a special case of cyclostationary processes, where the autocorrelation function (hence the power) is independent of the time t; it follows that $R_{z_i}(\tau) \equiv 0$ if $i \neq 0$.

The HPSDs of the circuit are illustrated in Figure 2.16. Since the input noise $i(t)$ is stationary, its only nonzero HPSD is the stationary component $S_{i_0}(f)$, assumed to be unity in the frequency band $[-B/2, B/2]$, as shown. From equation 2.25 applied to the first mixer, *three* nonzero HPSDs (S_{x_0}, S_{x_2} and $S_{x_{-2}}$, shown in the figure) are obtained for $x(t)$. These are generated by shifting the input PSD by $\pm f_0$ and scaling by $1/4$; in contrast to the naïve analysis, the stationary HPSD is not the only spectrum used to describe the upconverted noise. From equation 2.27, it is seen that the ideal bandpass filter propagates the three HPSDs of $x(t)$ unchanged to $y(t)$. Through equation 2.25, the second mixer generates five nonzero HPSDs, of which only the stationary component $S_{o_0}(f)$ is shown in the figure. This is obtained by scaling and shifting not only the stationary HPSD of $y(t)$, but also the cyclostationary HPSDs, which in fact contribute an extra $1/4$ to the lobe centered at zero. The average output noise (the shaded area under $S_{o_0}(f)$) equals $3/8$ of the input noise.

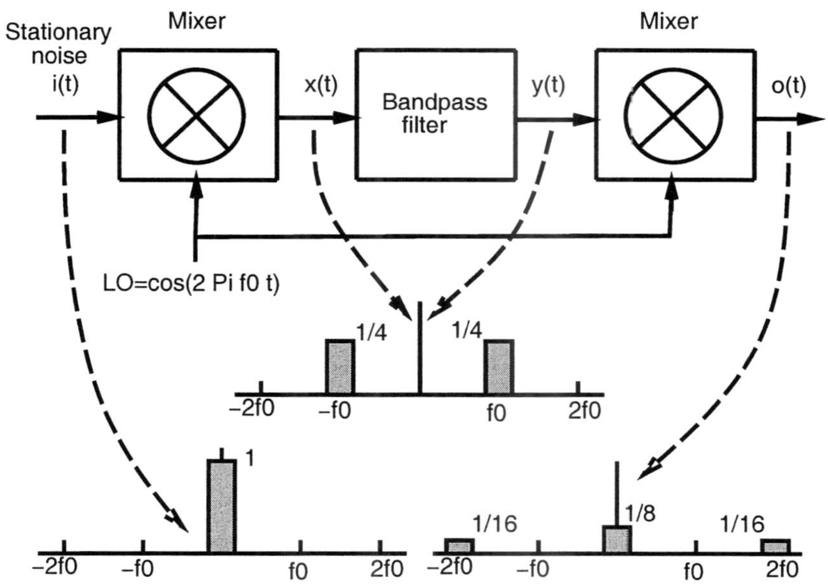

Figure 2.15 Mixer-filter-mixer circuit: naïve analysis.

We now sketch the general procedure for analyzing mixing noise in circuits. The noise sources within a circuit can be represented by a small additive term $Au(t)$ to (2.1), where $u(t)$ is a vector of noise sources, and A an incidence matrix capturing their connections to the circuit. (2.1) is first solved for a (quasi-)periodic steady state in the absence of noise, and then linearized as in (2.20), to obtain

$$C(t)\,\dot{x} + G(t)\,x + A\,u(t) = 0 \tag{2.28}$$

where $x(t)$ represents the small-signal deviations due to noise. Equation 2.28 describes a linear periodically time-varying (LPTV) system with input $u(t)$ and output $x(t)$. The system can be characterized by its time-varying transfer function $H(t, f)$.

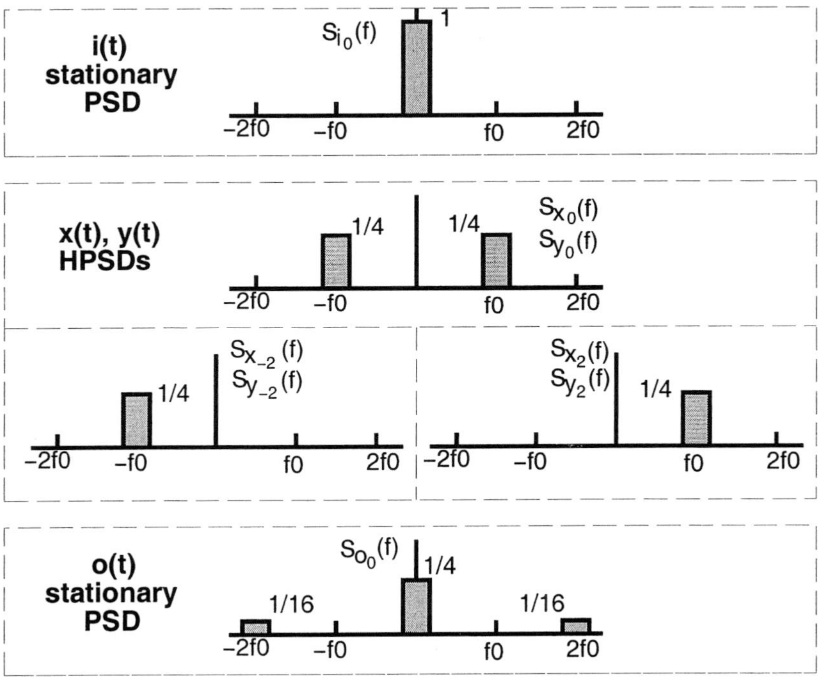

Figure 2.16 HPSDs of mixer-filter-mixer circuit.

This transfer function is periodic in t and can be expanded in a Fourier series similar to equation 2.24. We denote the Fourier components (*harmonic transfer functions*) by $H_i(f)$, .

Since $u(t)$ and $x(t)$ are vectors, their autocorrelation functions are *matrices* $R_{zz}(t,\tau) = E\left[z(t)z^T(t+\tau)\right]$, consisting of auto- and cross-correlations. Similarly, the HPSDs $S_{z_i}(f)$ are also matrices. It can be shown [36] that the HPSD matrices of y and u are related by:

$$\mathcal{S}_{xx}(f) = \mathcal{H}(f)\,\mathcal{S}_{uu}(f)\,\mathcal{H}^*(f). \tag{2.29}$$

Here $\mathcal{H}(f)$ (the *conversion matrix*) is the following block-structured matrix (f^k denotes $f + kf_0$):

$$\mathcal{H}(f) = \begin{pmatrix} & \vdots & \vdots & \vdots & \\ \cdots & H_0(f^1) & H_1(f^0) & H_2(f^{-1}) & \cdots \\ \cdots & H_{-1}(f^1) & H_0(f^0) & H_1(f^{-1}) & \cdots \\ \cdots & H_{-2}(f^1) & H_{-1}(f^0) & H_0(f^{-1}) & \cdots \\ & \vdots & \vdots & \vdots & \end{pmatrix} \tag{2.30}$$

$\mathcal{S}_{uu}(f)$ and $\mathcal{S}_{xx}(f)$ are similar to $\mathcal{H}(f)$: their transposes $\mathcal{S}_{zz}^T(f)$ have the same structure, but with $H_i(f^k)$ replaced by $S_{z_i}^T(f^k)$.

Equation 2.29 expresses the output HPSDs, contained in $\mathcal{S}_{xx}(f)$, in terms of the input HPSDs (contained in $\mathcal{S}_{uu}(f)$) and the harmonic transfer functions of the circuit (contained in $\mathcal{H}(f)$). The HPSDs of a single output variable $x_p(t)$ (both auto- and cross-terms with all other output variables) are available in the pth column of the central block-column of $\mathcal{S}_{xx}^T(f)$. To pick this column, $\mathcal{S}_{xx}^T(f)$ is applied to a unit vector E_{0p}, as follows ($\overline{}$ denotes the conjugate):

$$\mathcal{S}_{xx}^T(f)\,E_{0p} = \overline{\mathcal{H}}(f)\,\mathcal{S}_{uu}^T(f)\,\mathcal{H}^T(f)\,E_{0p} \tag{2.31}$$

Evaluating equation 2.31 involves two kinds of matrix-vector products: $\mathcal{H}(f)\,z$ and $\mathcal{S}_{uu}(f)\,z$ for some vectors z. Consider the latter product first. If the inputs $u(t)$ are stationary, as can be assumed without loss of generality [39], then $\mathcal{S}_{uu}(f)$ is block-diagonal. In practical circuits, the inputs $u(t)$ are either uncorrelated or sparsely correlated. This results in each diagonal block of $\mathcal{S}_{uu}(f)$ being either diagonal or sparse. In both cases, the matrix-vector product can be performed efficiently.

The product with $\mathcal{H}(f)$ can also be performed efficiently, by exploiting the relation $\mathcal{H}(f) = J^{-1}(f)\,\mathcal{A}$ [3]. $\mathcal{A}$ is a sparse incidence matrix of the device noise generators, hence its product with a vector can be computed efficiently. $J(0)$ is the harmonic balance Jacobian matrix [8] at the large-signal solution $x^*(t)$. $J(f)$ is obtained by replacing kf_0 by $kf_0 + f$ in the expression for the Jacobian. The product $J^{-1}z$ can therefore be computed efficiently using the fast techniques outlined in Section 2.1.2. As a result, equation 2.31 can be computed efficiently for large circuits to provide the auto- and cross-HPSDs of any output of interest.

As an example, a portion of the Lucent W2013 RFIC, consisting of an I-channel buffer feeding a mixer, was simulated using (2.31). The circuit consisted of about 360 nodes, and was excited by two tones — a local oscillator at 178 MHz driving the mixer, and a strong RF signal tone at 80 kHz feeding into the I-channel buffer. Two noise analyses were performed. The first analysis included both LO and RF tones (sometimes called a three-tone noise analysis). The circuit was also analysed with only the LO tone to determine if the RF signal affects the noise significantly. The two-tone noise simulation, using a total of 525 large-signal mix components, required 300MB of memory and for each frequency point, took 40 minutes on an SGI machine (200 MHz R10000 CPU). The one-tone noise simulation, using 45 harmonics, needed 70MB of memory and took 2 minutes per point.

The stationary PSDs of the mixer output noise for the two simulations are shown in Figure 2.17. It can be seen that the presence of the large RF signal increases the noise by about $1/3$. This is due to *noise folding*, the result of devices being driven into nonlinear regions by the strong RF input tone. The peaks in the two waveforms, located at the LO frequency, are due to up- and down-conversion of noise from other frequencies.

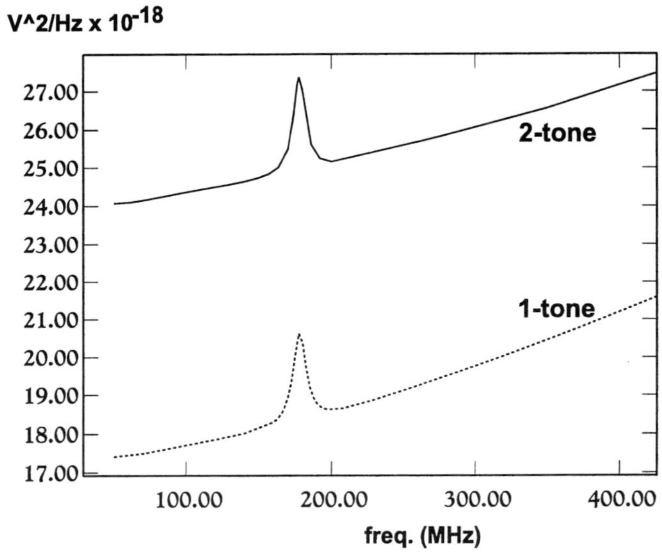

Figure 2.17 Stationary PSDs for the I-Q mixer/buffer circuit.

2.3.2 Phase Noise

Even small noise in an oscillator leads to dramatic changes in its frequency spectrum and timing properties, *i.e.*, to phase noise. This effect can lead to interchannel interference and increased bit-error-rates (BER) in RF communication systems. Another manifestation of the same phenomenon, jitter, is important in clocked and sampled-data systems: uncertainties in switching instants caused by noise can affect synchronisation.

Although a large body of literature is available on phase noise[3], treatments of the phenomenon from the design perspective have typically been phenomenological, *e.g.*, the well-known treatment of Leeson [41]. Most analyses have been based on linear time-invariant or time-varying approaches, which, though providing useful design guidelines, contain qualitative inaccuracies – *e.g.*, they can predict infinite noise power. Recently, however, the work of Kärtner [42] and Demir et al [40] have provided a more correct understanding of phase noise. Here, we sketch the approach in [40].

The starting point for phase noise analysis is the DAE given in equation (2.1), reproduced here for oscillators with no external forcing:

$$\dot{q}(x) + f(x) = 0. \tag{2.32}$$

We assume (2.32) to be the equations for an oscillator with an orbitally stable[4], nontrivial periodic solution, *i.e.*, an oscillation waveform $x_s(t)$. With small noise generators in the circuit, possibly dependent on circuit state, the equation becomes

$$\dot{q}(x) + f(x) = B(x)b(t), \tag{2.33}$$

where $b(t)$ now represents small perturbations.

When $b(t)$ is small, it can be shown [44] that the originally periodic oscillation $x_s(t)$ changes to

$$x(t) = x_s(t + \alpha(t)) + y(t), \tag{2.34}$$

where $y(t)$ remains small, but $\alpha(t)$ (a time/phase deviation) can grow unboundedly with time, no matter how small the perturbation $b(t)$ is (see Figure 2.18). For driven circuits (the mixing noise case) $\alpha(t)$ remains bounded and small and its effects can therefore be lumped into the $y(t)$ term. This is the difference illustrated in Figure 2.12, Figure 2.13 and Figure 2.14. The underlying reason for this difference is that oscillators by their very definition are phase-unstable, hence phase errors build up indefinitely.

Furthermore, it can be shown that $\alpha(t)$ is given by a nonlinear scalar differential equation

$$\dot{\alpha} = v_1^T(t + \alpha(t))B(x_s(t + \alpha(t)))b(t), \tag{2.35}$$

where $v_1(t)$ is a periodic vector function dubbed the Perturbation Projection Vector (PPV). The PPV, which is characteristic of an oscillator in steady state and does not depend on noise parameters, is an important quantity for phase noise calculation. Roughly speaking, it is a "transfer function" that relates perturbations to resulting time or phase jitter in the oscillator. The PPV can be found through only a linear time-varying analysis of the oscillator around its oscillatory solution, and simple techniques to calculate it using HB or shooting are available [45].

---

[3] [40] contains a list of references.

[4] see, *e.g.*, [43] for a precise definition; roughly speaking, an orbitally stable oscillator is one that eventually reaches a unique, periodic waveform with a definite magnitude.

In general, (2.35) can be difficult to solve analytically. When the perturbation $b(t)$ is white noise, however, it can be shown that $\alpha(t)$ becomes a Gaussian random walk process with linearly increasing variance ct, where c is a scalar constant given by

$$c = \frac{1}{T} \int_0^T v_1^T(t) B(x_s(t)) B^T(x_s(t)) v_1(t)\, dt, \tag{2.36}$$

with T the period of the unperturbed oscillation.

This random walk stochastic characterization of the phase error $\alpha(t)$ implies that:

1. the average spread of the jitter (mean-square jitter) increases *linearly* with time, with cT being the jitter per cycle.
2. the spectrum of the oscillator's output, *i.e.*, the power spectrum of $x_s(t+\alpha(t))$, is *Lorentzian*[5] about each harmonic. For example, around the fundamental (with angular frequency $\omega_0 = 2\pi/T$ and power P_{fund}), the spectrum is

$$S_p(f) = P_{\text{fund}} \frac{\omega_0^2 c}{\frac{\omega_0^4 c^2}{4} + (2\pi f - \omega_0)^2}. \tag{2.37}$$

   This means that the spectrum decays as $1/f^2$ beyond a certain knee distance away from the original oscillation frequency and its harmonics, as is well-known for white noise in oscillators [41]. The $1/f^2$ dependence does not, however, continue as $f \to 0$, *i.e.*, close to and at the oscillation frequency; instead, the spectrum reaches a finite maximum value.
3. the oscillator's output is a *stationary* stochastic process.

The Lorentzian shape of the spectrum also implies that the power spectral density at the carrier frequency and its harmonics has a finite value, and that the total carrier power is preserved despite spectral spreading due to noise. Equation (2.35) can also be solved for coloured noise perturbations $b(t)$ [44], and it can be shown that if $S(f)$ is the spectrum of the coloured noise, then the phase noise spectrum generated falls as $S(f)/(f - f_0)^2$, away from f_0.

Numerical methods based on the above insights are available to calculate phase noise. The main effort is calculating the PPV. Once it is known, c can be calculated easily using (2.36) and the spectrum obtained directly from (2.37). The PPV can be found from the time-varying linearization of the oscillator around its steady state. Two numerical methods can be used to find the PPV. The first calculates the time-domain monodromy (or state-transition) matrix of the linearized oscillator explicitly, and obtains the PPV by eigendecomposing this matrix [40]. A more recent method [45] relies on simple postprocessing of internal matrices generated during the solution of the operator's steady state using HB or shooting, and as such, can take advantage of the fast techniques of Section 2.1.2. The separate contributions of noise sources, and the sensitivity of phase noise to individual circuit devices and nodes, can be obtained easily.

[5] A Lorentzian is the shape of the squared magnitude of a one-pole lowpass filter transfer function.

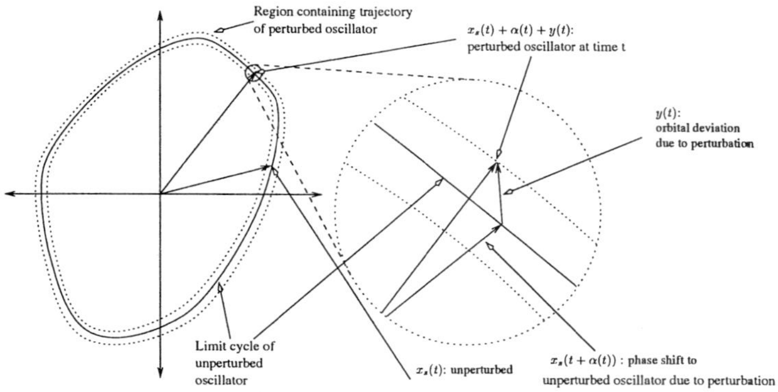

Figure 2.18 Oscillator trajectories.

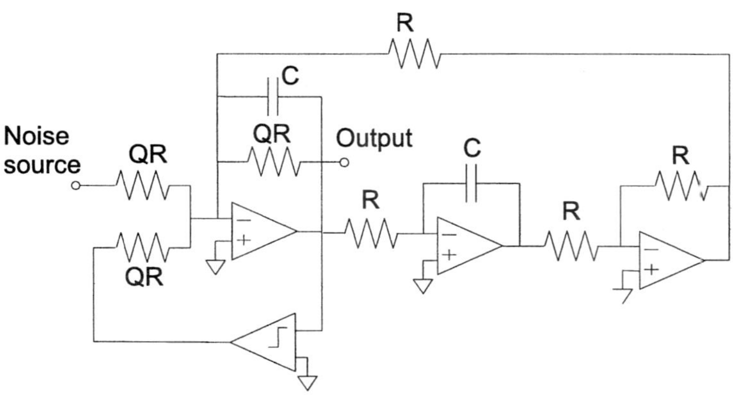

Figure 2.19 Oscillator with a band-pass filter and a comparator [46].

As an example, the oscillator in Figure 2.19 consists of a Tow-Thomas second-order bandpass filter and a comparator [46]. If the opamps are considered to be ideal, it can be shown that this oscillator is equivalent (in the sense of the differential equations that describe it) to a parallel RLC circuit in parallel with a nonlinear voltage-controlled current source (or equivalently a series RLC circuit in series with a nonlinear current-controlled voltage source). In [46], authors breadboarded this circuit with an external white noise source (intensity of which was chosen such that its effect is much larger than the other internal noise sources), and measured the PSD of the output with a spectrum analyzer. For $Q = 1$ and $f_o = 6.66$ kHz, a phase noise characterisation of this oscillator was performed to yield the periodic oscillation waveform $x_s(t)$ for the output and $c = 7.56 \times 10^{-8}$ sec^2.Hz. Figure 2.20 shows the PSD of the oscillator output and Figure 2.21 shows the spectrum analyzer measurement. Figure 2.22 shows a blown up version of the PSD around the first harmonic. The single-sideband phase noise spectrum is in Figure 2.23. The oscillator model that was simulated has two state variables and a single stationary noise source. Figure 2.24 shows a plot of the periodic nonnegative scalar (essentially the squared magnitude of the PPV):

$$v_1^T(t)B(x_s(t))B^T(x_s(t))v_1(t) = (v_1^T(t)B)^2, \tag{2.38}$$

where B is independent of t since the noise source is stationary. Recall that c is the time average of this scalar that is periodic in time.

The value of c can also be obtained relatively accurately in this case using Monte-Carlo analysis (in general, however, Monte-Carlo based on transient simulations can be extremely time-consuming and also inaccurate, and should be avoided except as a sanity check). The circuit was simulated with 10000 random excitations and averaged the results to obtain the mean-square difference between the perturbed and unperturbed systems as a function of time. Figure 2.25 illustrates the result, the slope of the envelope of which determines c. The Monte-Carlo simulations required small timesteps to produce accurate results, since numerical integration methods easily lose accuracy for autonomous circuits.

2.4 CONCLUSIONS

The explosion of the market of wireless digital telecommunications (GSM, DECT, GPS, WLAN, ...) requires low-cost transceiver front-ends with high performance and low power consumption. These front-ends contain RF circuits and baseband analog and digital circuits. An accurate and efficient simulation of such front-ends is beyond the capabilities of SPICE . For example, an analysis of a circuit that is excited by a digitally modulated signal and with parts operating in the GHz range and parts operating at baseband, cannot be simulated efficiently with SPICE. Furthermore, aspects such as mixing noise and phase noise require new advanced analysis techniques. These techniques, together with efficient simulation approaches have been addressed in this chapter. Their use enables an accurate and efficient analysis and simulation of the most important aspects that are of interest in the design of wireless transceiver front-ends.

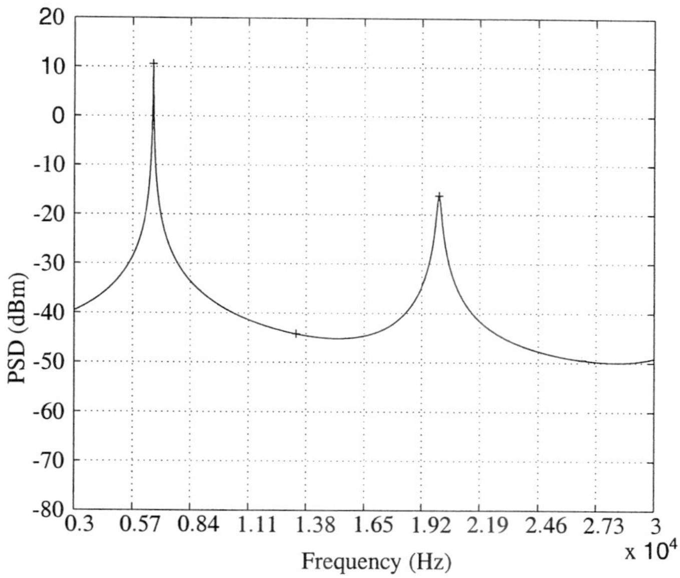

Figure 2.20 Phase noise characterisation for the oscillator in 2.19: computed PSD (4 harmonics).

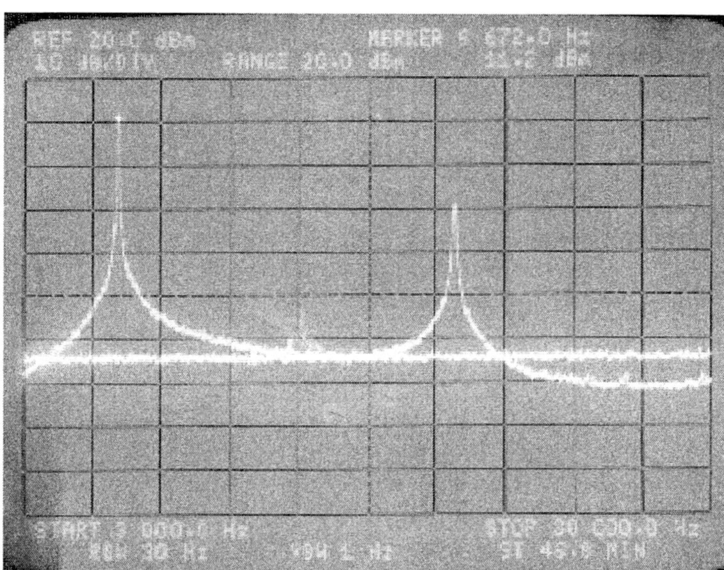

Figure 2.21 Phase noise characterisation for the oscillator in 2.19: spectrum analyzer measured PSD (in dBm) [46].

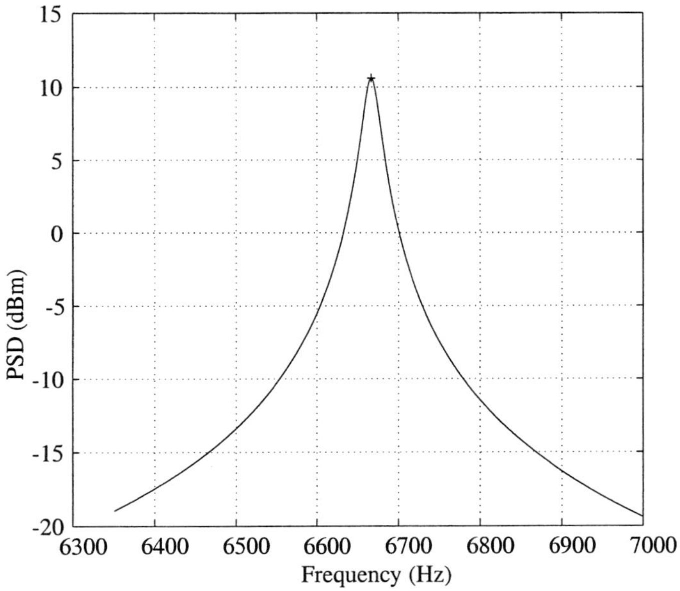

Figure 2.22 Phase noise characterisation for the oscillator in Figure 2.19: computed PSD (first harmonic).

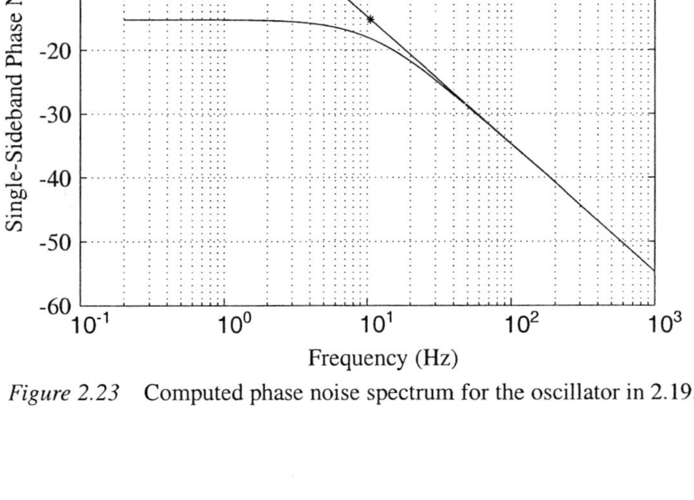

Figure 2.23 Computed phase noise spectrum for the oscillator in 2.19.

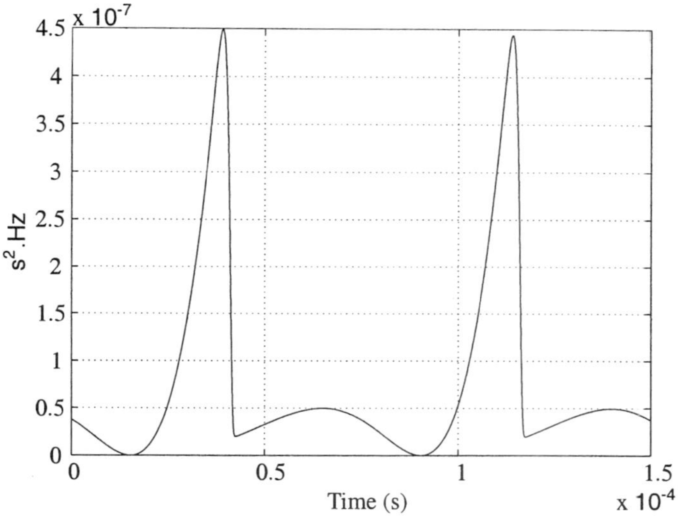

Figure 2.24 Phase noise characterisation for the oscillator in 2.19: $(v_1^T(t)B)^2$.

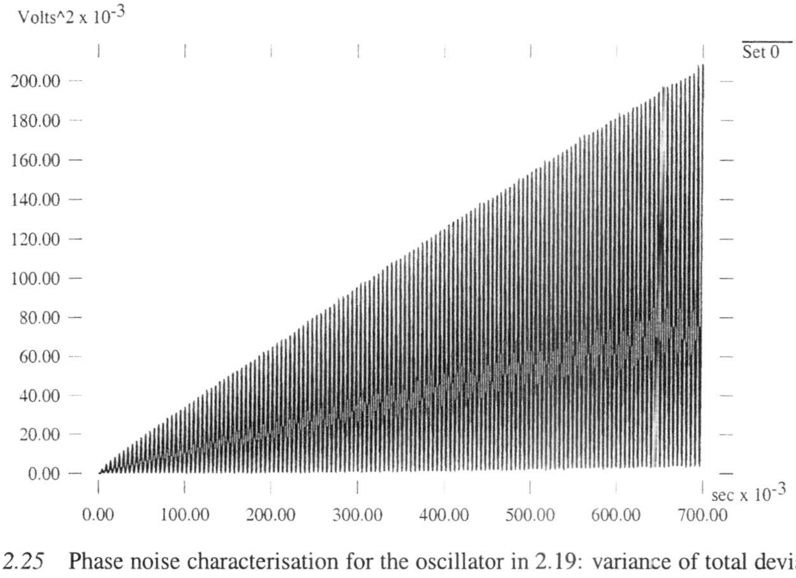

Figure 2.25 Phase noise characterisation for the oscillator in 2.19: variance of total deviation (Monte-Carlo method).

References

[1] K.S. Kundert, J.K. White, and A. Sangiovanni-Vincentelli. *Steady-state methods for simulating analog and microwave circuits*. Kluwer Academic Publishers, 1990.

[2] M.S. Nakhla and J. Vlach. A piecewise harmonic balance technique for determination of periodic responses of nonlinear systems. *IEEE Trans. Ckts. Syst.*, CAS-23:85, 1976.

[3] S.A. Haas. *Nonlinear Microwave Circuits*. Artech House, Norwood, MA, 1988.

[4] V. Rizzoli and A. Neri. State of the art and present trends in nonlinear microwave CAD techniques. *IEEE Trans. MTT*, 36(2):343–365, February 1988.

[5] R.J. Gilmore and M.B. Steer. Nonlinear circuit analysis using the method of harmonic balance – a review of the art. Part I. Introductory concepts. *Int. J. on Microwave and Millimeter Wave CAE*, 1(1), 1991.

[6] Markus Rösch. *Schnell Simulation des stationären Verhaltens nichtlinearer Schaltungen*. PhD thesis, Technischen Universität München, 1992.

[7] R. Mickens. *Oscillations in Planar Dynamic Systems*. World Scientific, 1995.

[8] R.C. Melville, P. Feldmann, and J. Roychowdhury. Efficient multi-tone distortion analysis of analog integrated circuits. In *Proc. IEEE CICC*, pages 241–244, May 1995.

[9] D. Long, R.C. Melville, et al. Full chip harmonic balance. In *Proc. IEEE CICC*, May 1997.

[10] W.H. Press, S.A. Teukolsky, W.T. Vetterling, and B.P. Flannery. *Numerical Recipes – The Art of Scientific Computing*. Cambridge University Press, 1989.

[11] T.J. Aprille and T.N. Trick. Steady-state analysis of nonlinear circuits with periodic inputs. *Proc. IEEE*, 60(1):108–114, January 1972.

[12] S. Skelboe. Computation of the periodic steady-state response of nonlinear networks by extrapolation methods. *IEEE Trans. Ckts. Syst.*, CAS-27(3):161–175, March 1980.

[13] A. Nayfeh and B. Balachandran. *Applied Nonlinear Dynamics*. Wiley, 1995.

[14] R. Telichevesky, K. Kundert, and J. White. Efficient steady-state analysis based on matrix-free krylov subspace methods. In *Proc. IEEE DAC*, pages 480–484, 1995.

[15] P. Feldmann, R.C. Melville, and D. Long. Efficient frequency domain analysis of large nonlinear analog circuits. In *Proc. IEEE CICC*, May 1996.

[16] Y. Saad. *Iterative methods for sparse linear systems*. PWS, Boston, 1996.

[17] R.W. Freund. Reduced-order modeling techniques based on Krylov subspaces and their use in circuit simulation. Technical Report 11273-980217-02TM, Bell Laboratories, 1998.

[18] R.W. Freund. Reduced-order modeling techniques based on Krylov subspaces and their use in circuit simulation. *Applied and Computational Control, Signals, and Circuits, Volume 1*, pages 435–498, 1999.

[19] J. Kevorkian and J.D. Cole. *Perturbation methods in Applied Mathematics*. Springer-Verlag, 1981.

[20] E. Ngoya and R. Larchevèque. Envelop transient analysis: a new method for the transient and steady state analysis of microwave communication circuits and systems. In *Proc. IEEE MTT Symp.*, 1996.

[21] H.G. Brachtendorf, G. Welsch, R. Laur, and A. Bunse-Gerstner. Numerical steady state analysis of electronic circuits driven by multi-tone signals. *Electrical Engineering (Springer-Verlag)*, 79:103–112, 1996.

[22] J. Roychowdhury. Efficient methods for simulating highly nonlinear multi-rate circuits. In *Proc. IEEE DAC*, 1997.

[23] O. Narayan and J. Roychowdhury. Multi-time simulation of voltage-controlled oscillators. In *Proc. IEEE DAC*, New Orleans, LA, June 1999.

[24] J. Phillips. Model Reduction of Time-Varying Linear Systems Using Approximate Multipoint Krylov-Subspace Projectors. In *Proc. ICCAD*, November 1998.

[25] J. Roychowdhury. Reduced-order modelling of time-varying systems. *IEEE Trans. Ckts. Syst. – II: Sig. Proc.*, 46(10), November 1999.

[26] L.T. Pillage and R.A. Rohrer. Asymptotic waveform evaluation for timing analysis. *IEEE Trans. CAD*, 9:352–366, April 1990.

[27] X. Huang, V. Raghavan, and R.A. Rohrer. AWEsim: A program for the efficient analysis of linear(ized) circuits. In *Proc. ICCAD*, pages 534–537, November 1990.

[28] E. Chiprout and M.S. Nakhla. *Asymptotic Waveform Evaluation*. Kluwer, Norwell, MA, 1994.

[29] P. Feldmann and R.W. Freund. Efficient linear circuit analysis by Padé approximation via the Lanczos process. *IEEE Trans. CAD*, 14(5):639–649, May 1995.

[30] P. Feldmann and R.W. Freund. Reduced-order modeling of large linear subcircuits via a block Lanczos algorithm. In *Proc. IEEE DAC*, pages 474–479, 1995.

[31] P. Feldmann and R.W. Freund. Circuit noise evaluation by Padé approximation based model-reduction techniques. In *Proc. ICCAD*, pages 132–138, November 1997.

[32] A. Odabasioglu, M. Celik, and L.T. Pileggi. PRIMA: passive reduced-order interconnect macromodelling algorithm. In *Proc. ICCAD*, pages 58–65, November 1997.

[33] W. Gardner. *Introduction to Random Processes*. McGraw-Hill, New York, 1986.

[34] T. Ström and S. Signell. Analysis of Periodically Switched Linear Circuits. *IEEE Trans. Ckts. Syst.*, CAS-24(10):531–541, Oct 1977.

[35] M. Okumura, H. Tanimoto, T. Itakura, and T. Sugawara. Numerical Noise Analysis for Nonlinear Circuits with a Periodic Large Signal Excitation Including Cyclostationary Noise Sources. *IEEE Trans. Ckts. Syst. – I: Fund. Th. Appl.*, 40(9):581–590, Sep 1993.

[36] J. Roychowdhury, D. Long, and P. Feldmann. Cyclostationary noise analysis of large RF circuits with multitone excitations. *IEEE J. Solid-State Ckts.*, 33:324–336, Mar 1998.

[37] V. Rizzoli, F. Mastri, and D. Masotti. General Noise Analysis of Nonlinear Microwave Circuits by the Piecewise Harmonic-Balance Technique. *IEEE Trans. MTT*, 42(5):807–819, May 1994.

[38] A.R. Kerr. Noise and loss in balanced and subharmonically pumped mixers: Part 1 – Theory. *IEEE Trans. MTT*, MTT-27:938–943, Dec 1979.

[39] J. Roychowdhury and P. Feldmann. A new linear-time harmonic balance algorithm for cyclostationary noise analysis in RF circuits. In *Proc ASP-DAC*, pages 483–492, 1997.

[40] A. Demir, A. Mehrotra, and J. Roychowdhury. Phase noise in oscillators: a unifying theory and numerical methods for characterization. *IEEE Trans. Ckts. Syst. – I: Fund. Th. Appl.*, 47:655–674, May 2000.

[41] D.B. Leeson. A simple model of feedback oscillator noise spectrum. *Proceedings of the IEEE*, 54(2):329, February 1966.

[42] F. Kärtner. Analysis of white and $f^{-\alpha}$ noise in oscillators. *International Journal of Circuit Theory and Applications*, 18:485–519, 1990.

[43] M. Farkas. *Periodic Motions*. Springer-Verlag, 1994.

[44] A. Demir. Phase Noise in Oscillators: DAEs and Colored Noise Sources. In *Proc. ICCAD*, pages 170–177, 1998.

[45] A. Demir, D. Long and J. Roychowdhury. Computing Phase Noise Eigenfunctions Directly from Steady-State Jacobian Matrices. In *Proc. ICCAD*, November 2000.

[46] A. Dec, L. Toth, and K. Suyama. Noise analysis of a class of oscillators. IEEE Transactions on Circuits and Systems, 1998.

3 FAST: AN EFFICIENT HIGH-LEVEL DATAFLOW SIMULATOR OF MIXED-SIGNAL FRONT-ENDS OF DIGITAL TELECOM TRANSCEIVERS

Gerd Vandersteen, Piet Wambacq, Stéphane Donnay, Wolfgang Eberle

IMEC, Heverlee, Belgium

E-mail: [Gerd.Vandersteen, wambacq, donnay, eberle]@imec.be

Yves Rolain

Vrije Universiteit Brussel, Belgium

E-mail: Yves.Rolain@vub.ac.be

Cost-effective implementations of wireless transceivers must have a small size and a low power consumption. To this purpose the degree of integration should be increased. This requires new front-end architectures and will result into ICs which combine analog and digital circuits. The development of front-end architectures is best supported by simulations at the architectural level. Current methodologies and corresponding tools suffer from common drawbacks, such as lack of accuracy, long simulation times, etc. A new methodology has been developed for efficient simulation, at the architectural level, of mixed-signal front-ends of digital telecom transceivers. The efficient execution is obtained using a local multirate, multicarrier signal representation together with a dataflow simulation scheme that dynamically switches to the most efficient signal processing technique available. The methodology has been implemented in the program FAST (<u>F</u>ront-end <u>A</u>rchitecture <u>S</u>imulator for digital <u>T</u>elecom applica-

P. Wambacq et al. (eds.),
Low-Power Design Techniques and CAD Tools for Analog and RF Integrated Circuits, 43–59.

tions). Simulation examples show both excellent runtimes and a high accuracy for realistic front-end architectures.

During the architectural design of the digital blocks of a telecom transceiver, the performance of the complete telecom link is often measured with bit-error-rate (BER) simulations. An optimal design of a complete transceiver requires simulations that take into account the signal degradations caused by the analog front-end blocks. Analog blocks are most often simulated at the transistor level. However, this high level of detail yields a low simulation efficiency compared to the digital blocks, which are most often simulated at a higher level. The simulation efficiency improves if the analog blocks are simulated at a higher abstraction level with dedicated simulation techniques. This chapter describes such technique.

Efficient numerical computations can take advantage of the technique of *vector processing*, which handles data at different timepoints in large vectors instead of processing single data points. This makes it possible to take full advantage of the capabilities of the processor, as is done for example in MATLAB [1]. This technique is useful for front-end simulations at the architectural level, as long as the architectures do not contain any feedback. Although front-ends in essence are feedforward structures, they can contain several feedback paths, which is the case e.g. with a phase-locked loop or an automatic gain control. In feedback loops the signals need to be calculated on a timepoint-by-timepoint or *sample-by-sample* basis. This approach is far less efficient than vector processing.

The most straightforward technique to simulate analog front-ends that contain RF circuits as well as baseband circuits is a SPICE-like time-domain simulation approach that solves a set of nonlinear differential equations with numerical integration (using a non-equidistant timestep). This approach is not efficient for wireless systems where RF frequencies are in the GHz range and the baseband frequencies in the MHz or even in the kHz range. Indeed, this approach requires a timestep that is small enough in order not to introduce aliasing by the sampling of the waveforms. In this way the timestep is upper bounded by the RF signals and lower bounded by the period of the lowest frequency in the simulation. This yields very long simulation results.

A harmonic balance approach does not suffer from this large difference of frequencies and it is sometimes used for system-level simulations of analog front-ends [2]. However, harmonic balance methods are only good at simulating nonlinear circuits with periodic signals that can be described by a small number of sinusoidal tones and their harmonics. A digitally-modulated signal, however, cannot be represented accurately in this way.

The problem of large differences in operating frequency of the front-end blocks is often solved with a complex lowpass signal representation [3]. This is used in tools such as COSSAP [4] and SPW [5] to co-simulate RF blocks and digital blocks with a dataflow approach. With the complex lowpass representation, only in–band distortion is considered for the nonlinear blocks by modeling these blocks with AM/AM and AM/PM characteristics. Modeling in–band distortion only, however, yields inaccurate results when two nonlinearities are cascaded. Indeed, out–of–band distortion generated by the first nonlinear block can be transformed into in-band distortion by the second nonlinearity. New front-end architectures that are investigated to increase the degree of integration, such as zero–IF [6], low–IF [7] or wideband IF double conversion architectures [8], generally contain more cascade connections of active (and hence nonlinear) blocks than classical superheterodyne architectures. Further, in some cases (e.g. in I/Q modulators) PM/AM and PM/PM conversion should be taken into account [9] in addition to AM/AM and AM/PM conversion.

The circuit envelope approach [11, 12] or envelope transient analysis [13] solves the problem of large differences in operating frequency by performing successive harmonic balance analyses. In order to take into account the dynamic effects on the modulation, a time-domain numerical integration method is used to compute the influence of one harmonic balance analysis onto the other. This implies that the original harmonic balance equations are augmented with a transient term. With Ptolemy [11,12] it is even possible to couple different envelope simulation processes. However, since the envelope method is based on the harmonic balance method, it suffers from the same drawbacks when performing system level simulations:

- a large memory usage which is proportional to number of carriers times the number of nodes. This is especially a problem for the simulation of strongly nonlinear behavior.

- a global definition of the simulation frequencies and the number of harmonics. This implies that the simulator cannot take advantage of the fact that some signals in the signal path can be represented by a subset of these simulation frequencies.

Furthermore, for its numerical integration in the time domain the envelope simulator uses a common timestep for all blocks. Hence, it cannot take advantage of a change in the signal bandwidth.

In this chapter a simulation methodology is described that maintains the simulation efficiency of a complex lowpass representation while out-of-band distortion can be modeled as well. Further, the frequencies of the modulated carriers are local quantities, instead of global ones. Also, the timestep is local, in the sense that it is not common for all blocks. The methodology is demonstrated in a dataflow-type simulator called FAST (**F**ront-end **A**rchitecture **S**imulator for digital **T**elecom applications). For the simulation of analog blocks no sets of nonlinear differential network equations need to be solved. Analog continuous-time blocks that are typically described in the frequency domain with linear transfer functions, are automatically translated into a discrete-time model with digital filters. In general, the high-level models of analog blocks can be accurate although their evaluation should not involve iteration at each timepoint. This does not impose a severe restriction on the high-level models, as is explained in Section 3.2. Since FAST is a dataflow simulator it can be coupled with a simulator for digital blocks such as the digital simulation environment OCAPI [10]. This allows efficient bit-error-rate simulations of a complete telecom link with the inclusion of the signal degradations caused by the analog front-ends such as phase noise, nonlinear behavior, I/Q mismatches, ... The ability of taking into account out-of-band distortion with the simulation efficiency of a complex lowpass representation is achieved with a local multirate, multicarrier (MRMC) representation of signals: each signal in a front-end is considered as a set of one or more modulated carriers. These carriers are each represented with a complex lowpass model and with a — possibly different — timestep. The carriers are used locally. This means that carriers that are important at some place in the architecture are no longer considered at places where they are negligible. Also, the simulation timestep is local: it varies throughout the front-end according to the bandwidth of the modulated signals at a given place in the front-end. The MRMC signal representation is discussed in Section 3.3. In Section 3.4 different approaches are explained to compute the response of a front-end block to a MRMC signal.

The methodology explained in this chapter can also be used in other programs than FAST. However, the extra computational blocks that are required to use the MRMC signal representation are generated inside FAST. This is performed during the translation of the architecture into a computational graph, as explained in Section 3.5. Next, in Section 3.6 the scheduler of FAST is presented. This scheduler exploits the nature of a front-end architecture, which is basically a feedforward structure, possibly with some local feedbacks. A coupling with the digital simulation environment OCAPI [10] is explained in Section 3.7. Finally, in Section 3.8 runtime examples are presented on a 5 GHz WLAN receiver front-end (see Figure 3.1) that consists of a superheterodyne stage, followed by a zero-IF stage.

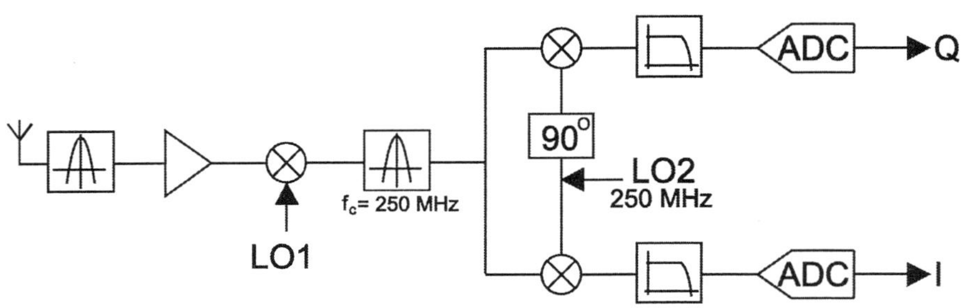

Figure 3.1 A simplified 5 GHz receiver front-end.

3.1 HIGH-LEVEL MODELS OF FRONT-END BLOCKS

FAST is a simulation program for front-end architectures that contain both analog and digital blocks that are described at a high level. Figure 3.2 shows a schematic front-end architecture with some representative blocks. The figure shows for each block some characteristics that are typically modeled in FAST.

Continuous-time blocks are modeled by linear terminal impedances and by a nonlinear controlled voltage or current source at the output, as shown in Figure 3.3. This source is controlled by the input voltage or current. In general the relationship between the controlling and the controlled quantity is a functional **H** that models linear (e.g. a transfer function) or weakly nonlinear behavior (e.g. a Volterra series). In the case of a Volterra series, the functional **H** can be represented by a combination of a small number of scale factors, linear transfer functions and static nonlinearities (multipliers, x^2, x^3) as described in [14]. These blocks can be efficiently evaluated in simulations.

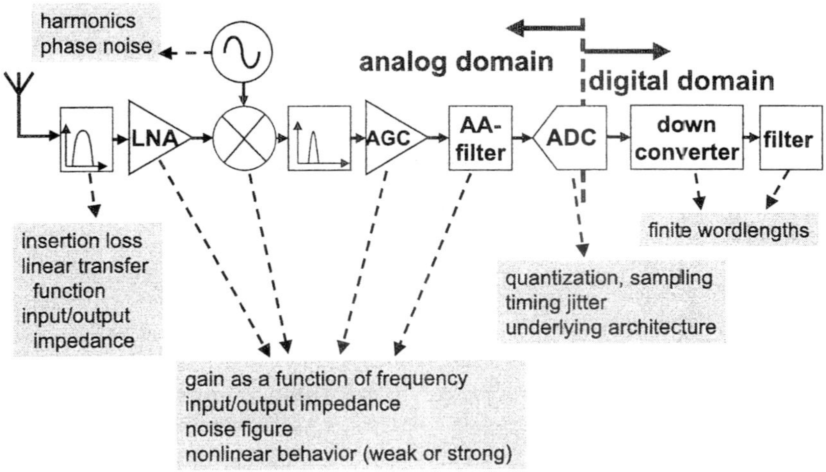

Figure 3.2 Front-end blocks and their characteristics that can be described at a high level.

Impedance loading of one block onto another one can be simulated efficiently when

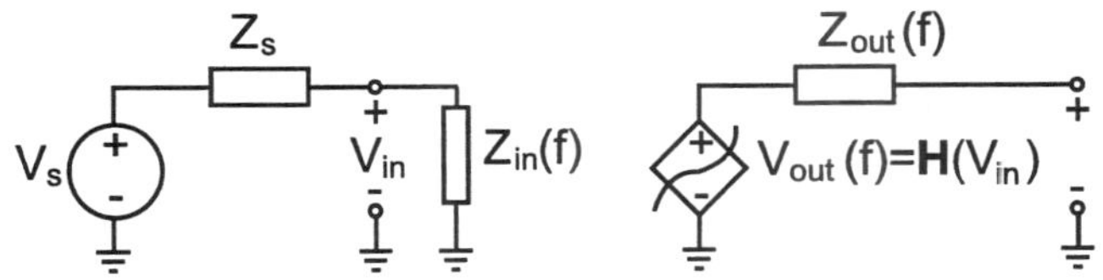

Figure 3.3 Unidirectional high-level model of a continuous-time block. A dual model in terms of admittances and a controlled current source is possible as well.

working with unidirectional models with linear terminal impedances. In this way, the impedance loading effect can be solved explicitly (using simple voltage division) and can be computed using vector processing. If reverse transmission of nonlinear loading is involved, then sample-by-sample processing is required.

3.2 MULTIRATE MULTICARRIER (MRMC) REPRESENTATION OF SIGNALS

The starting point of the MRMC representation in FAST is the complex lowpass equivalent representation of a signal [3]. With this representation a bandpass signal $x(t)$ centered around a frequency f with amplitude and phase modulation has the form

$$x(t) = \frac{1}{2}[x_L(t) \cdot \exp(j2\pi\, ft) \;+\; x_L^*(t) \cdot \exp(-j2\pi\, ft)], \tag{3.1}$$

in which $x_L(t)$ is the complex lowpass signal representation of $x(t)$.

Whereas the complex lowpass representation of equation (3.1) only models one carrier, a practical signal often consists of different bandpass signals, e.g. the modulated carrier of interest and its (modulated) harmonics, or an interfering signal. These out-of-band signals can cause in-band distortion due to nonlinear behavior of a front-end block. In our signal representation out-of-band signals are taken into account. To this purpose, a signal $x(t)$ is considered as being composed of different bandpass signals, which are all represented by a complex lowpass representation $x_{Li}(t)$:

$$x(t) = \frac{1}{2}\sum_{i=1}^{n} [x_{Li}(t) \cdot \exp(j2\pi\, f_i t) \;+\; x^*_{Li}(t) \cdot \exp(-j2\pi\, f_i t)]. \qquad (3.2)$$

Here f_i is the center frequency of the bandpass signal represented by $x_{Li}(t)$. Each $x_{Li}(t)$ is represented as a discrete-time signal with a different timestep T_{si} that is large enough such that $x_{Li}(t)$ is represented without aliasing (see Figure 3.4).

When an MRMC signal is fed into a nonlinearity, then extra modulated carriers are generated. However, not all of these carriers are equally important. Several bands or sub-bands can be neglected since they will finally not yield a significant contribution to the in-band distortion when they are fed into a subsequent nonlinearity. FAST can neglect unimportant frequency bands by using information of a short test simulation with a limited number of input samples. This test simulation does not use yet the MRMC representation.

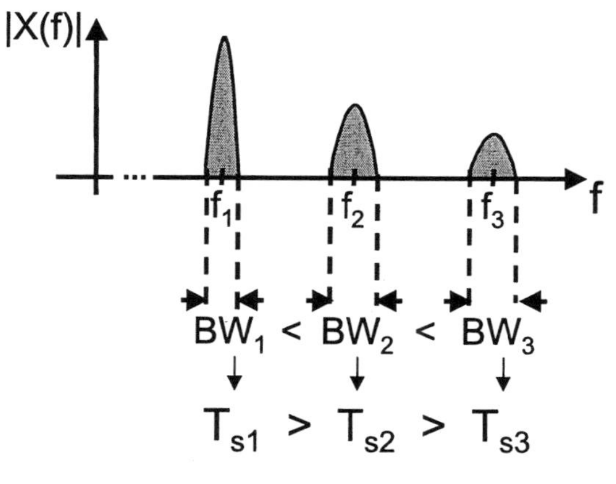

Figure 3.4 Spectral representation of an MRMC signal that consists of three carriers, each having their own bandwidth BW_i ($i = 1, 2, 3$) and simulation timestep T_{si} with $T_{si} \leq 1/BW_i$.

At the different places in the front-end the waveforms computed in this test simulation are inspected and the modulated carriers which are smaller than a user-specified threshold are indicated to be negligible in the final simulation. Also, from this simulation the carrier frequency of each non-negligible modulated carrier is determined, together with the bandwidth of its complex lowpass representation. Further, the total number of carriers can be reduced by combining bandpass signals that are sufficiently close to each other.

3.3 COMPUTATION OF THE RESPONSE TO AN MRMC SIGNAL

When computing the output of a block described by a linear transfer function, FAST processes the different components of an MRMC signal individually. Nonlinear blocks, however, combine the different bands of the MRMC signal to produce their output. To compute the output correctly, the timesteps for the different components of the MRMC signal are first made equal to a common timestep T_{common} that is large enough to represent the widest band in the complete output signal without aliasing. This is accomplished by an interpolation operation. After the computation of the output signal with one single timestep, an MRMC representation is constructed from this signal. Changes in the timestep that are required in this construction are realized by the insertion of interpolators or decimators.

FAST has different methods to compute the output of a nonlinear block, namely the combinatorial approach (Section 3.4.1) and the Fourier transform approach (Section 3.4.2). The type of nonlinearity (weak, strong) and the number of components in the MRMC signal determine which method is the most efficient. We limit the discussion here to static nonlinearities since weak dynamic nonlinearities can be separated into linear transfer functions and static nonlinearities [14].

3.3.1 Combinatorial approach

For a small number of carriers that is fed into a nonlinearity of a low degree, the output of the nonlinearity can be computed by raising the sum of carriers, as given in equation (3.2), to the appropriate power and consider the different terms, each of which considers to a contribution to a carrier at some frequency. For example, assume that an MRMC signal with n carriers is fed into a static nonlinearity of degree m, namely x^m, then the output of this nonlinearity contains terms of the form

$$\prod_{i=1}^{n} K_{k_i,l_i}[(x_{Li}(t))^{k_i} \cdot (x_{Li}^*(t))^{l_i} \cdot \exp(j2\pi(k_i - l_i) \cdot f_i \cdot t)], \tag{3.3}$$

and the complex conjugate term at the corresponding negative frequency. In this equation the indices k_i, l_i are non–negative such that

$$\sum_i (k_i + l_i) = m \tag{3.4}$$

Further, the integer coefficient K_{k_i,l_i} is precomputed using combinatorics [15]. From (3.3) it is seen that the modulation of the carrier at the frequency $(k_i\text{-}l_i)f_1+ \ldots +(k_i\text{-}l_i)\, f_n$ can be computed by multiplying complex baseband signals or their complex conjugate. As an example, assume that a third-order nonlinearity x^3 is excited by three modulated carriers that are harmonically related:

$$in(t) = \frac{1}{2}\sum_{i=1}^{3} [x_{Li}(t) \cdot \exp(j2\pi\, i f_0 t) \; + \; x_{Li}^*(t) \cdot \exp(-j2\pi\, i f_0 t)]. \tag{3.5}$$

The carrier at the output of the nonlinearity at the frequency f_0 is then given by

$$\text{response at } f_0 = \text{Re}\left[6x_{L1} \cdot x_{L3} \cdot x_{L3}^* + 3x_{L1}^2 \cdot x_{L1}^* + 6x_{L1} \cdot x_{L2} \cdot x_{L2}^* + 3x_{L1}^{*2} x_{L3} + 3x_{L2}^2 x_{L3}^*\right]. \quad (3.6)$$

This combinatorial approach is only feasible for a small number of carriers and a low degree of the nonlinearity since the number of contributions that need to be considered at each output carrier increases exponentially with the number of carriers and the degree of nonlinearity [15].

3.3.2 Fourier transform approach

In a signal that consists of several modulated carriers, such as the one given in equation (3.2), each carrier can be represented by a high-frequency sinusoidal signal with a slowly varying amplitude and phase. At each discrete timepoint $t_p = p.T_{common}$ the set of modulations $x_{L1}(t_p), \ldots, x_{Ln}(t_p)$ on the carriers are used to compute a time-domain representation of the quickly varying part of the signal, assuming that the carriers are not modulated. This time-domain representation is further in the text designated as an equivalent time-domain representation in order to make a distinction with the time-domain representation of the complex lowpass signals. The equivalent time-domain representation is computed by first taking an inverse discrete Fourier transform (IDFT) at each timepoint t_p. Next, a static nonlinearity can be applied to each individual equivalent time domain sample. Finally, the modulation on the different carriers at the output of the nonlinearity is computed by taking a DFT of the equivalent time-domain representation of the output of the nonlinearity. Notice that in contrast with the previous method, this approach can also be used for strong nonlinearities.

When the number of carriers at the input of the nonlinearity or the number of carriers of interest at the output is large, then FAST replaces the IDFT and the DFT by an IFFT and an FFT, respectively.

3.4 CONSTRUCTION OF A COMPUTATIONAL GRAPH

Before a FAST simulation, the description of a front-end architecture is first translated into a so-called computational graph. The vertices of this directed graph are referred to as computational nodes. They represent functions that operate on data and return data. Examples are: an FIR filter, an IIR filter, a static nonlinearity, a summer, an (I)FFT block, a (I)DFT block, a convolution block, a quantizer, an interpolator, a decimator,

different types of generator blocks (e.g. a Gaussian noise generator, a waveform generator, ...) The directed edges of the computational graph represent the data that is exchanged between the blocks using buffers. The optimal buffer sizes are determined during the construction of the graph.

Every front-end block gives rise to one or more nodes in the computational graph, the actual number of nodes depending on the number of carriers, the carrier frequencies and the sampling frequencies of the MRMC input and output signals of the block. For example, a block described by a linear transfer function and with n carriers at its input is translated into n parallel IIR or FIR filters, each with a complex transfer function. The translation step further generates the interpolators and decimators that perform the changes of the timesteps.

3.5 SCHEDULING AND EXECUTION

After the setup of the computational graph, the computations are scheduled and executed. During execution, the different computational nodes perform vector processing as much as possible. In this way, the architecture of processor is exploited optimally. Further, the execution scheme is flexible such that both synchronous and asynchronous nodes can be included. An example of the use of asynchronous nodes is the situation where the clock frequency of a digital part and the sampling frequencies of the continuous-time blocks are incommensurate.

A dynamic dataflow execution scheme has been chosen as the fundamental execution scheme. Groups of synchronous nodes – i.e. nodes that can be scheduled prior to the execution– are introduced as a subset of this dynamic execution scheme to increase the computational efficiency. The buffers in between the nodes operate in a blocking input / blocking output mode. This ensures that the amount of memory used remains fixed, in contrast with other methods such as blocking input / non-blocking output, where one needs to take special precautions to guarantee that the amount of memory used remains limited [16]. In order to make vector processing possible, it is necessary that the buffers in between the nodes can contain a sufficient number of tokens. Careful management of the buffers makes it possible to read and write directly into these buffers without additional data movement.

Feedforward parts of an architecture are computed with vector processing whereas feedback loops are computed on a sample-by-sample basis. The scheduler automatically switches the calculations between vector and sample-by-sample processing. Introducing vector processing in a dynamic way is done using a scheduler which is built around a priority queue with four levels of priority:

1. nodes that are not ready for execution
2. nodes not in a feedback loop and ready for sample-by-sample execution
3. nodes in a feedback loop and ready for sample-by-sample execution
4. nodes ready for vector processing.

Every node is marked in advance whether it is in a feedback loop or not. This information -together with the number of input tokens and the number of free output tokens- is used when determining the priority level of the node. Executing the node with the highest priority guarantees that vector processing will be used as much as possible. Nodes outside a feedback loop either wait until vector processing is possible or no nodes in any feedback loop can be executed.

3.6 COUPLING OF FAST WITH THE DIGITAL SIMULATION ENVIRONMENT OCAPI

The FAST simulator described in this chapter, is in essence a dataflow simulator. In this way, the digital blocks of a transceiver could be simulated with FAST at the dataflow level. However, we opted for a co-simulation with the digital simulation environment OCAPI [10], since this environment supports different abstraction levels for the digital blocks, as well as a path down to VHDL or Verilog.

Since both OCAPI and FAST are based on the same dataflow scheduling along with generalized firing rules, the coupling interface has to implement queue management and data type adaptation only. To maximize simulation speed and minimize computing platform dependencies, an interface C++ class EKOCAPI with direct access to both its OCAPI and/or FAST I/O queues was preferred over a memory pipe, system pipe, or file-based solution.

Maximum independence between OCAPI and FAST partitions is achieved by slaving all OCAPI schedulers as subprocesses under a FAST master scheduler in a hierarchical way. Both OCAPI and FAST partitions are developed and tested as stand-alone applications before the system integration. This preserves the localities of the dataflow scheduling in every partition leading to a simpler system-level schedule. At instantiation time, the EKOCAPI class defines the connections between OCAPI and FAST partitions. At runtime, i.e. during the simulation, it handles I/O queue management including multirate adaptation at the partition boundaries.

The result is a distributed, hierarchical scheduling with lean communication at partition boundaries only, which translates into a low coupling overhead.

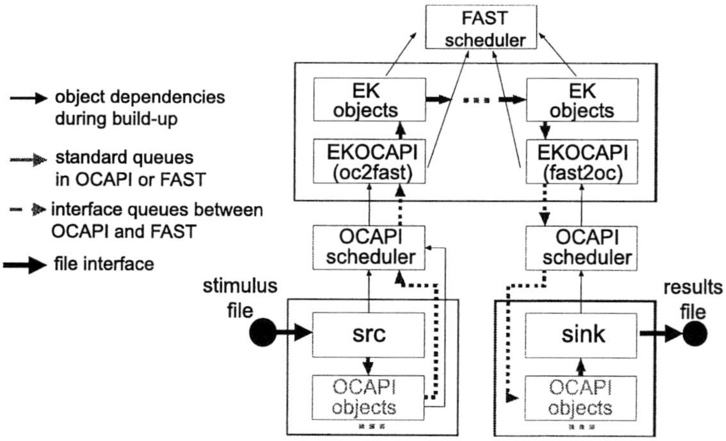

Figure 3.5 Coupling connects the underlying object and scheduler classes of OCAPI and FAST, respectively, by an interface object, which defines the connection during instantiation and handles the inter-scheduler communication during simulation.

An example for a forward chain topology (Fig. 3.5) shows how the stimulus is read from a file, preprocessed by an OCAPI partition, handed over to FAST being further processed, then passed to another OCAPI partition for post-processing and the output finally being written to a results file. This scenario is typically found in end-to-end link simulations of a transmitter and receiver over a channel. The coupling mechanism handles also other relevant topologies, including bidirectional or feedback communication between OCAPI and FAST partitions. For this case, initialization methods are foreseen in both tool classes to support cyclo-static dataflow.

3.7 RUNTIME EXAMPLES

The design methodology discussed in the previous sections is implemented in the program FAST with a Microsoft Visual C++ version 5 compiler. The runtimes reported in this section are for a Pentium II 266MHz processor.

3.7.1 Influence of the buffer size on the CPU time

The runtimes of some basic computational nodes, given in Table 3.1, illustrate the advantage of vector processing over sample-by-sample processing.

Table 3.1 CPU time in μs for a single timepoint of some basic computational nodes as a function of the size of the buffer before and after the nodes. A size of 1 corresponds to sample-by-sample processing.

buffer size	1	4	16	64	256	1024	4096
Gauss. Noise	1.01	0.75	0.69	0.67	0.66	0.68	0.68
Waveform	0.35	0.09	0.03	0.02	0.007	0.008	0.018
FIR 4 taps	2.7	0.78	0.25	0.12	0.08	0.07	0.082
FIR 16 taps	3.0	1.12	0.65	0.34	0.25	0.23	0.23
FIR 64 taps	4.1	2.2	1.7	1.6	0.67	0.36	0.38
FIR 256 taps	8.6	6.6	6.0	6.0	5.9	1.8	0.53
FIR 1024 taps	25.7	24.1	23.5	23.5	23.6	23.2	6.4

The gain is only marginal for the Gaussian noise generator since the calling overhead is marginal compared to the computation itself.

The efficiency of the waveform generator on the other hand increases up to a factor 50. The computational speed of the FIR filters also increases considerably with the buffer size. This is due to a reduction of the calling overhead and due to a change in the method for the computation of the convolution when this is appropriate. For example, for a FIR filter with 1024 taps, the overlap-save method is used from a buffer size of 4096 and this results in a simulation speed that is four times higher than with a time-domain convolution sum.

Even when a feedback loop is present, such as in Figure 3.6, vector processing can still be used to speed up the computations. In Figure 3.6 the source produces time samples and the sink consumes the data at the output of the $\Sigma\Delta$ ADC without any further operation. The outputs of the blocks outside of the feedback loop are computed with vector processing, whereas the feedback loop is processed on a sample-by-sample basis.

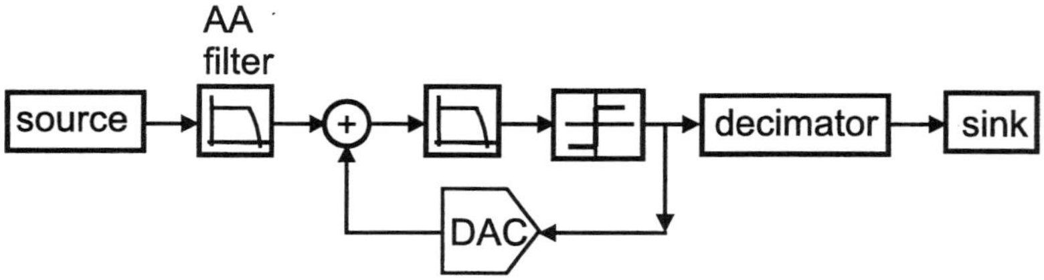

Figure 3.6 A $\Sigma\Delta$ ADC, preceded by an anti-alias filter and followed by a sink that consumes the output data of the ADC.

The filter inside the modulator, the filter of the decimator and the anti-alias filter are each approximated by a FIR filter with 9 taps. Table 3.2 again illustrates that large buffer sizes speed up the computations. If the anti-alias filter is approximated by a FIR filter with 256 taps, which is more realistic, then the speed advantage of vector processing is even more pronounced since more computations are performed outside the feedback loop.

The example of the $\Sigma\Delta$ ADC further illustrates the speed advantage of a speed advantage of static scheduling compared to dynamic scheduling [16].

Table 3.2 CPU time (μs) for one iteration (= processing of one time sample) in the example of the sigma-delta ADC. The time is averaged over the different blocks of the modulator.

buffer size	1	4	16	64	256
synchronous dataflow	0.50	0.33	0.27	0.24	0.23
dynamic dataflow	1.31	0.81	0.66	0.58	0.57

3.7.2 Processing of nonlinear blocks

From the two methods to compute the output of nonlinearities, the combinatorial approach and the Fourier transform approach, the latter becomes more efficient when the number of carriers at the input and/or output is high. This is illustrated with the next experiments. For an evaluation of the combinatorial approach FAST computes the output of a third-order nonlinearity to a sum of three modulated carriers at f_0, $2*f_0$ and $3*f_0$. The output carriers of interest are at the same frequencies. The computation of this response involves the computation of 14 contributions, each consisting of a product of three carriers (or their complex conjugate). The CPU times for this computation, given in Table 3.3, again illustrate that vector processing speeds up the calculations. However, for a large buffer size the execution time increases slightly due to the fact that the data no longer fits into the cache memory of the processor.

Table 3.3 CPU time per timepoint to compute the output of a third-order nonlinearity with the combinatorial method.

buffer size	1	4	16	64	256	1024	4096
CPU time (μs)	45	14.4	7.37	5.42	5.62	6	8.29

The Fourier transform approach can use either the fast Fourier transform or the discrete transform.

For a comparison of these two approaches, one could measure the CPU time of the three consecutive steps that are used in the Fourier transform approach:

1. Computation of an IFFT or IDFT.

2. Computation of the output of the nonlinearity (time domain).

3. Computation of an FFT or DFT.

Since step 2 is the same for the two transforms, we can concentrate on the CPU times for the (I)FFT and the (I)DFT.

Table 3.4 illustrates the advantage of the (I)FFT over the (I)DFT when the number of carriers increases: if the input consists of more than 4 carriers, then the use of the IFFT method is more efficient. Also, if the number of carriers of interest at the output of a nonlinearity is larger than four, then the (I)FFT is more efficient than the (I)DFT.

For the example of the third-order nonlinearity with three carriers at its input and three carriers of interest at the output, the combinatorial approach is slightly more efficient than the Fourier transform approach. With an optimal buffer size the CPU time obtained with the former approach is no more than 6 μs per timepoint according to Table 3.3. The CPU time with the Fourier transform approach which does not vary much with the buffer size at the input, is about 7.5 μs per timepoint.

Finally, the CPU time per timepoint for a simulation of the complete receiver front-end of Figure 3.1 equals 9.78 μs. The input at the antenna is an OFDM signal with 256 carriers, each with a QAM 16 modulation. The carrier frequency f_c is 5.25 GHz. In addition to the input signal, two other waveform generators have been used for the two local oscillators (LO1 and LO2). These generators produce a sinusoidal signal with phase noise. Their frequency is $f_{LO1} = 5$ GHz and $f_{LO2} = 0.25$ GHz. The LNA is a static nonlinearity, described by a polynomial of order three that relates the output to the input. The coefficients of this polynomial are related to the intercept points [14].

Table 3.4 CPU times (μs) for one FFT/DFT or IFFT/IDFT operation. These are used in the Fourier transform approach to compute the output of a nonlinearity. The horizontal axis is the number of spectral inputs, the vertical axis is the number of real data points of the FFT.

# input carriers →	1		4		8	
# data points ↓	FFT	DFT	FFT	DFT	FFT	DFT
16	2.63	0.95	3.18	3.08	3.76	5.69
64	9.23	3.44	10.3	10.1	10.8	18.8
256	40.0	13.1	40.4	39.1	41.3	73.6
1024	254	118	253	351	257	666
4096	1275	511	1273	1414	1275	2576

The three mixers in this example are also described with a third-order polynomial, but now as a function of two inputs, the local oscillator input and the RF input. All filters in the front-end are elliptic filters. The RF bandpass filter has order three, the bandpass filter at 250 MHz order four, and the lowpass filters order six. Finally, the analog-to-digital converters (ADC) are ideal samplers that quantize the signal with a given resolution.

Prior to simulation, the front-end architecture is translated into a computational graph, as shown in Figure 3.7.

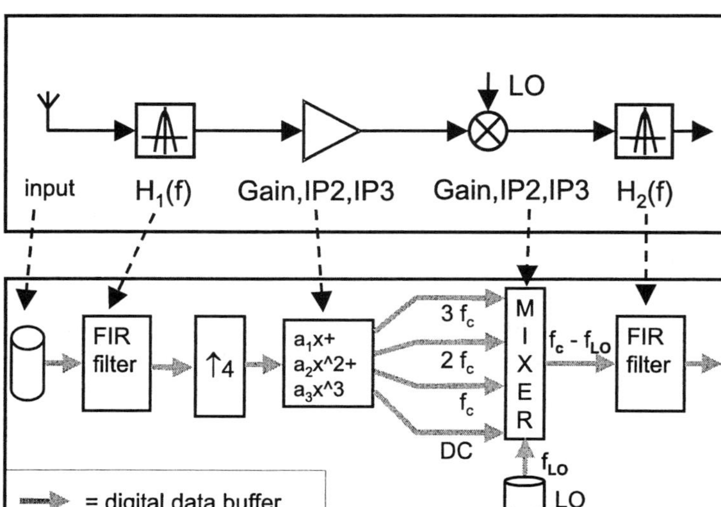

Figure 3.7 Translation of (part of) the front-end of Figure 1 into a computational graph.

The RF bandpass filter and the bandpass filter centered around $f_c - f_{LO1}$ are each translated into an FIR filter with 32 and 128 taps, respectively. The two lowpass filters in Figure 3.1 are translated into two FIR filters with 128 taps.

The signals before and after the RF bandpass filter only contain one carrier. This is represented at baseband, where it is oversampled twice. The nonlinearities of the LNA and the mixer generate extra modulated carriers and they widen the frequency band of interest. A 4-fold interpolator, placed in front of the LNA, changes the simulation timestep to allow a correct computation of the spectral regrowth and the extra carriers. The polyphase filter that accomplishes this interpolation has 128 taps.

The output of the LNA consists of four modulated carriers, one at baseband, and one at f_c, $2f_c$ and $3f_c$, respectively. The mixer's second- and third-order nonlinearities combine these carriers with the noisy local oscillator signal. At the output of the mixer only the frequency component at $f_c - f_{LO1}$ is considered. The other carriers are strongly suppressed by the subsequent bandpass filter. Since the number of carriers is not large anywhere in the front-end, the responses of the nonlinearities are computed with the combinatorial approach. At the output of the IF mixers that downconvert the I- and the Q-signal to baseband, only the baseband signal is taken into account.

The actual simulation executes the generated computational graph. This simulation has been performed both with FAST and with MATLAB, in which exactly the same computational graph has been implemented. The MATLAB code was written such that all operations are performed in a vectorized way. Hence, MATLAB runs as fast as possible. The length of the signals is such that calling overhead in MATLAB is negligible. No additional optimization has been performed to optimize the FAST or the MATLAB code. Both FAST and MATLAB use the same input. The outputs of MATLAB and FAST are equal within the numerical precision. The CPU times in MATLAB per simulation timestep and per carrier symbol are $446\mu s$ and $56\mu s$, respectively. This is a factor 10 slower than with FAST. If no MRMC signal representation were used in this example, then the simulation time for this example increases with a factor of 70.

3.8 CONCLUSIONS

High-level simulations are essential in the design of architectures of mixed-signal front-ends of digital telecom transceivers. In this chapter a methodology, implemented in a program called FAST, has been described for efficient dataflow simulations of such front-ends at the architectural level. The high-level models of the front-end blocks are such that iterations for one timepoint are avoided during their evaluation. The high simulation efficiency is obtained using a local multirate, multi-carrier (MRMC) signal representation. This is an extension of the complex lowpass representation of one carrier. The latter approach is less accurate than the MRMC approach, since this one can take into account out-of-band distortion while maintaining the high simulation efficiency of working with complex lowpass representations. The high simulation efficiency is also due to a translation, prior to simulation, of the architecture description, to a computational graph, that can be efficiently evaluated. Finally, the calculations are speeded up by using vector processing of data as much as possible. Sample-by-sample processing is used locally when feedback loops are present. Runtime examples show an execution time for a receiver front-end of 9.8 μs per timepoint of the modulating signals on a Pentium II 266MHz processor. This high efficiency allows an efficient co-simulation of the front-ends with the digital blocks of a transceiver in order to determine bit-error-rates of a complete telecom link, including the analog front-end. This co-simulation has been demonstrated here with a coupling of FAST with the digital simulation environment OCAPI.

References

[1] Communication toolbox of MATLAB/ SIMULINK, http://www.mathworks.com/products/communications/ .

[2] I. Vassiliou and A. Sangiovanni-Vincentelli, "A frequency-domain, Volterra series-based behavioral simulation tool for RF systems," *Proc. IEEE Custom Integrated Circuits Conference*, pp. 21-24, 1999.

[3] Jeruchim, Balaban and Shanmugan, "Simulation of Communication Systems," Plenum, 1992.

[4] COSSAP of Synopsis, http://www.synopsis.com/products/dsp/cossap_ds.html .

[5] SPW of Cadence, http://www.cadence.com/software/cierto/

[6] A. Rofougaran *et al.*, "A 1 GHz CMOS RF front-end IC for a direct-conversion wireless transceiver," *IEEE J. Solid-State Circuits,* Vol. 31, pp. 880-889, July 1996.

[7] J. Crols and M. Steyaert, "A single-chip 900 MHz CMOS receiver front-end with a high-performance low-IF topology," *IEEE J. Solid-State Circuits,* Vol. 30, pp. 1483-1492, Dec. 1995.

[8] J. Rudell *et al.*, "A 1.9-GHz wide-band IF double conversion CMOS receiver for cordless telephone applications," *IEEE J. Solid-State Circuits,* Vol. 32, pp. 2071-2088, Dec. 1997.

[9] J. Chen, D. Feng, J. Philips and K. Kundert, "Simulation and modeling of intermodulation distortion in communication circuits," *Proc. IEEE Custom Integrated Circuits Conference*, pp. 5-8, 1999.

[10] P. Schaumont *et al.*, "A programming environment for the design of complex high speed ASICs", *Proceedings of DAC*, pp. 315-320, June 1998.

[11] J.L. Pino and K. Kalbasi, "Cosimulating synchronous DSP applications with analog RF circuits," *Proc. 32nd Annual Asilomar Conference on Signals, Systems, and Computers*, Nov. 1998.

[12] HP-ADS of Hewlett-Packard, http://www.tm.agilent.com/tmo/hpeesof/products/ads/adsoview.html .

[13] E. Ngoya and R. Larchevèque, "Envelop transient analysis: a new method for the transient and steady state analysis of microwave communication circuits and systems," *Proc. IEEE MTT-S*, pp. 1365-1368, 1996.

[14] P. Wambacq and W. Sansen, "Distortion analysis of analog integrated circuits," *Kluwer Academic Publishers*, 1998.

[15] R. Sea, "An algebraic formula for amplitudes of intermodulation products involving an arbitrary number of frequencies," *Proc. IEEE*, pp. 1388-1389, Aug. 1968.

[16] Ptolemy of the University of Berkeley, http://ptolemy.eecs.berkeley.edu .

4 EFFICIENT HIGH-LEVEL SIMULATION OF ANALOG TELECOM FRONTENDS

Piet Vanassche, Georges Gielen, Willy Sansen

ESAT-MICAS,
Katholieke Universiteiten Leuven, Belgium

E-mail: Piet.Vanassche@esat.kuleuven.ac.be

During the last decade, the telecommunication market is experiencing a tremendous growth. This includes both a growth in the number of consumers as in the number of applications. DECT, GSM, ADSL, VDSL GPRS, UMTS, Bluetooth, etc. are just a few examples of evolving technologies that try to cope with the increasing demand for higher data rates. This makes it a very attractive and very competitive market, which requires short time to market in introducing new products. In order to deal with these requirements, designers are in need of efficient tools for evaluation (simulation) of their telecommunication systems at all levels of design abstraction. This chapter describes a technique that is particularly well suited for simulation of the frontend part of these telecom systems at architectural level.

The algorithm being presented works in the time-domain and is based upon the use of complex damped exponential basis functions. This means that all known and unknown signals are modeled as

$$x(t) = \sum_k X_k e^{z_k t} = \sum_k X_k e^{(p_k + j\omega_k)t} \tag{4.1}$$

This approach tries to exploit both the characteristics of the telecommunication signals and those of the frontend building blocks in order to accelerate sytem-level evaluation of the frontend with real-life telecom signals being applied at its inputs. One of the most important characteristics of the signals occurring at the different nodes in a telecom transceiver is the fact that they contain components with widely varying time constants. The exponential basis tries to deal with this. As for the frontend characteristics, we assume the system to behave weakly nonlinearly. This restriction of weak nonlinearity can be justified by the fact that telecommunication frontends are essentially designed to behave linearly (in its most general, time-varying, sense). The

P. Wambacq et al. (eds.),
Low-Power Design Techniques and CAD Tools for Analog and RF Integrated Circuits, 61–81.

nonlinear behavior is parasitic and hence suppressed as much as possible. This makes telecom frontends weakly nonlinear in nature.

This chapter is made up out of four major parts.

Section 4.1: Here we describe the simulation problem from a general point of view. Choices and trade-offs involved in the construction of a simulation algorithm are discussed. It is also shown how system and signal characteristics influence algorithmic performance. This general framework is then used to classify some existing simulation algorithms. Particular attention is payed to the problems encountered in applying these simulation algorithms to telecommunication frontends and how these problems are dealt with using the exponential approach.

Section 4.2: Here, we investigate the efficiency of the complex damped exponential basis in modeling the high-frequency digitally modulated signals encountered in telecom frontends. It is shown that with relatively few basis functions, signal models can be constructed that are valid over a long range of time, allowing for a much larger simulation time step than possible using traditional methods.

Section 4.3: This section discusses the details of the time-domain simulation algorithm based upon the complex damped exponential signal model. This includes dealing with linear systems (both time-invariant and time-varying), weakly nonlinear systems and finding the appropriate values of the exponents z_k in 4.1.

Section 4.4: The simulation approach is applied to a DCS1800 receiver system with a 1.8GHz modulated GMSK signal at the input. Comparison is made with Spice-like integration techniques.

This chapter is ended with some conclusions.

4.1 SITUATING THE EXPONENTIAL APPROACH WITHIN A GLOBAL FRAMEWORK FOR SIMULATION ALGORITHMS

In this first section, we treat the simulation problem from a general point of view. The basic ideas behind simulation algorithms are outlined together with the choices and trade-offs involved. It is explained how system and signal characteristics influence the performance of a particular algorithm. This general framework is then used to discuss some existing simulation algorithms. Particular attention is paid to the problems encountered when applying these simulation algorithms to telecommunication frontends. Finally, based upon this framework, the complex damped exponential approach, which is discussed in more detail in the subsequent sections, is briefly situated among existing algorithms.

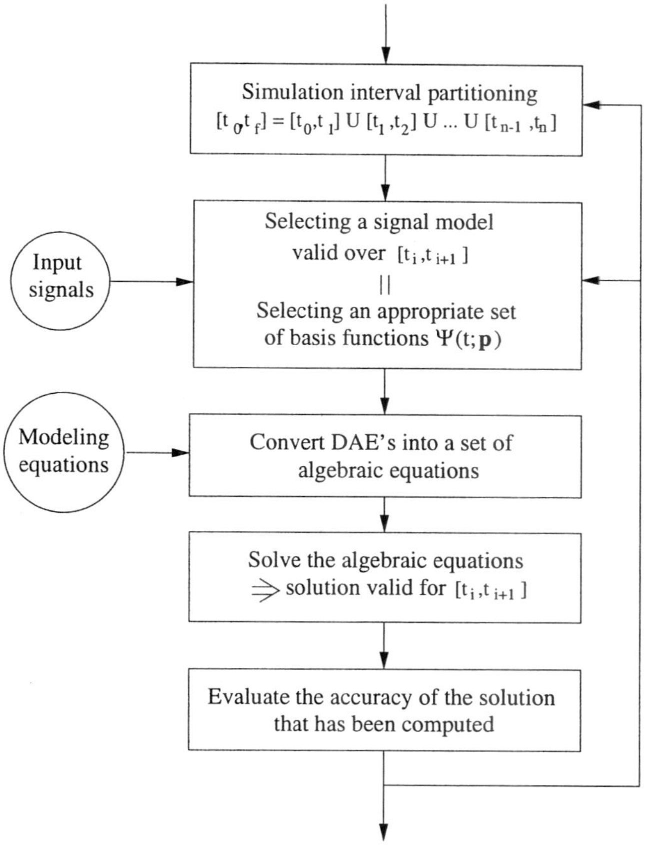

Figure 4.1 Global structure of a simulation algorithm.

4.1.1 Basic ideas behind simulation

Interpreted in a purely mathematical fashion, simulation can be defined as solving a set of (nonlinear) differential algebraic equations (DAE's)

$$\mathbf{F}_{DAE}\left(\frac{d\mathbf{x}}{dt}, \mathbf{x}, \mathbf{x}_{in}, t\right) = 0 \tag{4.2}$$

where $\mathbf{x}(t) = \begin{bmatrix} x_1(t) & \cdots & x_N(t) \end{bmatrix}^T$ and $\mathbf{x}_{in}(t)$ is defined similarly. These are respectively the unknown signals and the input signals. The solution needs to be determined over a given range $[t_0, t_f]$ and for a given set of initial (or boundary) conditions. Looking at algorithms for solving these kinds of problems, we observe that most of them possess a similar structure. This structure is illustrated in Fig.4.1. The figure clearly illustrates how the global algorithm can be decomposed into several subalgorithms. Different simulation algorithms provide different implementations for one or more of these subalgorithms. In what follows, we briefly discuss each of these subalgorithms and their impact upon the overall algorithmic performance.

The first two steps of a simulation algorithm involve the selection of a time interval partitioning strategy and a signal modeling strategy. The first subalgorithm partitions the overall simulation interval $[t_0, t_f]$ into a number of subintervals $[t_k, t_{k+1}]$ within which a solution is easier to compute with sufficient accuracy. The basic idea behind interval partitioning is that the signal variations become easier to model, and hence easier to compute, when the time interval (i.e. the timestep) is small and vice versa. The second subalgorithm selects a set of differentiable basis functions $\Psi_{i,j}(t)$ for $j = 1, \ldots, J_i$ such that the signals $x_i(t)$ can be modeled as

$$x_i(t) = \sum_{j=1}^{J_i} X_{i,j} \Psi_{i,j}(t) \tag{4.3}$$

within the subintervals $[t_k, t_{k+1}]$ and this with sufficient accuracy. This signal modeling step turns equation (4.2), which is essentially a problem in an infinite number of variables, into the computation of the finite number of unknowns $X_{i,j}$. From a computational point of view, this results in a complexity given by

$$C_{[t_0,t_f]} = N_T \times \overline{C}_{[t_k,t_{k+1}]} \tag{4.4}$$

where N_T is the number of subintervals and $\overline{C}_{[t_k,t_{k+1}]}$ is the average computational complexity involved in computing a solution over one of the subintervals. The latter complexity strongly depends upon the choice of the signal models. This choice determines the total number of unknown variables $X_{i,j}$ which need to be computed per subinterval $[t_k, t_{k+1}]$, and hence the size of the systems of nonlinear and linear equations which need to be solved further on. This number equals $\sum_{i=1}^{N} J_i = N \times J$ where N is the total number of unknown signals $x_i(t)$ and J is the average number of modeling functions per node. In order to keep this number as low as possible, *it is important to select basis functions which correspond as closely as possible to the characteristics of the signals. These models may differ from node to node.*

As the interaction between the first two steps in Fig.4.1 is concerned, there are two limit strategies that can be pursued:

1. Fix the complexity (the number of modeling functions) per subinterval and adapt the size of the subintervals $[t_k, t_{k+1}]$ (the time step) in order to meet the required accuracy. This is a strategy applied in Spice-like algorithms.

2. Fix the size of subintervals (the time step) and increase the number of modeling functions in order to achieve sufficient accuracy. Harmonic balance, for example, applies this second approach.

It is however also possible to choose a strategy in between these two extremes.

Once the signal models have been constructed, they can be substituted into (4.2) in order to convert the DAE's into a set of purely algebraic equations

$$\mathbf{F}_{AE}(X_{i,j}) = 0 \tag{4.5}$$

that can be solved for the unknowns $X_{i,j}$. Note that this needs to be done for each subinterval $[t_k, t_{k+1}]$ separately. The complexity involved in solving these algebraic equations is mainly determined by the cost for evaluation of the left hand side of (4.5) and the corresponding Jacobian matrix

$$\mathbf{J_F} = \left[\frac{\partial \mathbf{F}_{AE}}{\partial X_{i,j}}\right] = \begin{bmatrix} \frac{\partial F_{AE,1}}{\partial X_{1,1}} & \cdots & \frac{\partial F_{AE,1}}{\partial X_{N,J}} \\ \vdots & \ddots & \vdots \\ \frac{\partial F_{AE,N\times J}}{\partial X_{1,1}} & \cdots & \frac{\partial F_{AE,N\times J}}{\partial X_{N,J}} \end{bmatrix}$$

on the one hand and the cost for solving linear systems involving this same Jacobian on the other hand. This can be expressed as

$$\overline{C}_{[t_k,t_{k+1}]} \sim (a_1 C_{\mathbf{F}} + a_2 C_{\mathbf{J_F}} + a_3 C_{lin}) \tag{4.6}$$

The right hand side of the latter equations often (but not always) tends to be dominated by the cost C_{lin} for solving the linear systems. This cost depends superlinearly (cubically in the worst case) upon the number of unknowns, or

$$C_{lin} \sim (N \times J)^{\alpha} \tag{4.7}$$

where $1 \leq \alpha \leq 3$ depending upon the sparsity of the Jacobian matrix and the type of linear solver being used. This implies that the computational complexity per subinterval $[t_k, t_{k+1}]$ could seriously increase if we increase the number of modeling functions. However, if this increase in the number of modeling functions allows us to substantially increase the time step (decrease the number N_T of subintervals in which the overall simulation interval $[t_0, t_f]$ is partitioned), the overall complexity as determined by (4.4) might still decrease, despite the increased value of $\overline{C}_{[t_k,t_{k+1}]}$. This hence implies a trade-off between the number of subintervals and the computational complexity involved in finding a solution over one subinterval. Besides the selection of the basis functions $\Psi_{i,j}(t)$, this trade-off is an important parameter that can be used in the performance optimization of simulation algorithms as will be shown in section 4.4 This optimum strategy will be seen to depend upon the system characteristics (like the degree of nonlinearity, etc.).

4.1.2 An overview of some existing simulation algorithms

In what follows, some simulation algorithms, relevant with respect to the simulation of analog telecom frontends are briefly reviewed. They are situated within the framework shown in Fig.4.1. Time step control, signal modeling strategy and other possible techniques to speed up simulations are discussed together with their main drawbacks. Particular attention is paid to the problems encountered in applying these simulation algorithms to telecommunication frontends.

Spice: Traditional Spice-like solvers [21] typically select a 2nd-order polynomial model and decrease the size of the subintervals (time step) in order to ensure the accuracy of the signal model over $[t_k, t_{k+1}]$. This is often combined with adaptive stepsize control in order to obtain the largest timestep possible. In [23] an extension is presented which allows you to make a trade-off between timestep and polynomial order. Also, since Spice-like solvers typically tend to spend a great deal of time on the nonlinear function evaluations, some commercial implementations try to speed this up by using table lookup methods combined with interpolation. Finally, circuit partitioning strategies are applied, in order to break up a large system into a number of smaller ones. From a computational point of view, this reduces the component C_{lin} in (4.6). This follows from (4.7) by observing that it requires less time to solve M systems of N_b nodes than one large system of $N = M \times N_b$ nodes.

Drawbacks: When dealing with telecom systems and their associated IF and RF modulated signals, these approaches always lead to a huge number N_T of subintervals (small timestep), resulting in long simulation times. The basic reason for this is that the polynomial basis is not suited for modeling high-frequency modulated signals.

Harmonic balance: Harmonic balance [5] algorithms avoid partitioning of the simulation interval by selecting only one subinterval $[t_0, t_f] = [0, T]$ where T is the signal period. The timestep is hence maximal. The signals are modeled using the harmonic functions $e^{j2\pi kt/T}$, which allows efficient modeling of high-frequency, periodic signals. The number of harmonics is chosen in a way to make the signal model valid over the entire simulation interval $[0, T]$. In case of strongly nonlinear systems, this however leads to a large number of harmonics, resulting in large linear systems to be solved. In this area, major improvements were achieved by applying Krylov subspace techniques [13, 1] instead of direct methods in solving the linear systems involved. This allowed to reduce the exponent α in (4.7) to sometimes as low as 1. Also, some variants [8, 9] have been presented which first apply a transformation of the time variable $\hat{t} = \lambda(t)$ such that the number of harmonic basis functions $e^{j2\pi k\hat{t}/T}$ necessary to model signals with respect to $\hat{t}$ is smaller than the number that would be needed for modeling with respect to t. Looking at this in another way, the above is equivalent to saying that $e^{j2\pi k\lambda(t)/T}$ is a more efficient set of modeling functions than $e^{j2\pi kt/T}$. This is actually also the idea behind other transformation of variables techniques [10].

Drawbacks: All of these methods only allow to find the steady-state solution of nonlinear circuits with periodic inputs. No transients can be taken into account and it is not possible to model the digitally modulated signals that serve as the actual inputs to telecommunication systems.

Circuit envelope: Circuit envelope [14] supports transient simulations of high-frequency modulated signals in a way that can be seen as combining the polynomial and the harmonic basis, using $t^m e^{j2\pi nt/T}$ as basis functions. It is often combined with some kind of adaptive stepsize control.

Drawbacks: The algorithm uses a global signal model, meaning that the basis functions are the same for all signals in the system. This implies that the simulator cannot take advantage of the sometimes widely different signal characteristics at different nodes in a frontend, something which is typical in telecom applications (high frequency input, low frequency output or vice versa). This global model leads to an unnecessary increase in the number of basis functions per subinterval $[t_k, t_{k+1}]$ and hence in the number of unknowns, complicating the solution

4.1.3 The complex damped exponential approach

The complex damped exponential [18] approach tries to incorporate the advantages of the previous methods while avoiding the drawbacks in dealing with the simulation of weakly nonlinear telecom frontends at architectural level. The approach makes use of a complex damped exponential signal model $\Psi(t) = e^{(p+j\omega)t} = e^{zt}$, which allows for efficient modeling of high-frequency modulated telecommunication signals. Compared to harmonic balance, the introduction of the damping factor p makes it possible to perform simulations in the time-domain and this for non-periodic input signals. The simulation algorithm is based on a Volterra series expansion approach. It selects the necessary basis functions at runtime, avoiding the global signal model used in the circuit envelope method. The simulation stepsize can be used as a parameter to optimize simulation performance. The algorithm also allows to compute wanted and unwanted signals separately. This, together with the natural relationship between the exponential signal model and frequency content, also greatly facilitates analysis of the results afterwards.

4.2 THE COMPLEX DAMPED EXPONENTIAL BASIS AND ITS SIGNAL MODELING CAPABILITIES

One of the most important properties of telecommunication signals is the fact that they often contain greatly different time constants, especially when dealing with RF-applications. A set of basis functions that has the natural ability to deal with this property is the complex damped exponential basis. In its most general form, a signal $x(t)$ is modeled as

$$x(t) = \sum_k X_k t^{n_k} e^{z_k t} \tag{4.8}$$

with $n_k \in N$ and $z_k = p_k + j\omega_k \in C$. This signal model shows great resemblances with the harmonic basis $\left\{e^{jn2\pi f_0 t}\right\}$. The latter is actually a subset of the former. There are however some important differences. By adding the damping factor p_k, it becomes *possible to model real-life telecom signals* like OFDM, GMSK, etc. The harmonic basis on the other hand only allows for periodic input signals. It also becomes *possible to perform simulations in the time domain and to take transients into account.* In what follows, we limit ourselves to the subset of (4.8) for which $n_k = 0$.

In order to get an idea of the modeling efficiency of the exponential basis with respect to telecommunication signals, we compare it with the results obtained using a polynomial basis. For testing purposes, we use a GMSK modulated bitstream [11]. Similar results can be obtained for other kinds of modulation strategies. The GMSK signal can be written as

$$\begin{aligned} s(t) &= Re\left[e^{j2\pi f_0 t} \times (\cos(\Phi(t)) + j\sin(\Phi(t)))\right] \\ &= Re\left[e^{j(2\pi f_0 t + \Phi(t))}\right] \end{aligned}$$

RMS error	J_{GMSK}	N_T/T
-40 dB	4	1
-60 dB	6	1

Table 4.1 Number of exponentials J_{GMSK} and number of subintervals N_T per symbol necessary to model a GMSK modulated input signal.

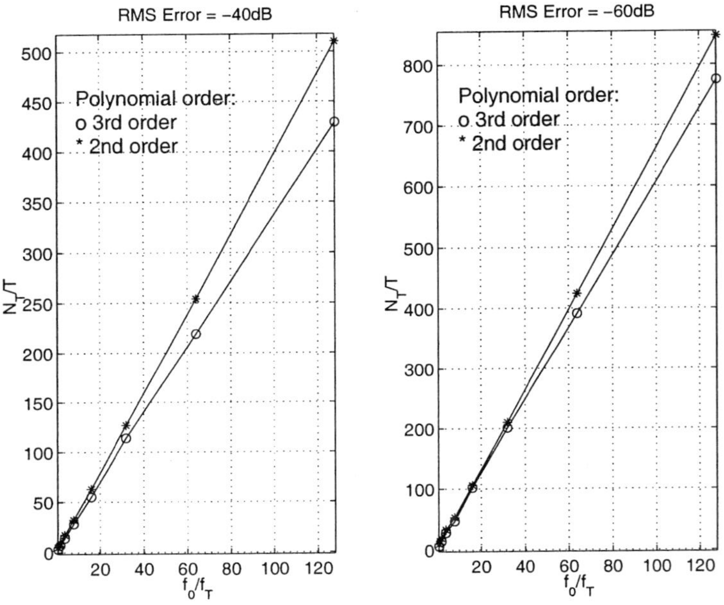

Figure 4.2 Necessary number of subintervals for polynomial fit of a modulated GMSK signal.

where $\cos(\Phi(t))$ and $\sin(\Phi(t))$ are respectively the inphase and quadrature signal components. The GMSK symbol rate equals $f_T = 1/T$. For both signal modeling approaches, exponential and polynomial, the number of modeling functions J_{GMSK} and the number of simulation subintervals N_T were computed for varying values of the carrier frequency f_0. All polynomial fits are performed using a least-squares approach, while the exponential fitting was done using the HTLS algorithm as described in [19]. All fits were performed for RMS errors equalling respectively -40dB and -60dB. This error was computed as the difference between the input samples and the resulting fit. The results of the exponential fit are summarized in Table 4.1 where *all of the numbers are independent of the modulation frequency* f_0. Fig. 4.2 shows the resulting (normalized) number of subintervals N_T/T for the corresponding polynomial fits using 2nd- and 3rd-order polynomial models. As is to be expected, N_T increases (decreased timestep) about linearly with the frequency f_0. The complex-

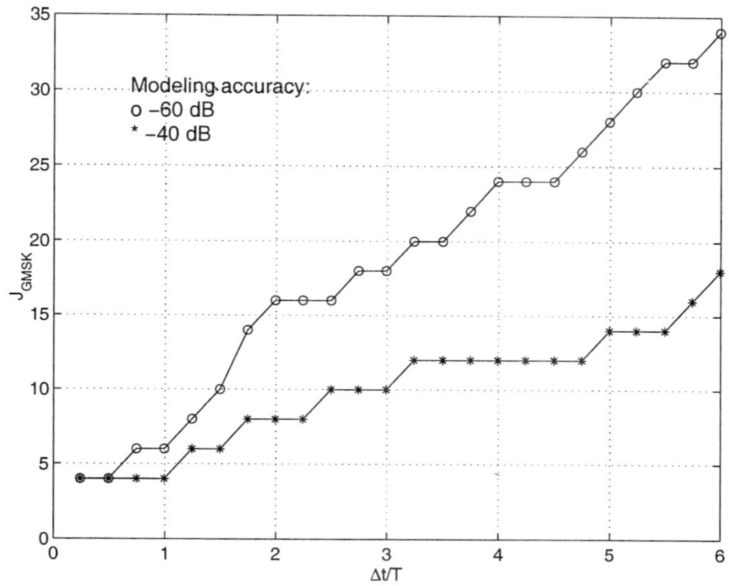

Figure 4.3 Number of modeling exponentials versus the length of the modeling interval.

ity of the corresponding simulation algorithm will hence also grow linearly with f_0. Since the exponential basis does not suffer this drawback, it clearly provides a much more efficient model than polynomials, especially when f_0/f_T gets large (which is typically the case).

A second experiment demonstrates that also when using exponentials, there is a trade off possible between the length of the timestep (length of the modeling subinterval $[t_k, t_{k+1}]$) and the complexity of the signal model within $[t_k, t_{k+1}]$, as was discussed in section 4.1. Using the same GMSK signal (with $f_0 = 10 \cdot f_T$), Fig. 4.3 plots the number of modeling exponentials versus the normalized timestep $\Delta t/T$. This is done for two different values for the accuracy of the resulting fit. Again, these numbers are independent of the carrier frequency f_0. This figure clearly illustrates how it is possible to increase the timestep (decrease the number N_T of subintervals) by increasing the number of basis functions (modeling complexity). Which choice of Δt is optimal depends upon the system degree of nonlinearity as will be demonstrated in section 4.4.

As a final note, it is worth mentioning that one could argue that the exponential method uses both X_k and z_k in (4.8) to obtain a good fit, effectively doubling the number of unknowns that have to be computed. The exponents z_k can however be determined on beforehand, based upon the knowledge of the input signals and their harmonics. Another approach would be to determine them using the results of some short trial simulations. Doing so allows us to *incorporate the input signal characteristics into the simulation algorithm.* Stated in another way, the basis functions $e^{z_k t}$ are chosen, before starting the actual simulation, for them to be best suited for modeling the signals which arise when a given set of input signals is applied. This can be compared with harmonic balance where the functions $e^{j2\pi kt/T}$ are chosen because they are very well suited to model T-periodic functions.

4.3 A SIMULATION APPROACH USING COMPLEX DAMPED EXPONENTIALS

Having demonstrated the efficiency of the complex damped exponential basis in modeling telecom signals, we now briefly outline a simulation approach for weakly nonlinear systems using this basis. The algorithm is absolutely stable by construction and avoids the necessity of a global signal model. The restriction of weak nonlinearity can be justified by the fact that telecommunication frontends are essentially designed to behave linearly (in its most general, time-varying, sense). The nonlinear behavior is parasitic and hence suppressed as much as possible. This makes telecom frontends weakly nonlinear in nature. In what follows, we also assume that the telecom system can be modeled as an interconnection of linear blocks and static (memoryless) nonlinearities. This also poses no great restrictions, especially when dealing with high-level models, i.e. models used in simulations at the architectural level.

4.3.1 Elementary arithmetics using exponentials

Before going into the details of more involved algorithms for dealing with systems of linear and nonlinear equations, we first deal with some elementary operators being applied to signals in exponential form. The operators being considered are

$$\left\{ \times, \int_0^t, \frac{d}{dt} \right\} \tag{4.9}$$

The effect of these operators on signals of the form (4.8) is given by:

$$A_1 t^{n_1} e^{z_1 t} \times A_2 t^{n_2} e^{z_2 t} = A_1 A_2 t^{n_1+n_2} e^{(z_1+z_2)t} \tag{4.10}$$

$$\int_0^t \tau^n e^{z\tau} d\tau = \sum_{k=0}^{n} \frac{n!}{k!} \frac{(-1)^{n-k}}{z^{n+1-k}} t^k e^{zt} - n! \frac{(-1)^n}{z^{n+1}} \tag{4.11}$$

$$\frac{d}{dt}\left(t^n e^{zt}\right) = n t^{n-1} e^{zt} + z t^n e^{zt} \tag{4.12}$$

Looking at these equations, we see the simplicity in computing the coefficients of the output signals in dealing with the most important operators. This is another argument in favor of this set of basis functions.

4.3.2 Algorithms for dealing with linear systems

Linear systems are important by themselves, for example in modeling gain blocks and filters. They also are one of the basic building blocks in the modeling of weakly nonlinear systems. In what follows we treat algorithms for dealing with both linear time-invariant (LTI), such as LNA's or filters, and linear periodically time varying (LPTV) system blocks, such as mixers, etc.

Linear time-invariant systems In case of an LTI system, the system equations are given by

$$\frac{d\mathbf{x}}{dt}(t) = \mathbf{A}\mathbf{x}(t) + \mathbf{B}v_{in}(t) \tag{4.13}$$
$$\mathbf{y}(t) = \mathbf{C}\mathbf{x}(t) + \mathbf{D}v_{in}(t) \tag{4.14}$$

combined with some initial condition $\mathbf{x}(0) = \mathbf{X}_0$. Here $\mathbf{A} \in \Re^{N\times N}$ with system poles z_k. Without loss of generality, we assume $v_{in} = \sum_{k=0}^{n} v_k t^k e^{zt}$.

The algorithm used for solving the equations (4.13)-(4.14) is, like any integration scheme, based on the fact that it is possible to generate a "template" for the solution that is in the form of equation (4.8). This leaves only the coefficients X_k to be computed. We illustrate this procedure for the case where the signal pole z does not coincide with one of the system poles z_k and when none of the system poles z_k is degenerate. Other cases are a little bit more involved, but proceed along the same lines.

Under the assumptions mentioned above, it is well known that $\mathbf{x}(t)$ takes on the form

$$\mathbf{x}(t) = \sum_{k=0}^{n} \mathbf{X}_k^v t^k e^{zt} + \sum_{k=1}^{N} \mathbf{X}_k^A e^{z_k t} \tag{4.15}$$

where the $\mathbf{X}_k^v$ and $\mathbf{X}_k^A$ have to be determined. Substituting this in (4.13) and equating the coefficients belonging to the same basis functions yields:

$$z_k \mathbf{X}_k^A = \mathbf{A}\mathbf{X}_k^A \tag{4.16}$$
$$z\mathbf{X}_k^v + (k+1)\mathbf{X}_{k+1}^v = \mathbf{A}\mathbf{X}_k^v + \mathbf{B}v_k \tag{4.17}$$
$$z\mathbf{X}_n^v = \mathbf{A}\mathbf{X}_n^v + \mathbf{B}v_n \tag{4.18}$$

Equation (4.17) can be solved for $\mathbf{X}_k^v$ through downward recursion starting from (4.18). This is always possible because of the assumption that z does not coincide with one of the system poles. Equation (4.16) learns that $\mathbf{X}_k^A$ is proportional to the eigenvector $\mathbf{V}_k$ corresponding to the system pole z_k, or $\mathbf{X}_k^A = c_k \mathbf{V}_k$. The proportionality factors can be determined out of the initial condition $\mathbf{x}(0) = \mathbf{X}_0$ which yields:

$$\sum_{k=1}^{N} \mathbf{X}_k^A + \mathbf{X}_0^v = \mathbf{V}\mathbf{c} + \mathbf{X}_0^v = \mathbf{X}_0 \tag{4.19}$$

where $\mathbf{V} = [\mathbf{V}_1, \dots, \mathbf{V}_N]$ and $\mathbf{c} = [c_1, \dots, c_N]^T$.

From a computational point of view, the easiest way to solve (4.16)-(4.19) it by using the eigenvalue decomposition of $\mathbf{A} = \mathbf{V}\mathbf{D}\mathbf{V}^{-1}$. Substituting $\mathbf{X}_k^A = \mathbf{V}\mathbf{Y}_k^A$, $\mathbf{X}_k^v = \mathbf{V}\mathbf{Y}_k^v$ and $\mathbf{B} = \mathbf{V}\overline{\mathbf{B}}$ in the equations above then results in:

$$z_k \mathbf{Y}_k^A = \mathbf{D}\mathbf{Y}_k^A \tag{4.20}$$
$$z\mathbf{Y}_k^v + (k+1)\mathbf{Y}_{k+1}^v = \mathbf{D}\mathbf{Y}_k^v + \overline{\mathbf{B}}v_k \tag{4.21}$$
$$z\mathbf{Y}_n^\nu = \mathbf{D}\mathbf{Y}_n^v + \overline{\mathbf{B}}v_n \tag{4.22}$$

which is trivial to solve.

Linear periodically time varying systems In order to derive an approach for the solution of LPTV systems, we start from the basic input-output relation for linear systems

$$\frac{d\mathbf{x}}{dt}(t) = \mathbf{A}(t)\,\mathbf{x}(t) + \mathbf{B}(t)\,v_{in}(t) \tag{4.23}$$

$$\mathbf{y}(t) = \mathbf{C}(t)\,\mathbf{x}(t) + \mathbf{D}(t)\,v_{in}(t) \tag{4.24}$$

For a given initial value $x(0) = \mathbf{X}_0$, the solution $\mathbf{y}(t)$ is given by

$$\mathbf{y}(t) = \mathbf{C}(t)\,\Phi(t,0)\,\mathbf{X}_0 + \mathbf{C}(t)\int_0^t \Phi(t,\tau)\,\mathbf{B}(t)\,v_{in}(t)\,dt + \mathbf{D}(t)\,v_{in}(t) \tag{4.25}$$

where for an Nth-order system $\Phi(t,\tau) = \Phi(t,0)\,\Phi^{-1}(\tau,0) \in \Re \times \Re \rightarrow \Re^{N\times N}$. Here the kth-column of $\Phi(t,0)$ represents the solution to (4.23) with $v_{in}(t) = 0$ and $\mathbf{X}_0 = [0,\ldots,0,1,0,\ldots,0]^T$ where the 1 is at the kth position.

For periodic systems, where $\mathbf{A}(t)\,,\mathbf{B}(t)\,,\mathbf{C}(t)$ and $\mathbf{D}(t)$ are T-periodic functions, it can be shown that [7]:

$$\Phi(t,0) = \mathbf{U}(t)\,e^{\mathbf{W}t} \tag{4.26}$$

where $\mathbf{W} \in \Re^{N\times N}$ and $\mathbf{U}(t) \in \Re \rightarrow \Re^{N\times N}$ is T-periodic. Substituting (4.26) into (4.25) and using $\mathbf{C}(t)\,\mathbf{U}(t) = \sum_m \mathbf{C}_m e^{jm\omega_0 t}$, $\mathbf{U}^{-1}(t)\,\mathbf{B}(t) = \sum_n \mathbf{B}_n e^{-jn\omega_0 t}$ with $\omega_0 = 2\pi/T$, one obtains

$$\mathbf{y}(t) = \sum_{m,n} e^{jm\omega_0 t}\mathbf{C}_m \left(e^{\mathbf{W}t}\mathbf{X}_0 + \int_0^t e^{\mathbf{W}(t-\tau)}\mathbf{B}_n \left(e^{-jn\omega_0\tau} v_{in}(\tau)\right) d\tau\right) \tag{4.27}$$

This can be seen as a downconversion of the input signal, followed by an LTI filtering operation, the result of which is upconverted again. Both up- and downconversion are trivial to implement when $v_{in}(\tau)$ is represented with respect to an exponential basis. An algorithm for LTI signal processing was already presented earlier in this section. Note that use of (4.27) allows efficient evaluation on how information moves from one frequency band to another and by which amount unwanted signal bands are suppressed.

As a final note, we mention that the matrix $\mathbf{W}$ and the coefficient matrices $\mathbf{B}_n$ and $\mathbf{C}_m$ can be determined through a number of simulations solving (4.23) an (4.24) over the time interval $[0,T]$ for different initial conditions. This can be done efficiently using classical Spice-like integration algorithms.

4.3.3 Algorithms for dealing with weakly nonlinear systems

In most cases, a telecom frontend is designed to behave in a (possibly time-varying) linear way, including amplification, up- or downconversion and filtering. Any intermodulation of the signal with itself or with neighbouring signals is often undesired and hence has to be suppressed as much as possible. This makes many frontends weakly nonlinear in nature.

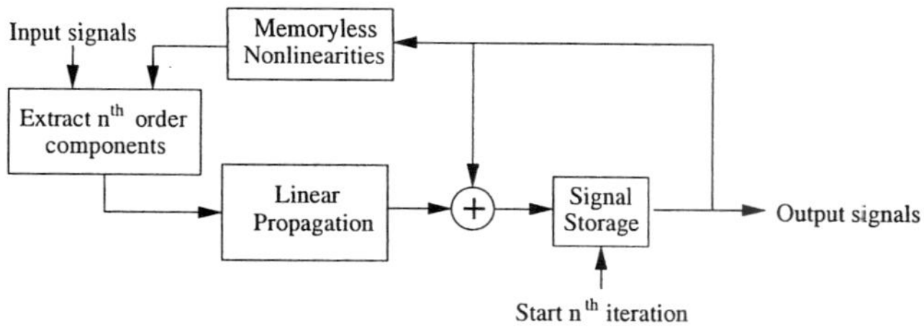

Figure 4.4 Volterra series expansion iterations

In order to simulate the behavior of these weakly nonlinear systems, it was chosen to use an approach based upon a runtime Volterra series expansion [22, 20]. It is assumed that a system can be modeled as an interconnection of linear systems and memoryless weak nonlinearities, approximated through their (multivariate) Taylor expansion. The components of the output signal $y(t) = y_1(t) + y_2(t) + y_3(t) + \ldots$ are computed in increasing order of nonlinearity, where $y_1(t)$ represents the linear (1st order) components, $y_2(t)$ the 2nd order components, and so on. A basic procedure to do so is given by the following steps:

1. Propagate the input signals through the (possibly time-varying) linear part of the system imposing the appropriate initial conditions. This results in the first-order components $y_1(t)$ of the states and the output signals.

2. Compute the second-order components at the outputs of the memoryless nonlinearities. Those are made up out of the product of 2 first-order components, like $y_1(t) \times z_1(t)$ or $y_1(t)^2$. Propagate these components through the linear part of the system. This results in the second order components $y_2(t)$ of the states and the output signals.

3. Compute the third-order components at the outputs of the third-order nonlinearities. Those can be made up out of both first and second order components of the state and output signals. $y_1(t) \times y_2(t)$ or $y_1(t)^3$ are examples of such third order components. Propagate these components through the linear part of the system. This results in the third order components $y_3(t)$ of the states and the output signals.

4. Continue this procedure until the desired order of nonlinearity has been reached.

This procedure is pictured in Fig. 4.4.

This Volterra series expansion is computationally efficient as long as the order of nonlinearity does not become too high. There is, however, one possible pitfall. Straightforward application of the strategy above will result in about N^n exponentials for representation of the nth-order distortion components, where N is the number of exponentials used to model the input signals. In order to avoid this problem, both the inputs to and outputs of the nonlinearities should be simplified. This means that they are represented using a smaller number of exponentials. Pruning the exponentials with negligible power contribution or a sample and refit procedure (see section

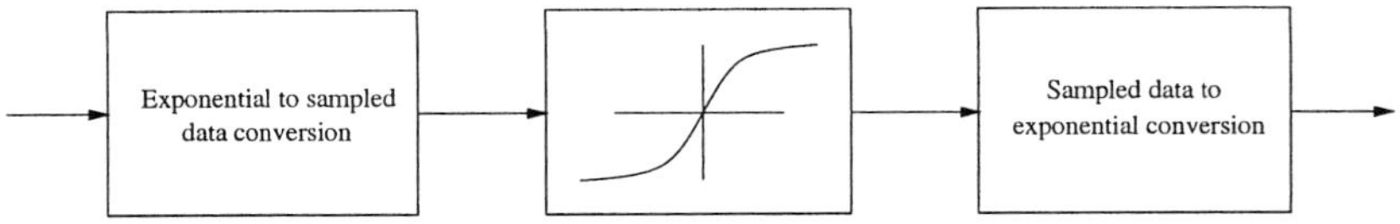

Figure 4.5 Dealing with strongly nonlinear blocks

4.3.4) are possible ways to do so. This introduces some error, which, however, if small enough, is acceptable since the nonlinearities themselves are often small compared to the desired linear signal.

A second way in which this Volterra series expansion can be made more efficient, is by only computing those components of the series expansion which are of interest. For example, if one is not interested in the high-frequency components at the output of a downconverion mixer (because they are known to very small, or because they are filtered out by some subsequent lowpass filter), they can simply be ignored during computation. It is up to the user to specify which frequency bands to compute at which node.

Besides being quite efficient in simulating weakly nonlinear systems, this Volterra series expansion has another significant advantage over other algorithms. This lies in the fact that the simulation results obtained, are, by construction, decomposed into wanted (1st-order) and unwanted (2nd-order, 3rd-order, ...) components. This greatly facilitates analysis of the results by the designer afterwards since it allows him or her to identify that part of the system behaviour causing most of the trouble in meeting the overall system specifications.

4.3.4 Conversion between sampled data and exponential representation

We conclude this section on exponential-based signal processing with a discussion of the procedure for conversion from a sampled-data to an exponential representation. This algorithm can be used in many places in the simulation of telecom frontends. It can be used to perform the necessary signal conversions in order to deal with strongly nonlinear blocks, as is illustrated in Fig. 4.5. Here, the input signal is sampled in the time domain before it is fed into the strong nonlinearity. The output is then converted back again to an exponential representation to do further processing. Another application is the reduction of the number of exponentials to represent a given signal. This can be done through sampling of the original signal followed by a refit in which the allowed number of exponentials is limited. The algorithm can also be used to model the input signals of a certain block when these input signals are given in a sampled-data form.

Fitting a set of data samples $y_n = y(n\Delta t), n \in [0, \ldots, N-1]$ to an exponential representation is finding a set of z_k, c_k with $k \in [1 \ldots K]$ such that

$$y_n = \sum_{k=1}^{K} c_k e^{z_k n \Delta t} + e_n = \sum_{k=1}^{K} c_k \zeta_k^n + e_n \tag{4.28}$$

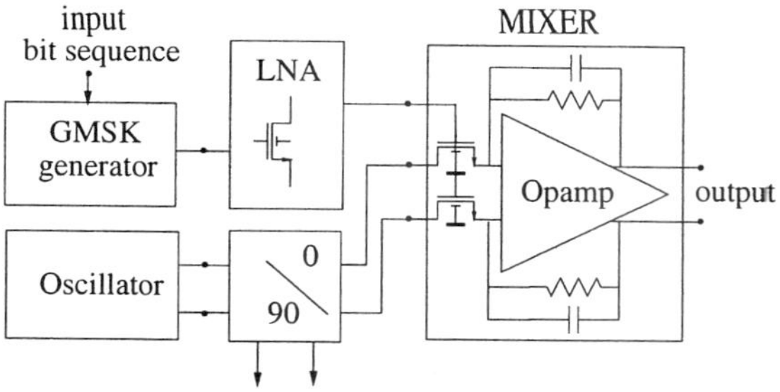

Figure 4.6 Schematic of the DCS1800 receiver.

where $\zeta_k = e^{z_k \Delta t}$ and the fit error e_n can be treated as numerical noise. Note that the samples must be taken equidistantly. Algorithms for solving this problem have for instance been developed for use in NMR spectroscopy. A good overview of the available algorithms is given in [3]. The algorithm implemented in our environment is based on an accelarated version of the state space method as described in [19]. This algorithm allows to compute the minimum number of exponentials necessary to model the input data for a given accuracy. No discretization of the frequency axis is necessary as is the case with the Gabor transform. This makes it very well suited to deal with signals for which the ratio of the time constants is non-rational. These HTLS based algorithms also perform slightly better than linear prediction (LP) fitting [3] procedures as far as accuracy is concerned.

4.4 EXPERIMENTAL RESULTS

In order to demonstrate the efficiency of our approach, we present the results from the analysis of a low-IF DCS1800 receiver [15] using the simulation approach discussed above. All of the algorithms are implemented in Matlab™. The results were obtained on a Sun ULTRA30. Both CPU times and flop counts are presented.

The DCS1800 receiver is depicted in Fig. 4.6. The RF input is a GMSK signal according to the GSM specifications, with a symbol length $T = 3.7\mu s$ at a carrier frequency somewhere around the 1.8 GHz. The low-noise amplifier (LNA) is represented with a 12th-order small-signal model extracted from a Spice netlist. For the mixer opamps, a two-stage macromodel was constructed where each stage is modeled using a transconductance plus a resistive and capacitive load. The transconductances are nonlinear such that the mixer has an IIP3 of 14dBm. The oscillator is taken to be an ideal cosine. Note that Fig. 4.6 only shows the inphase mixer, but the total system contains a quadrature mixer as well.

In a first step, an exponential model of the input signal is computed. This is done by fitting the complex baseband equivalent $e^{j\Phi(t)}$, the result of which is shown in Fig. 4.7. It takes about 0.1 seconds of CPU time and 1 Mflops (per symbol) to compute them. After upconversion, which has now become a trivial operation, this results in an exponential model for the high-frequency modulated output signal $\cos(2\pi f_c t + \phi(t))$. Two different models were constructed, one wich is accurate up to -40dB and one up to -60 dB. As was already illustrated in Fig. 4.3, the higher accuracy is at the cost of

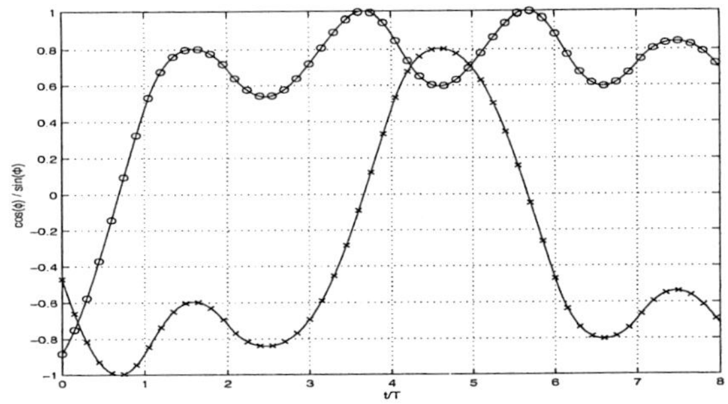

Figure 4.7 $\cos(\phi)$ and $\sin(\phi)$ as a function of time. The solid lines represent the sampled-data values. The x-marks and circles represent samples of the exponential fit of respectively $\cos(\phi)$ and $\sin(\phi)$.

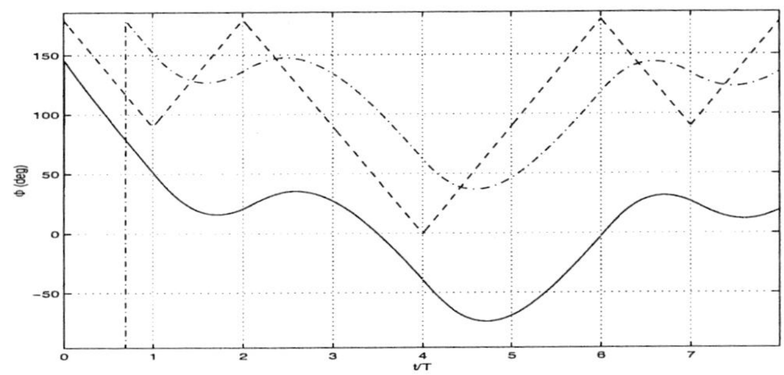

Figure 4.8 Input phase before (dashed line) and after (dashdot line) the Gaussian GMSK filter and the output phase extracted out of the mixer output signals (solid line).

a more complex model (more exponentials). Since these models need to be computed only once (they can be reused for different simulations), the corresponding CPU times and flop counts are not taken into account in the simulation times below.

Next, the input signal is propagated through the receiver, while varying the stepsize Δt of the simulation subintervals. This means that the exponential signal models at each node must be valid over a range Δt. This is repeated for both the mixer opamp nonlinearities turned on and off and for both levels of modeling accuracy for the input signals. When computing the nonlinear system behavior, the high-frequency components (at 3.6 GHz and higher) are neglected, since their power is negligible. In order to avoid explosion of the number of exponentials, signals are simplified using a sample and refit procedure. Fig. 4.8 shows the input phase and the output phase, as extracted from the outputs of the inphase and quadrature mixers. The resulting CPU times and flop counts per symbol period T are shown in Fig. 4.9 and Fig. 4.10 respectively. These figures illustrate how the overall simulation time can be optimized by selecting an optimum value for the timestep Δt. They also show how this optimum depends upon the properties of the system being simulated. For simulation of the ideal system behavior only, the timestep Δt can be taken very large. The resulting decrease

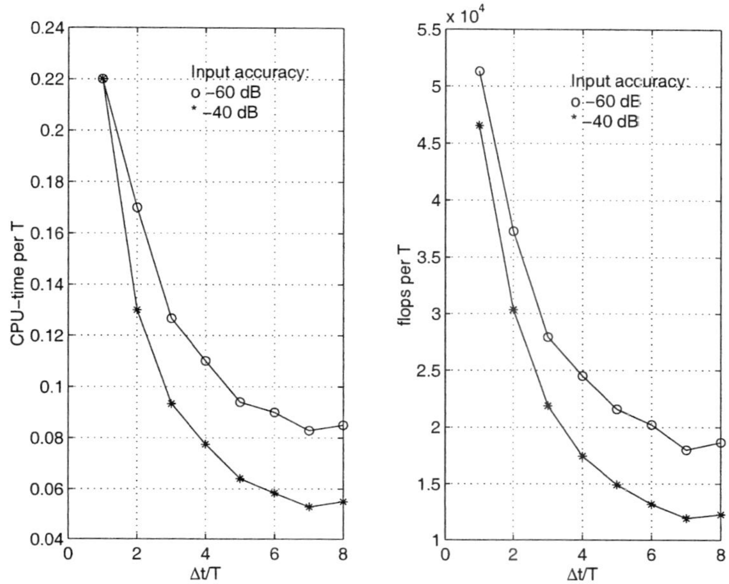

Figure 4.9 CPU times and flop counts per symbol period T versus the stepsize Δt when the mixer opamp behaves strictly linearly.

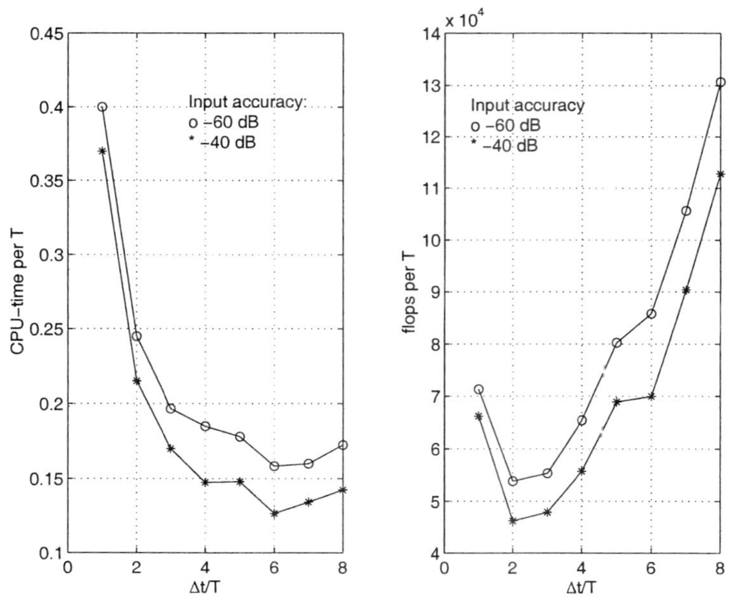

Figure 4.10 CPU times and flop counts per symbol period T versus the stepsize Δt when the mixer opamp contains nonlinear behavior.

f_0/f_T	CPU gain	Flop gain	Acc./Dist. (dB)
4	8.0	8.2	-17
8	6.0	6.4	-20
16	11.3	12.4	-22
32	20.7	22.9	-23
64	42.4	44.9	-23
128	73.4	70.0	-25

Table 4.2 Comparison between classical polynomial based integration methods and an exponential approach.

in the number of simulation subintervals is more important than the increased modeling complexity (number of exponentials) per subinterval. When the opamp behaves nonlinearly, the simulation algorithm generates extra exponentials to model the nonlinear signal components. For large Δt, this increase in signal modeling complexity becomes dominant and the overall simulation time starts to increase. In this case it is interesting to decrease the timestep compared to the case where the opamp behaves linearly. The difference in Fig. 4.10 between CPU times and flop counts is due to the extra overhead in Matlab function calls and memory management. Note that in both the linear and nonlinear case, the timestep Δt is still a multiple of the period T of the GMSK symbols.

In a second experiment, the exponential approach is compared to the SPICE-like Matlab integration method *ode15s*, which is a variable-order, adaptive-stepsize method suited for solving stiff problems. The comparison was performed by applying a modulated GMSK signal to an opamp-RC lowpass filter (similar to the mixer in Fig. 4.6, but with the MOSFETS replaced by their (unmodulated) resistance, opamp nonlinearities included). The timestep for the algorithm based upon complex damped exponentials was chosen to be T, the length of one GMSK symbol. This step is independent of the carrier frequency f_0. The experiment is repeated for increasing values of the modulation frequency f_0. Table 4.2 presents the gain in both CPU time and flop count, obtained by using the exponential approach. The last column compares the energy contained in the difference between both methods and the energy contained in the signal distortion components (smallest relevant signal components). It is clearly seen that the performance gain increases about linearly with f_0/f_T, the ratio of the carrier frequency over the signal frequency. This is because of the increasing timestep of the *ode15s* integration method. The behavior of the CPU and flop gains for low modulation frequencies are explained by the fact that in these cases the time step of the *ode15s*-routine is determined by the impulse responses of the amplifier, and not by the frequency content of the input signal.

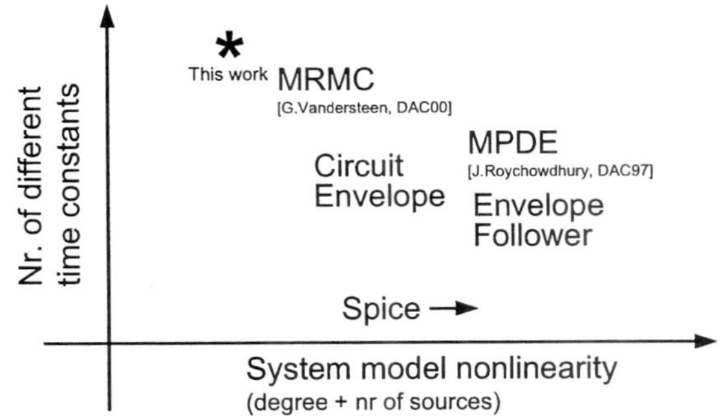

Figure 4.11 Situating complex damped exponential based simulation and some other algorithms as a a function of the characteristics of the (input) signals and system models involved.

4.5 CONCLUSIONS

The importance of telecommunication systems justifies the research towards algorithms specifically targeted to increase simulation speed by incorporating telecom system and signal properties into the algorithm. This chapter presented such an approach based upon a complex damped exponential signal model that incorporates the typical properties of digitally modulated telecom signals, like their many different time constants, in a natural way. This allows to construct simple signal models (containing only a few basis functions) that are valid over a long time interval. This results into a significant increase of the simulation time step, speeding the simulations up. This exponential signal model was combined with a runtime Volterra series expansion into an algorithm which is particularly well suited for the simulation of weakly nonlinear (telecom) systems. The algorithm is absolutely stable and avoids the necessity of a global signal model (i.e. it allows node independent signal representations). The algorithm also allows to compute wanted and unwanted signals separately. This, together with the natural relationship between the exponential signal model and frequency content, greatly facilitates analysis of the results by the designer.

As a final conclusion, we summarize the properties of the algorithm here described by situating it's optimal use as a function of the characteristics of the (input) signals and system models involved. This is as shown in Fig. 4.11 where, for sake of comparison, we also include a (qualitative) positioning of some other algorithms that allow for transient simulation. Again, as was already mentioned, this stresses that the algorithm works best for weakly nonlinear systems with signals and models containing largely different time constants. This is typical for a lot of telecommunication applications.

References

[1] P.Brown and Y.Saad, "Convergence Theory for Nonlinear Newton-Krylov Algorithms", Technical Report UCRL-102434-REV-1, Lawrence Livermore National Laboratory, 1992

[2] P.Feldmann, B.Melville and D.Long, "Efficient Frequency Domain Analysis of Large Nonlinear Analog Circuits", In *Proc. CICC*, pp. 21.6.1-21.6.4, 1996

[3] De Beer and Van Ormondt, "Analysis of NMR data using time-domain fitting procedures", In *In-vivo Magnetic Resonance Sectrocospy I: Probeheads, Radiofrequency Pulses, Spectrum Analysis* (M. Ruding and J. Seelig, Eds), pp. 201-248, Springer-Verlag, Berlin/Heidelberg, 1992

[4] G.H.Golub and C.F.Van Loan, *Matrix Computations,* 3rd ed., The Johns Hopkins University Press, Baltimore and London, 1996

[5] K.Kundert and A.Sangiovanni-Vincentelli, "Simulation of Nonlinear Circuits in the Frequency Domain", In *Trans. on CAD*, Vol. CAD-5, No.4, pp. 521-535, October 1986

[6] K.Kundert, J.K.White and A.Sangiovanni-Vincentelli, *Steady-State Methods for Simulating Analog and Microwave Circuits,* Kluwer Academic Publishers, Dordrecht, 1990.

[7] R.Lamour, R.März and R.Winkler, "How Floquet-Theory Applies to Differential-Algebraic Equations", Humboldt-Univ, Inst. für Mathem., Preprint 96-15, 1996

[8] O.J.Nastov and J.K.White, "Time-Mapped Harmonic Balance", In *Proc. DAC*, Session 36, 1999

[9] O.J.Nastov and J.K.White, "Grid Selection Strategies for Time-Mapped Harmonic Balance Simulation of Circuits with Rapid Transitions", In *Proc. CICC*, Session 2, 1999

[10] O.Narayan and J.Roychowdhury, "Multi-Time Simulation of Voltage-Controlled Oscillators", In *Proc. DAC*, Session 36, 1999

[11] J.G.Proakis, *Digital Communications*, McGRAW-HILL, 1995

[12] J.Rowchowdhury, "Efficient Methods for Simulating Highly Nonlinear Multi-Rate Circuits", In *Proc. DAC,* pp. 269-274, 1997

[13] Y.Saad, *Iterative Methods for Sparse Linear Systems,* International Thomson Publishing, 1996

[14] D.Sharrit, "New Method of Analysis of Communication Systems", *MTTS'96 WMFA: Nonlinear CAD Workshop*, June 1996

[15] M.Steyaert et al., "A 2V CMOS Cellular Transceiver Front-End", In *Proc. of the ISSCC*, Session 16, 2000

[16] R.Telicheveski et al., "Efficient Steady-State Analysis based on Matrix-Free Krylov-Subspace Methods", In *Proc. DAC*, Session 28, 1995

[17] R.Telichevesky et al., "Fast Simulation Algorithms for RF Circuits", In *Proc. CICC*, pp. 21.1.1-21.1.8, 1996

[18] P.Vanassche, G.Gielen and W.Sansen, "Efficient Time-Domain Simulation of Telecom Frontends Using a Complex Damped Exponential Signal Model", *In Proc. DATE2001*, Session 4A, Munich, 2001

[19] S.Van Huffel et al., "Algorithm for Time-Domain NMR Data Fitting Based on Total Least Squares", *Journal of Magnetic Resonance*, A 100, pp. 228-237, 1994

[20] I.Vassiliou and A.Sangiovanni-Vincentelli, "A Frequency-Domain, Volterra Series-Based Behavioral Simulation Tool for RF Systems", In *Proc. CICC*, pp. 2.5.1-2.5.4, 1999

[21] J.Vlach and K.Singhal, *Computer Methods for Circuit Analysis and Design,* Van Nostrand Reinhold, 1983

[22] P.Wambacq and W.Sansen, *Distortion Analysis of Analog Integrated Circuits*, Kluwer Academic Publishers, 1998

[23] B.Yang and J.Philips, "A multi-interval Chebyshev collocation method for efficient high-accuracy RF circuit simulation", *In Proc. DAC*, Session 11, 2000

5 HIGH-LEVEL POWER ESTIMATION OF ANALOG FRONT-END BLOCKS

Erik Lauwers, Georges Gielen

ESAT-MICAS
Katholieke Universiteit Leuven, Belgium
E-mail: [Erik.Lauwers, Georges.Gielen]@esat.kuleuven.ac.be

When talking about high-level design techniques for telecom front ends or mixed-signal integrated systems in general, many different aspects are involved. These include: architecture selection, simulation, behavioural modeling, synthesis, performance modeling, ... This chapter focuses on performance modeling and more precisely on the estimation at a high level of abstraction of the power consumed by an analog building block of a telecom front end.

First, high-level design in general and the relation between power estimators and high-level design will be highlighted: where do power estimators come into play and why. This also leads to a good definition of what a power estimator is. Next, before elaborating on how a power estimator can be constructed, a few reflections about the accuracy of power estimators are given. This is followed by examples of fundamental estimators and their limitations. Finally a few practical examples are given.

5.1 SYSTEM DESIGN OF TELECOM FRONT ENDS

The design of analog telecom system front-ends is a difficult task. Therefore, the system-level architecture and building block specifications are normally defined by a very experienced designer. Even then, the new architecture is usually based on an old existing design in which some modifications have been done that take into account any extra requirements or a new technology. The system specifications are typically refined into the specifications of the subblocks in the same way or with the help of a spreadsheet and eventually some intuitive relations. This approach however does not guarantee an optimal solution, and although the system will probably work, changing only one parameter in the system may jeopardize the resulting performance.

To better illustrate the problem, take a look at *Figure 5.1* which depicts a simple receiver front end. The total available noise-plus-distortion budget is divided over the

P. Wambacq et al. (eds.),
Low-Power Design Techniques and CAD Tools for Analog and RF Integrated Circuits, 83–105.

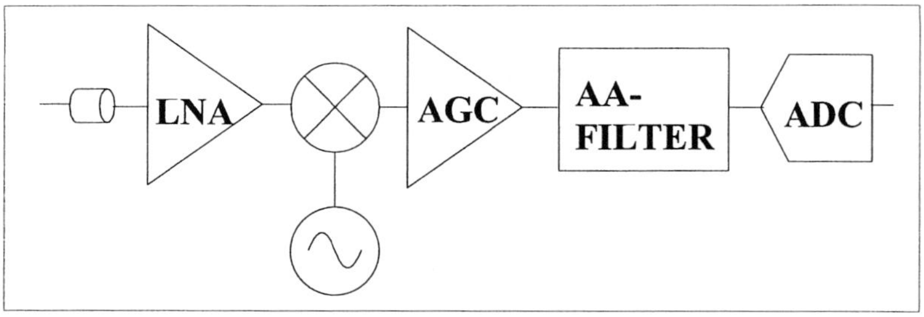

Figure 5.1 Example: a simple receiver front end

different building blocks by an experienced designer and the design is well on course to meet the preset time to market. Suddenly, a competitor presents (on a conference) a similar receiver but with better specifications, for example 1 bit more accuracy in the analog to digital converter. What are the options left to rapidly change the design to meet the new standard market specifications? The answer is left to the imagination of the reader.

To support the analysis of such problems, efforts are being done to construct CAD tools that formalize and speed up the system design phase and this while minimizing a given cost function which in most cases is power minimization. In the next paragraphs, the design cycle is refined and the problems that occur are discussed in order to converge to the how and why of analog power estimators.

5.1.1 Localising the digital and the analog part

Many integrated systems interact with the environment through analog signals while their core consists of digital electronics. This implies that part of the system, usually the interface to the environment, consists of analog electronics. This is not different for telecom systems. Also, this means that during the design of the system, a division has to be made between what will be done digitally and what will be done using analog parts. Traditionally this partitioning is done based on experience and heuristics. Once this is done, the analog and digital parts are developed and simulated separately and then brought together, eventually on one die but sometimes on separate dies, to form the complete mixed-mode integrated circuit. To be able to simulate the total system, mixed-mode simulators are developed. It is worth noting that once the initial division is made, there is no further interaction between the digital and the analog part. This accentuates the importance of a right choice at this level. An example can make this easier to understand. Imagine a system that uses a downconversion mixer to modulate the signal from a frequency suited for transport through the application medium, for example a coax cable, to the baseband. The signal is first converted from the carrier frequency to an intermediate frequency and then modulated a second time to the baseband. Both mixing operations are either done in the analog domain, or only one is done in the analog domain and the second one from IF to baseband is done digitally or in the digital time-discrete domain (*Figure 5.2*).

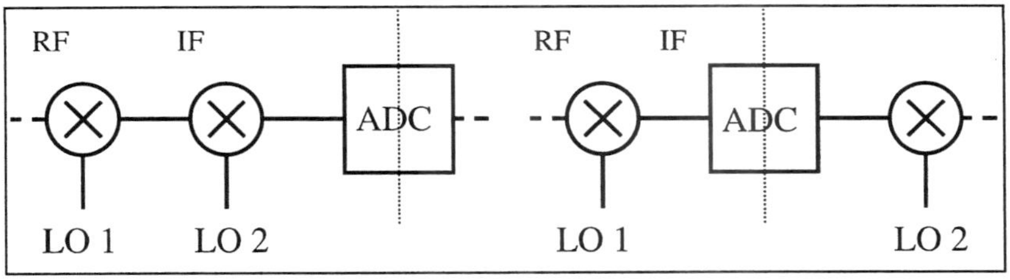

Figure 5.2 Two possible solutions: 2 analog mixers or 1 analog and 1 digital

This choice however greatly affects the system characterisation. Doing both down-conversions with analog mixers requires more analog design effort because of the extra phase noise, distortion, ... that have to be taken into account. On the other hand, if one conversion is done digitally, then there is a need for a faster analog-to-digital converter (ADC) and more digital electronics are required. This means that there is in this system a trade off between extra analog electronics and thus more power consumption or less analog and more digital electronics but also with extra ADC power consumption due to the higher requirements.

Two important fundamental steps of automated system design are present at this level of the design: architecture selection and simulation.

5.1.2 Refining the analog part

Once the trade-off has been made between analog and digital operations on a high level of abstraction, there is a second trade-off to be made. When zooming in on the analog domain, it can be understood that this part also consists of many functional blocks. Although the total transfer function of the analog part is fixed, because of the application of the system under design and the division in an analog and a digital part, the way to achieve this transfer function is not yet definitive. There are still many degrees of freedom internally that can be exploited. The analog system architecture can vary, but even when that is fixed degrees of freedom do exist. As a simple example, one could think of the overall gain. This gain can be generated in one single block or can be built up throughout consecutive blocks. The aim would be to distribute the gain in such a way as to optimise a certain parameter such as total system power consumption or area. Another example is the division of the total allowed distortion among different blocks (*Figure 5.3*). If the combination of a low-distortion buffer followed by a sample and hold with less stringent distortion requirements consumes less power than another configuration but with the same total distortion, then that might be a better solution (of course, in general, there are still other considerations to be taken into account).

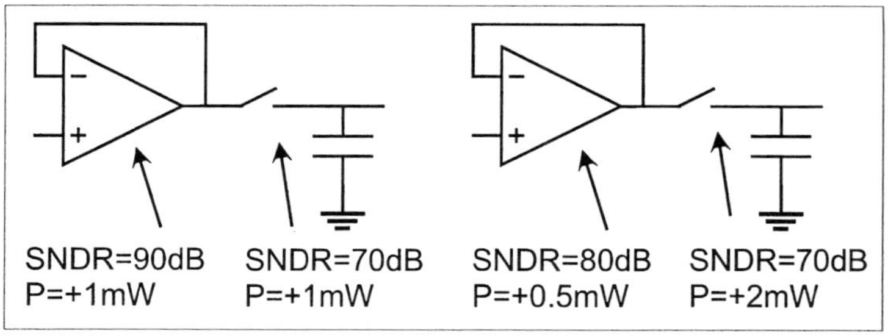

Figure 5.3 Power versus accuracy trade-off

5.1.3 Conclusions from current system design and possible improvements

The conclusion to be drawn from all the above is that designing a system, or at least specifying a system is not that trivial. A first major improvement would be to be able to specify the total system, including digital and analog electronics, globally. This means considering everything before making any split up. In order for this approach to work, good models are needed for all the usable building blocks. These models should describe the ideal functionality of the block in terms of key performance parameters such as (amongst many others) bandwidth, gain, frequency shift, ... Furthermore, they should describe or contain the most important signal degradation mechanisms which are the cause of the non-ideal behaviour of the building block and which are characteristic for that block and for the given application. This clearly leaves some place for interpretability: what non-idealities have to be taken into account for a first-order system design on a high level of abstraction and which not? This question has to be addressed but not here where the focus is on power estimators.

Even with these models available, it is still impossible to design a total system. This is because there is always more than one solution for achieving the wanted signal manipulation. What is needed is a cost function that has to be optimised in order to compare the different alternatives. Normally, a system has to be designed for minimal total cost, which includes minimal area, minimal power consumption, maximal yield, etc .... These extra requirements enable the design tool to construct the best design with the given constraints and the available models. Nowadays, certainly the minimal power consumption constraint is very important due to increasing mobile telecommunication applications and the ever-larger market of portable computing. In addition, as CMOS deep submicron technology enables smaller minimal transistor dimensions, more transistors are integrated on one chip while keeping the same total size, causing problems for heat dissipation through the packaging. And of course environmental arguments can also be given. This means that a clear need for power estimators exists.

5.2 ANALOG POWER ESTIMATORS

This paragraph focuses on the requirements for analog power estimators and on how they can be constructed. A good definition for a power estimator is: A power estimator is a function that returns an estimated value for the power consumed by a functional block when given some relevant input specifications for that block without knowing the detailed implementation of the block. Note that this is a general definition valid for analog as well as digital electronics. Detailed implementation can be a description in terms of smaller subblocks such as operational amplifiers but can in the worst case also be the transistor level.

5.2.1 High-level power estimation

Before discussing techniques to construct power estimators, a few reflections about high-level power estimators are given. Power estimators are useful for high-level design problems and have to be used in the first place for first-order system design. This is an important statement because it holds two of the most important messages regarding power estimators. The estimators must have as input parameters only high-level block parameters. Secondly, the accuracy of the estimated value with respect to the exact, finally measured power consumption of a real transistor-level implementation of the block must only be within a first-order range. Both statements will now be explained in more detail.

When designing a system on a (moderately) high level, a designer makes use of block diagrams. The different blocks are for example for a telecom front-end (*Figure 5.1*): analog to digital converters, low-noise amplifiers, filters, mixers, ... Along with each block the designer will define parameters that are important for the performance of the total system. For example, when a digital to analog converter (DAC) is followed in the system by a low-pass filter, then supposedly only the signal-to-noise ratio (SNR) of the DAC is important and not its spurious-free dynamic range. Other parameters can be provided but they are not essential for the system behaviour. Because complexity of the design grows with each extra parameter taken into account, it can easily be understood that the set of used parameters on the high system level has to be minimized. The minimized set is then what is here called: the high-level parameters. When looking at the possible input parameters for power estimators, this set can probably be even more minimized. Only those parameters have to be taken into account, which have a major influence on the power consumption, because a very high absolute accuracy of power estimation is neither achievable nor necessary as will be explained below. Examples of such (reduced) parameter sets can be found in the section with fundamental and practical examples.

Defining the required accuracy of a power estimator is a much more difficult task and can certainly be subject to many discussions. Of course, the higher the accuracy, the better. But to what extent does this has to be followed? Does the estimator has to predict the power within a 1% error margin? Or is 100% sufficient? Maybe, on a truly high level it would be better or at least more realistic to talk about orders of magnitude. The decisive factor is the amount of detail that can be used without compromising the high-level approach. This is best explained with an example. If one is designing a total telecom system, then the analog receiver front end is just a building block. It doesn't really matter at this point whether the estimated power consumption of an analog filter in it is exactly estimated and an estimation within a few orders of magnitude will probably do. If however the level of abstraction is the analog receiver front end, then one order of magnitude is already much. Of course, if only one filter implementation is considered, then 100% is even too much for a good estimator. For a high-level estimator, as a good rule of thumb, an estimate within an order of magnitude is considered to be good and around 100% is very good.

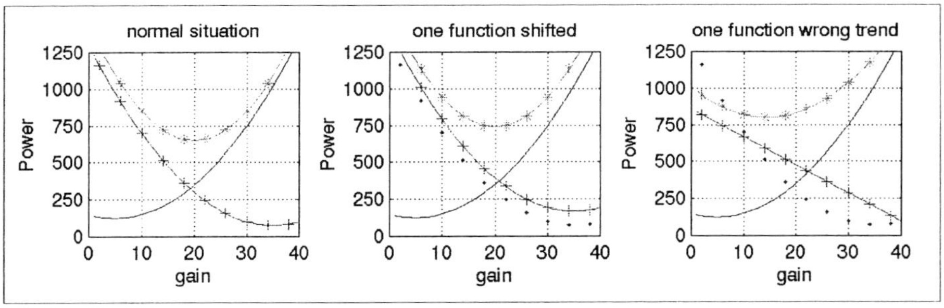

Figure 5.4 Tracking and accuracy example

Luckily another facet of the power estimator has to be considered. When an input parameter varies, the estimator has to correctly follow this change. A few intuitive examples can easily be understood. In a filter when the order increases, then so does the power consumption. Another example is taken from the design of high-speed data converters. If a higher number of effective bits is wanted, then the power consumption has to increase as well or the speed has to go down.

The relative accuracy with which the power estimate tracks this parameter Tracking as to be good. The simple reason being that when power estimators are used on a high system level, it is exactly this property that is the most important when comparing different architectural alternatives. Since one is looking for an optimal power consumption, parameters have to be varied and exchanged between different building blocks and even over the border of analog and digital design. Only if the parameter variations are well modelled by the estimator, trend analyses can be performed accurately. If an optimum exists, it is more important to know where it is than to know what the exact calculated value is. An example to illustrate this hypotheses is given in *Figure 5.4*. Imagine a system of two high-level blocks and with a fixed total gain. The power estimator of each block is a function of the gain of that block. If the gain of one is chosen, the other is fixed, so the total system power can be estimated. The power estimator for both blocks as a function of the gain of one of the blocks is sketched in the figure. On the left is the normal situation, when both estimators (normal curve and the '+' curve) are correct. An optimum can be found by looking at the sum of both (the '*' curve) around a gain of 20. The middle figure shows the situation when one of the estimators has the correct form but is shifted over a certain amount. The dots in this figure indicate the correct estimator. It can be seen that the location of the optimum is not shifted. The total amount of power is wrong but it is still a true optimum. In the figure on the right, the trend was badly chosen (linear instead of square dependence), and the result is that now the optimum is at a gain that is lower (around 15) and the real result is not optimal anymore.

So, the conclusion is that the absolute accuracy is more important for architecture exploration and that the tracking behaviour is more important for the distribution of specifications within one architecture. But both need to be reasonable in either case. For tracking it is harder to give orders of magnitude, but a good rule could be that the optimum should not deviate more than a certain percentage from the exact case. This is much harder to verify.

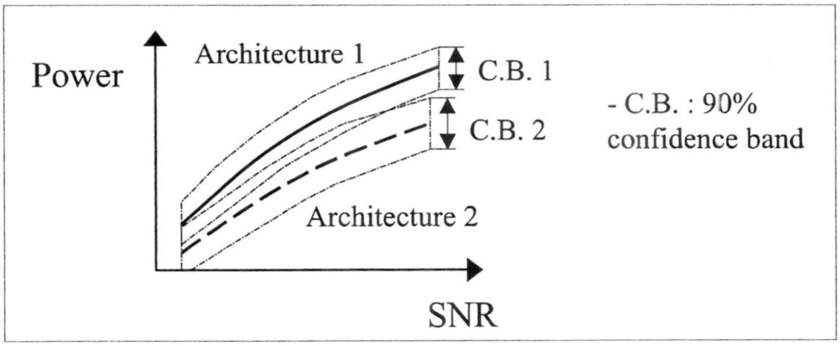

Figure 5.5 Statistical point of view regarding the accuracy of a power estimator

How do these inaccuracies have to be taken into account when designing a system on a high level of abstraction? Imagine (see *Figure 5.5*) that the power of the total system is estimated as a function of the signal-to-noise ratio (SNR). Two architectures are withhold and architecture two is better over the complete SNR range compared to architecture one. If now also the confidence interval consideration, it is better to state that architecture two is significantly better than architecture one with a certain significance level. This is a much safer statement, given the high level the design is being done at, and certainly more valuable when all the different system specifications have to be traded off against each other.

5.2.2 Construction of power estimators

After creating the framework for power estimators, now a more practical aspect is highlighted, namely the way of constructing power estimator functions. For obtaining the power estimator as a function of the input parameters, there are several possible approaches. Here the two basic approaches will be described which are both feasible and which both have their shortcomings and advantages. Other methods are possible but they typically are a combination of both basic approaches given below. The two methods are the bottom-up method and the top-down method.

5.2.2.1 First approach: bottom-up. In the bottom-up method a certain topology then from this exactly known schematic, equations are derived. In this way the power behaviour of the analog block is modelled. Topologies can be collected in a library of known designs. When a set of input parameters is given, a function must search the library for an existing design that matches them best. Another possibility is to design the different blocks once, model them and put these in a library. Then of course, the involved effort is much higher. The essence stays the same: at the base there is an implementation, described by the complete transistor netlist. See also *Figure 5.6*.

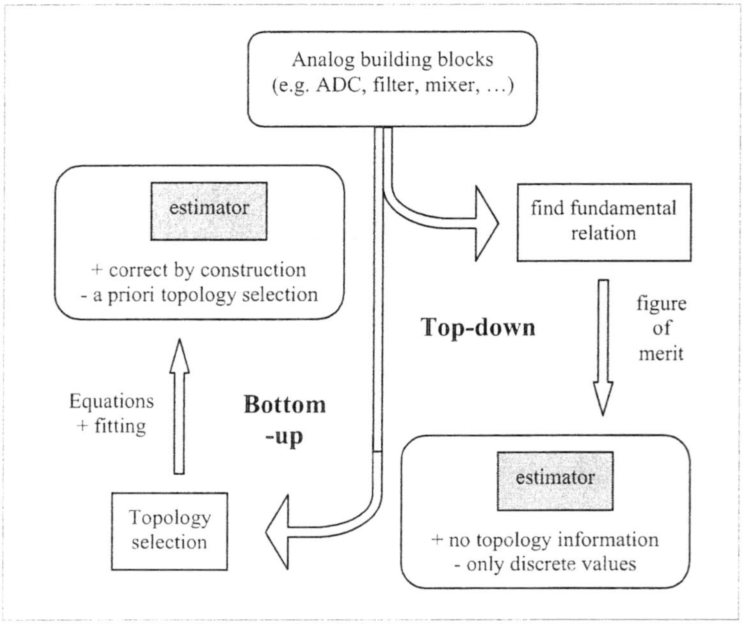

Figure 5.6 Overview of the top-down and bottom-up approaches for analog power estimation and their major properties

The most obvious disadvantage of this approach is that first a topology has to be chosen for each building block. This is fine if all possible topologies are included in the library, but unfortunately this is most often not the case. Including other topologies is to a certain practical extent possible but time consuming.

Also, the method inherently limits itself to discrete data points. This problem could be eased by making the entries in the library a little more continuous. For example, for one type of design several parameters could be varied over a certain range so that more accurate points can be found in terms of matching the parameters as closely as possible. Also, the design could be given in different technologies. The drawback of achieving these extended libraries is again that it takes more time to build them up correctly and that new entries, needed to stay up to date with the evolving technology, are more difficult. Additionally, when an existing design is taken and added to the library, it is never clear whether or not this design consumes the minimal achievable power or in other words, it is not sure whether or not it is an optimal design. This is not necessarily a problem since it can have other parameters that are very good and this might then relax the requirements on another building block in the system so that still a better minimal power consumption is obtained. The only way to not have this problem is to design the circuits and to model them yourself. Even with these estimation regions around each topology, it is never clear whether they overlap and thus whether an exact solution exists. Extrapolation between the regions is dangerous and has to be avoided.

As a last negative point, consider the case where all but one input parameter fits a certain high-level design requirement. It would be very tempting to just extrapolate the result of that design to the exact given parameters. This however is wrong because there is no indication whatsoever about the sensitivity of that particular design towards the power consumed as a function of the considered parameter.

Apparently this method has not much good in it. However one clear advantage exists that still gives a lot of credit to this method despite all the negative properties mentioned above. The advantage is that the models and estimators are exact and therefore are very feasible for real designs: they are correct by construction. Even the parameter tracking of the power consumption estimation of one topology can be made very accurately because the whole netlist is known and can be very well modelled.

To conclude, this method is a trade off between invested time on the one hand and completeness of the design space span of the estimator on the other hand. Accuracy is merely a problem of modeling than of the used method in this case.

5.2.2.2 Second approach: top-down. The second method, the top-down method, is much more optimal for real system-level design. For a certain analog function, for example analog to digital conversion, a fundamental relation is found that expresses the power consumption in terms of the high-level parameters. The set of input parameters does not have to be complete in the sense that not all design aspects have to be covered. Only those parameters that have a major impact on the power consumption have to be taken into account. Basically this means that the details have to be left out of the estimator, resulting unavoidably in a less accurate estimator. No assumptions are made regarding the topology of the building block leaving, all (and hopefully original) solutions open. This fits well in the idea that when making a real top down design, nothing is known in advance about the underlying, exact implementation in terms of transistors. The result is a set of simple equations that are very suitable for implementation in fast system exploration tools. See also *Figure 5.6*.

A drawback of this approach is that good accuracy of the models is often difficult to achieve because of the typical nature of analog design where one transistor more or less can have a great impact on the behaviour and specifications of the design. This also explains why it is more difficult to obtain a general, topology and technology independent model. The use of fitting parameters typically compensates this shortcoming towards different topological implementations of the same analog block. This approach still limits the use in system design because only discrete solutions can be used: every topology has its own fit values. This could be traded against the absolute accuracy by taking the average of the fit factors as a global fit factor for one analog function. The use of fit factors is unavoidable when designing truly top-down estimators because if the exact power relations are written down, then one has the exact implementation, and this of course means that the system is known and no optimisation can be done anymore. Finding the fit factors, however, is not a very time-consuming job and should not weigh up to the advantages mentioned before.

Another drawback is that in order to build up a good estimator, one first has to know and understand the considered building block very well in order to make the right simplifications. One has to first find out which details can be left out and which not. This takes a lot of time and is probably not always possible, at least not up to a practically correct enough estimator.

5.2.2.3 Illustration of both approaches. Before comparing both methods and drawing a conclusion, a simple example to illustrate each method is given. Also, as a link between system design and the construction of power estimators, in *Figure 5.7*, a small schematic overview is given. This schematic is typically used in analog synthesis tools [3]. The terms bottom-up and top-down are used in a similar way. The main similarity is that when doing the bottom-up part of the design cycle, all design parameters are exactly known. The difference is that first a top-down design step is performed to design an optimised circuit and that afterwards the bottom-up design step is used as verification. If the result is not satisfactory, an iteration is performed. Both steps are part of the design cycle, i.e. both steps have to be executed. For a power estimator this verification loop is usually not used and only one approach is used.

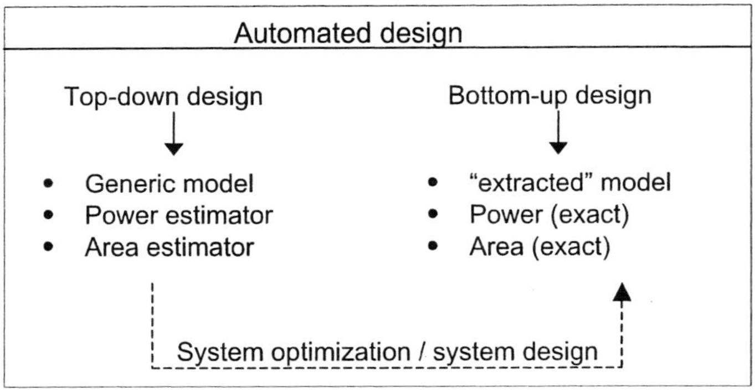

Figure 5.7 Automated design flow

The two approaches are now illustrated with two different examples. Both examples are taken from the field of high-speed analog to digital converters (ADCs).

The first example is taken from [1] and is an example of a bottom-up approach:

$$f_{\mathrm{in,max}} = \frac{2 \cdot \mathrm{BW}_{\mathrm{fold}}}{\pi \cdot \mathrm{F_F}} \tag{5.1}$$

In equation (5.1), $f_{\mathrm{in,max}}$, the maximum input signal frequency, is expressed as a function of the bandwidth of the folding pre-processing circuit and the folding factor F_F. It uses information specific to the used topology (F_F) to model the external specification $f_{in,max}$. It is clear that this equation is only valid for this exact topology and can only be used in system-level design tools if this restriction is not a problem. Likewise for this ADC one could derive a function expressing the power consumption as P=function(BW_{fold}, F_F, F_{sample}, Bits, F_{signal}, ...) but this takes time and has to be repeated for every (even little bit) different topology or technology.

The second example taken from [2] is an example of a top-down approach:

$$P_{ADC} = FOM \cdot DR \cdot SR \cdot 10^{12} \tag{5.2}$$

It expresses the power as a function of the dynamic range (defined as DR=2^N with N the number of bits), the sampling rate (SR) and a figure of merit (FOM). The FOM of several ADCs are also given in [2]. The equation contains no information about the topology of the ADC, the only fit parameter used is the FOM. When a new ADC is published, one can easily calculate the FOM. The estimator will always give discrete values and will in this way limit the effectiveness of the system-level design tool that uses it. No trend analysis is possible as interpolation between (completely) different designs is doubtful practice and probably leads to wrong results.

5.2.2.4 Conclusions. From an ideal point of view, the top-down approach leaves the system designer with the most freedom. Unfortunately, it is not sure whether or not for a given functional block exact fundamental relations regarding power consumption valid for all different designs can be found. Therefore, very often in practice a combination of both approaches will have to be used. Examples are given in the next section with practical power estimators. The reality is more like a meet-in-the-middle solution.

5.3 EXAMPLES OF FUNDAMENTAL RELATIONS USED TO ESTIMATE THE POWER CONSUMPTION

In the next two sections, examples are given that illustrate practical power estimators. In this section, the examples are limited to two existing estimators that express a fundamental relation valid for the analog circuit under consideration. The first one is valid for analog filters. The second one is a more general approach and specifically an example for a mixer is provided in more detail.

5.3.1 Analog integrated continuous-time filtering

The first example is taken from the field of analog continuous-time integrated filters. It can be found in [4], which is also a nice overview of the field. Based on the design of an optimal dynamic range filter, an expression for the total capacitance is given:

$$C_{tot} = \alpha \cdot \frac{kT \cdot Q}{V_P^2} \cdot \left(\frac{S}{N}\right)^2 \tag{5.3}$$

$$P = C_{tot} \cdot Vdd^2 \tag{5.4}$$

Combining equation (5.3) and (5.4) and assuming that V_P, the peak signal value, is proportional to the power supply voltage (Vdd), equation (5.5) is obtained.

$$P = \eta \cdot kT \cdot Q \cdot f_0 \cdot \left(\frac{S}{N}\right)^2 \tag{5.5}$$

It is an expression valid for a second-order filter section. F_0 is the filter's centre frequency, Q is the quality factor, k is Boltzmann's constant, T the absolute temperature and S/N is the maximum signal-to-noise ratio. The two constants α and η are heavily implementation dependent (filter order, specifications, topology, active elements, ...). This means that only rough comparisons (trend analysis) between very similar filters are possible. Comparison of different filters with one fixed constant results in absolute accuracy deviations of orders of magnitude. Useful rules of thumb for estimating the total capacitance can also be found in [5] and can be combined with equation (5.4).

5.3.2 Analog signal processing: mixers

In [6] a theoretical minimum power consumption is derived valid for any signal processing building block. The difference between the theoretical general value and the specific value for a certain analog function is described by an efficiency parameter.

$$\eta = \frac{P_{\min}}{P} = DR \cdot \frac{BW \cdot kT}{P} \tag{5.6}$$

$$\eta = \eta_i \cdot \eta_s \cdot \eta_p \cdot \eta_n \tag{5.7}$$

$P_{\min}$ is the theoretical minimum power consumption and P is the actual power consumption. The efficiency η is split into four parts (equation 5.7).

The intrinsic efficiency η_i is related to the building block type. The efficiency of a certain building block can, even with a theoretically perfect implementation, never be higher than this value. For a double balanced mixer for example η_i equals:

$$\eta_{i,mixer} = \left(\frac{G}{A_{amp}}\right)^2 \tag{5.8}$$

A double balanced mixer can be seen as an amplifier from which the gain can be varied between $+A_{amp}$ and $-A_{amp}$ and the output noise during operation is equal to the output noise of the mixer when used as an amplifier. The conversion gain G of a mixer is defined as the ratio between the amplitude of the wanted signal at the output and its amplitude at the input. The input-referred noise of the mixer is equal to the output-referred noise divided by the conversion gain G. When used as amplifier, this is A_{amp}. The ratio between the two gives the efficiency reduction.

The other efficiencies are less dependent on the performed function. η_n is the efficiency degradation due to excess noise sources. This accounts for all noise generated by devices different from the main dissipating elements which perform the wanted operation. η_p is the power efficiency with which the maximum output signal is generated and depends mainly on the class of the driver (A, AB, C). A second factor in η_p is the power needed to drive parasitic capacitances. Finally η_s is the efficiency due to the limited output swing:

$$\eta_s = \left(\frac{S}{Vdd}\right)^2 \tag{5.9}$$

This maximal output swing S is the signal level above which the distortion becomes too high.

5.3.3 Conclusions on fundamental estimators

Most estimators that depend on fundamental theoretical relations have in common that they use a lot of low-level information and bring this into account in the form of an efficiency parameter. Both here described estimators at least do. If a topology-independent estimator is to be found, an average of this parameter has to be used. The estimators that use fundamental relations and an average efficiency parameter usually offer an accuracy of (several) orders of magnitude which is not good enough for practical system design. Bringing into account more details fatally goes at the expense of the fundamental relations which are at the basis of the estimators. This brings us to the next section on practical examples of power estimators.

5.4 PRACTICAL HIGH-LEVEL POWER ESTIMATORS

Already one example was given in the previous section in equation (5.2). It is a nice example of a semi-fundamental estimator. Some relevant design parameters are combined and a figure of merit is used. However, for high-speed analog-to-digital converters (ADCs), a much better solution is possible. This is presented in the subsequent section. Next, in section 5.5.2, a power estimator for operational amplifiers using neural networks is explained. Finally, as a third example, the power consumption of analog continuous-time filters is analysed in section 5.5.3. This is to be compared with the fundamental relation of equation (5.5). A conclusion is given in section 5.5.4.

5.4.1 A power estimation model for high-speed CMOS ADCs

The estimator described here is valid for all topologies in the class of high-speed ADC converters [7]. These are flash topologies with or without folding and/or interpolating and pipelined architectures. Low-speed, high-accuracy architectures such as delta-sigma and two-step ADCs are not covered.

The used approach is a meet-in-the-middle strategy between top-down and bottom-up. The top-down approach is the first part of the derivation and is based on the fundamental relation:

$$P = \mathrm{Vdd} \cdot I = \mathrm{Vdd} \cdot f \cdot \mathrm{Charge} \tag{5.10}$$

This is of course a quite trivial equation, but the question is what frequency should be considered when talking about ADCs and then which charge is being transferred. A high-speed ADC always has two parts: comparators and (pre)processing circuitry (the digital decoding logic is considered to behave as the comparators). The comparator is clocked at F_{sample} (and reset every clock cycle) and the processing circuit varies at the frequency of the input signal. In pipelined ADCs also the sample and hold and the DAC are clocked at F_{sample} but the charges internally vary with the signal frequency amplitude. So it is better to split equation (5.10) in two parts as well:

$$P = P_{comparators@Fsample} + P_{processingcircuit@Fsignal} \tag{5.11}$$

The charge is stored on internal capacitances so that for each part of equation (5.11) the following is valid:

$$P = \mathrm{Vdd} \cdot \mathrm{freq} \cdot (\mathrm{Voltageswing}) \cdot C \tag{5.12}$$

The voltage swing for the comparators is always the full supply voltage (digital values). For the rest of the circuitry the swing depends on the signal swing, but if it is a good design it should be as large as possible and therefore approximately Vdd is taken which results in an expression of the following form for each part of equation (5.11):

$$P = \mathrm{Vdd}^2 \cdot C \cdot \mathrm{freq} \tag{5.13}$$

About the capacitances C, little can be said without going into topology details which has to be avoided to have a good high-level top-down estimator. Therefore, the equality is replaced by a proportionality and the capacitance is taken proportional to the technology's minimal transistor channel length L_{min} which yields for the total power estimator:

$$P \propto \mathrm{Vdd}^2 \cdot L_{min} \cdot (F_{sample} + F_{signal}) \tag{5.14}$$

Supposing that the capacitance scales with the technology is maybe a little strange and sometimes completely wrong depending on the capacitances used. The reasoning is that the thickness of the gate oxide is proportional to the minimal gate length, and that only for very deep submicron technologies this proportionality factor changes. In high-speed designs most capacitances are parasitic in the sense that they are transistor capacitances (mainly C_{GS}) and are thus also dependent on the technology.

However, this is far from being a good power estimator because nothing has been said about the accuracy in the derivation so far, and accuracy of course is an important factor in ADCs. It is clear from mismatch analyses that through the size of the devices the accuracy is varied [8]. Using larger devices results in a higher accuracy but also increases the total capacitance, thus limiting the speed and increasing the power consumption [9]. The accuracy is expressed here as the effective number of bits (ENOB) which is defined as shown in the well known [10] equation:

$$\mathrm{ENOB} = \frac{20 \cdot \log(\mathrm{SNDR}) - 1.76}{6.02} \tag{5.15}$$

The accuracy (5.15) is related to the size of the devices and in this way to the power (5.14). Of course, once that the device sizes become involved, topology details are needed. This yields the second part of the derivation which is a top-down part because specific circuit-related information is used. It was supposed that the obtained equation (5.14) divided by the power was somehow a function of the requested ENOB. To check this, a total of 22 designs has been looked up in the IEEE Journal of Solid State Circuits, and the 75 so obtained data points are put together in *Figure 5.8*. A linear regression line has been fitted through these data points. The correlation coefficient r is also given for this linear regression approximation and is 0.791. The mean square error equals 0.2405 and a 90% confidence band for the regression line is drawn.

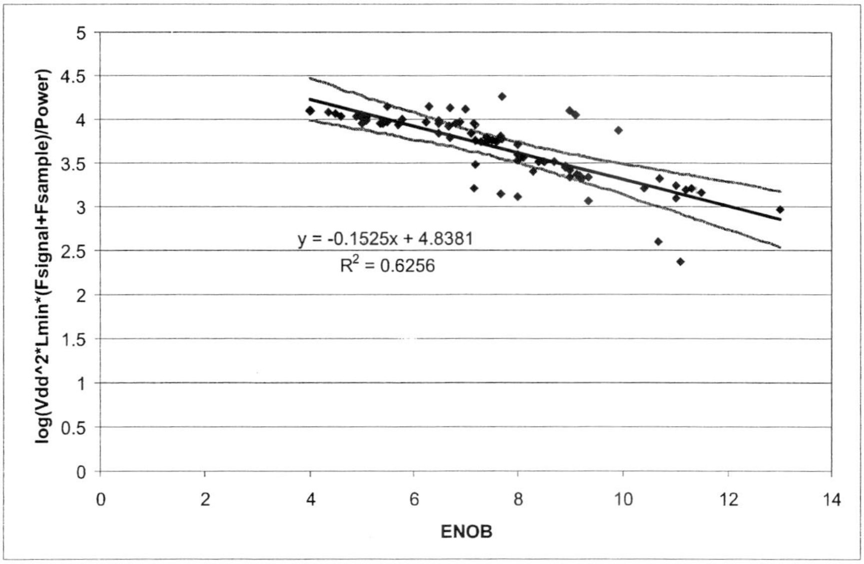

Figure 5.8 The result of the log of equation (5.14) divided by the power as a function of the ENOB, revealing a linear trend line and its 90% confidence band

From this regression line the final power estimator valid for the class of CMOS non-oversampling high-speed ADCs can be derived as:

$$P = \frac{Vdd^2 \cdot L_{min} \cdot (F_{sample} + F_{signal})}{10^{(-0.1525 \cdot ENOB + 4.838)}} \tag{5.16}$$

Of all the data points 85% or 19 out of the 22 designs taken into account fall within an absolute estimation accuracy factor of 2.2. A factor of 2.2 is about 1.44 times the sigma calculated from the linear regression analysis. This result is accurate enough for a system-level architectural design where a nominal value can be taken (the estimated value) and a certain margin on this value. A possible comment that will always remain, since it is related to the used data and therefore inherent to the method, is that the variance of the reported power dissipation of the set of samples is not known and thus not taken into account. Therefore, no absolute figure expressing the uncertainty of the estimator can be calculated. Only a relative figure, sigma, resulting from the regression fit was calculated to be 0.2405. Another way to prove the usefulness of the estimator is to take a design that was not used to derive the estimation function and to check the result or in other words by verification. Verification has proven the usefulness of the estimator. For example a design that expresses the ENOB as a function of fsample and fsignal is found in [11]. The estimated power consumption is calculated (5.16) to be 166mW and the published power consumption is 225mW or the relative error equals (166-225)/225=-0.26 which is about a factor 1.35 (well below the uncertainty margin of 2.2). The regression parameters can be updated by just adding new designs.

Despite the easy and straight forward construction of this power estimator, the method proved not to be universally usable to other classes of circuits, as will be illustrated for filters later on.

5.4.2 Power estimation using neural networks

The second example uses a bottom-up approach and is especially useful for showing that no golden solution exists for constructing power estimators but that many have been tried already.

In [13], a tool for automated estimation of the power is introduced. The method consists of the generation of a training set and a subsequent training of an artificial neural network (ANN) that takes as input the circuit's performance specifications and returns as output the power estimate. Use is made of the capability of neural networks of being able to model complex interactions without having to know the explicit relationship. The tool is demonstrated for a class AB operational amplifier.

The major problem is the generation of a good training set. In the reported approach [13], it is generated in two steps (*Figure 5.9*). In a first step, a grid is defined by the user to sample the parameter specification space of the analog block under consideration. From this grid random samples are selected and sorted. The samples are randomised to limit the amount of samples and thus computing time. The amount of samples determines the final accuracy of the estimator. Sorting is done using the distance from the origin of the n-dimensional specification space to make the subsequent calculations more efficient. Next, for each sample an optimal sizing is performed using an analog circuit design optimisation program OPTIMAN [14] and the resulting power (and area) is calculated. Once the training set is generated, the training of the neural network is done. Only 80% of the selected samples were used to train the ANN and

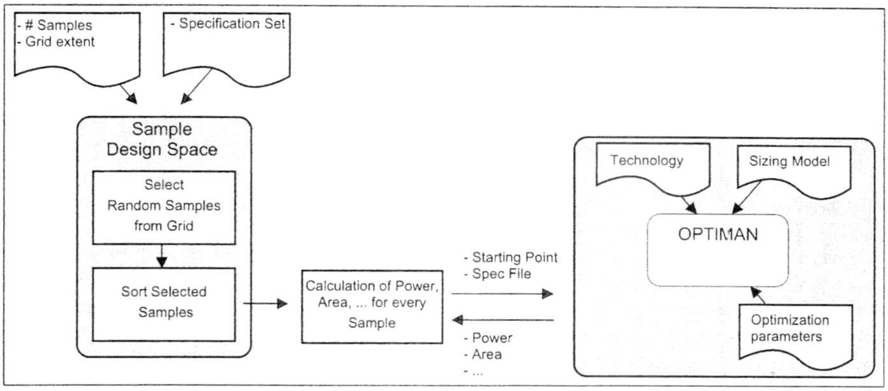

Figure 5.9 Generation of training set for Estemate

20% were used for verification. On these 20% an average error of 2% was obtained and the average evaluation time can be neglected (once the training has been done). So, for this estimator all the arguments mentioned in the paragraph on the bottom-up approach are valid and the conclusions stay the same.

5.4.3 High-level power estimation of analog continuous-time filters

A final example shows again the combination of the top-down and bottom-up approaches, but in this case, also topology information had to be used [15]. That is why the estimator is only suited for the class of high-speed OTA C or transconductance C filters. Using the same approach as for the high-speed ADCs of section 5.5.1 but then applied to analog filters, yielded inaccurate results [12]. Including topology information allowed to obtain very accurate results and turned out to be necessary for the case of filters.

The power consumed by an analog continuous-time (CT) OTA-C filter when given as input a limited set of high-level system parameters is estimated. Only three inputs are basically needed: the filter transfer function, the desired dynamic range (DR) and the maximal (differential) signal amplitude. The output is the power consumption needed to realize the given transfer function in a given technology. The estimation tool is divided in two major parts: the filter synthesis part and the OTA specification optimisation part (*Figure 5.10*). Linked to them are a filter topology library and an OTA model library. More information on both parts is given later in this section.

In this case, finding a good analog power estimator is finding the exact level of abstraction at which one can disconnect the topology information from all that comes below that abstraction level down to the transistor level. For the CT OTA-C filters, this level is found at the transconductor level. Finding the high-level specifications for the transconductors (gm and distortion) is done using the *filter synthesis* part. In the second step, the *OTA optimisation* part, optimisation techniques are used in com-

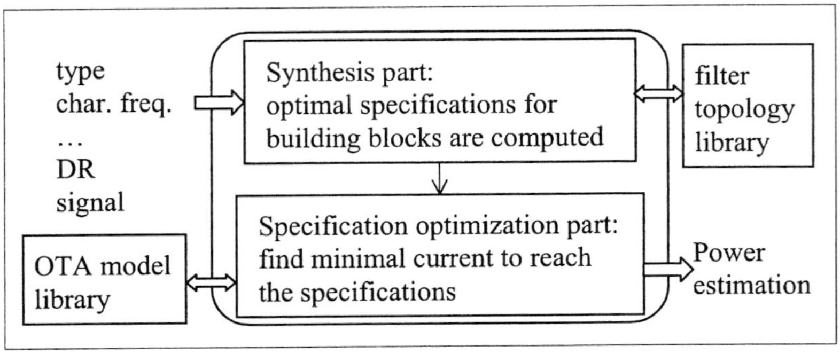

Figure 5.10 An overview of the estimation tool flow for analog filters

bination with behavioral models for the OTAs to find the minimally needed current to achieve the derived OTA specifications. Finding behavioral models that link the high-level specifications of the OTAs to the design parameters is difficult but significantly speeds up the estimation process. Other techniques would also be possible, such as table lookup and in-the-loop simulation.

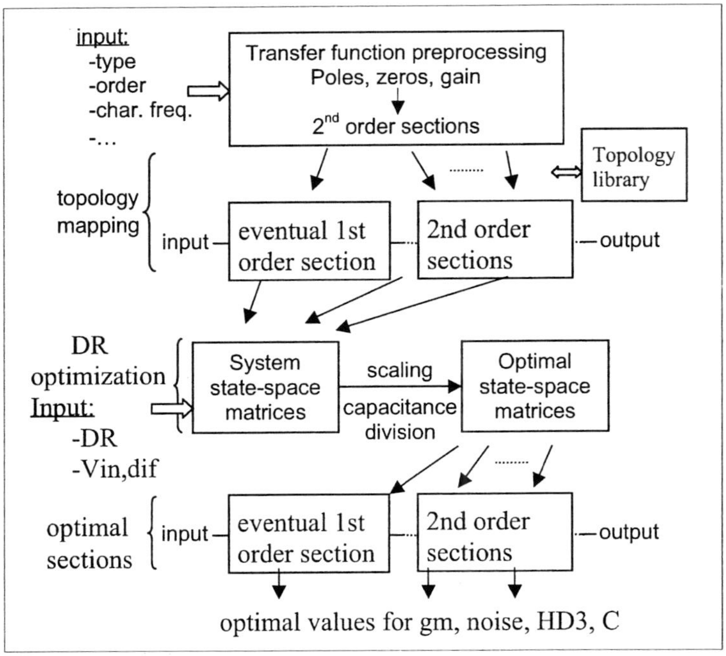

Figure 5.11 Synthesis workflow of the estimation tool

The synthesis workflow, illustrated in *Figure 5.11*, is as follows. On the top left part, the wanted filtering function is entered in the tool (for example a 4th-order elliptic bandpass filter). The filter transfer function is then split in a first-order stage, if the order is odd, and second-order stages. These stages are mapped on a biquad topology and for each section the state-space matrices are constructed. Next, the total system state-space matrices are calculated. Using the desired dynamic range and signal level from the input, the optimal system state-space matrices are then obtained after scaling and optimal capacitance distribution. The result is then broken up again into 2nd-order sections and the optimal values for the needed OTA specifications are derived. For more details about this workflow and the used mathematical expressions, the reader is referred to [15]. At the end of the synthesis step, the gm and distortion levels of each OTA are known and are given to an optimiser, which minimizes the current to reach these values in selected OTA implementations. Only the third-order harmonic distortion HD3 is taken into account because higher-order terms are typically smaller, the second-order term is cancelled out in optimal differential designs and the intermodulation product is obtained from the harmonic distortion term. This means that for gm and HD3, models have to be developed. To this end, five different OTA stages have been simulated in a 0.35μm CMOS process with varying design parameters and with an input voltage amplitude sweep. The design parameters are characteristic for each gm topology. Again, for more details on the modeling approach, the reader is referred to [15]. The result is an OTA model library containing the following expressions for each different OTA topology.

$$\begin{cases} gm = f\left(I_{DS}, V_{GT}, w, R\right) \\ HD3 = f\left(I_{DS}, V_{GT}, w, R\right) \end{cases} \tag{5.17}$$

The parameters I_{DS}, V_{GT}, w and R are key design parameters of the OTA topologies. The distortion model is a piecewise linear model and purely based on simulations. The transconductor model is based on hand formulas and fitted with simulations through a fit factor.

The optimisation flow is as follows (*Figure 5.12*). For a given OTA parameter set, the boundaries of the regions in the HD3 model are calculated. If the region is known, then the expression for HD3 is known for that set of parameters. With the expression for gm added, the constraint-based gradient search optimisation as implemented in Matlab^{TM} can start looking for that parameter set that yields the asked gm and HD3 values at minimal current I_{DS}. However, it is possible that the new set of parameters changes the model region of the HD3 model. So a kind of rule-based optimisation was programmed that notices when region oscillation occurs and handles it.

The output of the optimisation is a table for each OTA needed in the filter as shown in *Figure 5.13*. It indicates the current needed to achieve the "found" values. If these are within the specifications, an indication "good" is given. If the values are not attained but are within a certain margin, the tag "maybe" is given because this is an indication that a goal could not be attained. If completely wrong values come out, a "bad" tag is given to indicate that the optimiser could not converge.

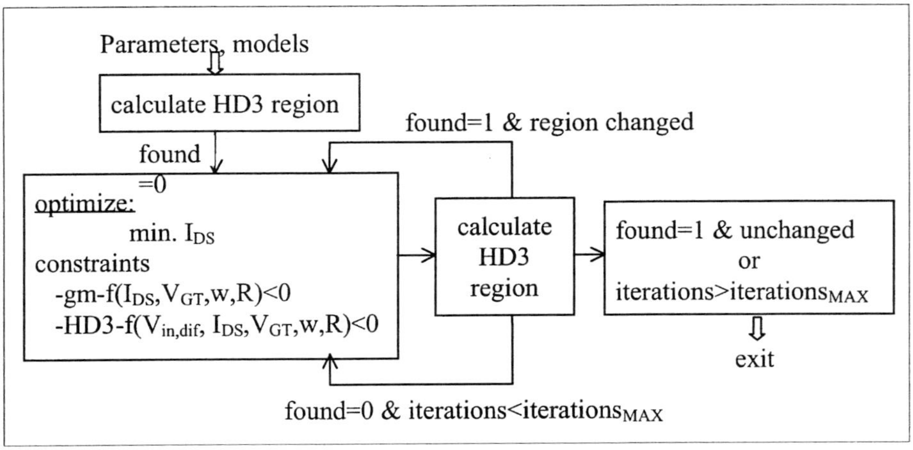

Figure 5.12 Optimisation flow

Type	Minimal current	goal gm	found gm	goal HD3	found HD3	indicator
SDP'	'1200.058'	'2.944e-07'	'0.0038001'	'55.9188'	'52.0009'	'maybe'
'SRDP'	'59.6275'	'2.944e-07'	'0.00013223'	'55.9188'	'55.9188'	'good'
'TOR'	'53.2937'	'2.944e-07'	'9.3392e-05'	'55.9188'	'55.9188'	'good'
'KRUM'	'1200.0196'	'2.944e-07'	'0.0035669'	'55.9188'	'54.6595'	'maybe'
'SIL'	'172.9208'	'2.944e-07'	'0.00034576'	'55.9188'	'55.9188'	'good'

Figure 5.13 Output table for one gm of a fictive filter example

It is typical that the gm values can easily be obtained and that the distortion specification is the limiting factor. This stresses the importance of including the distortion behavior of the OTAs in the power estimation of filters.

Two examples are given to illustrate the capabilities of the approach used in the estimation tool. Power estimation time for each is about 6 minutes using a SUN Sparc-30 running Matlab^{TM}.

Example 1: A 7th-order phase-equiripple 0.35μm CMOS LPF with a cut-off frequency of 70MHz was presented in [16]. It consumes 55mW, including the control circuitry, the DR is 42dB and the differential input swing is 400mV_{PP}. When given to ACTIF, the power estimate for high-speed OTA topologies is 34.9mW. This is quite good compared to the published 55mW, considering that the control circuitry is still missing in the estimator. Including tuning strategies in the estimator is subject of further work.

Example 2: A 5th-order 0.8μm CMOS LPF with a 4MHz cut-off frequency, a DR of 57.6dB, a $V_{in,dif}$ of 0.625V_{PP}, a total intermodulation distortion of 40 dB and a power consumption of 10mW was presented in [17]. This is a good example because the distortion level is lower than the DR and it is not a biquad implementation. It is in a different technology but one which also has a 3V supply voltage, which is the main limiting factor for distortion. If the correct distortion level is given to the filter power estimation tool, the outcome is that a total power consumption of about 3.5mW is needed for the most suited OTA topologies and about 10.6mW for the worst one, again without tuning taken into account. This is again close enough to the published power consumption of 10mW for a first-order estimation on a truly high level, even despite the fact that a different filter topology has been used in the real implementation. If, as an experiment, the filter has to be redesigned by increasing the distortion specification up to the DR, then a total power of 88.8mW is expected.

5.4.4 Conclusion about practical estimators

Three examples were given that reflect the main ideas of the two proposed approaches for the construction of power estimators. It seems that even in the case of a top-down approach, if a more or less accurate result is desired, including some information about the real underlying circuit is mandatory. In the case of the high-speed ADCs, the extra information comes in the form of measurement results. Using the topology information and adding some crude synthesis, adds the extra information needed to yield accurate results for the filter estimator. The neural-network approach uses the full topology information and hence is a case of a bottom-up estimation approach.

5.5 SUMMARY

First an overview of high-level system design was given to locate where and why power estimators come into play. The essence is that power estimators provide a criterion, minimal power consumption, that has to be optimised when comparing different architectural alternatives during system design. Next, a definition was given for an analog power estimator and requirements for a good estimator were discussed. The two main qualities are absolute accuracy and tracking accuracy. Then, two different approaches were presented and compared for the construction of power estimator functions: bottom-up and top-down. Next some theoretical estimator examples were discussed and their usefulness and shortcomings depicted. Finally, a few examples of practical power estimators were given, illustrating that in practice some kind of meet-in-the-middle strategy has to be used to generate good power estimates. The level at which topology-specific information has to be used varies from one analog circuit type to another, but inversely affects the generic properties of the estimator.

References

[1] A.G. Venes, R.J. van de Plassche, "An 80-mW, 8-b CMOS Folding A/D Converter with Distributed Track-and-Hold Preprocessing", Journal of Solid-State Circuits, pp. 1846-1853, December 1996

[2] F. Goodenough, "ADC's move to cut power dissipation", Electronic Design, January 9, 1995

[3] G. Gielen et Al., "An analog module generator for mixed analog/digital ASIC design", International journal of circuit theory and applications, Volume 23, pp 269-283, 1995

[4] Y. Tsividis, "Integrated continuous-time filter design – an overview", Journal of Solid-State Circuits, pp. 166-176, March 1994

[5] J.O. Voorman, "Continuous time integrated filters", IEEE Press, 1993

[6] J. Crols and M. Steyaert, "CMOS wireless tranceiver design", Kluwer academic publishers, Boston, June 1997

[7] E.Lauwers and G.Gielen, "A power estimation model for high-speed CMOS A/D convertors", Design Automation and Test in Europe – DATE 1999, pp.401-405

[8] M.Pelgrom, A.Duinmaijer, A.Welbers, "Matching properties of MOS transistors", Journal of Solid-State Circuits, pp. 1433-1439, October 1989

[9] P.Kinget, "Analog VLSI integration of parallel signal processing systems", PhD thesis, Katholieke Universiteit Leuven, May 1996

[10] R.J. van de Plassche, "Integrated anolog-to-digital and digital-to-analog convertors", July 1993

[11] I. Mehr, D. Dalton, "A 500 Msample/s 6-Bit Nyquist Rate ADC for Disk Drive Read Channel Applications", proceedings of ESSCIRC '98, pp.236-239.

[12] E.Lauwers and G.Gielen, "High-level power estimation functions for analog filters", proceedings of ProRISC99/IEEE workshop on Circuits, systems and signal processing

[13] G.Van der Plas, J.Vandenbussche, G.Gielen, W.Sansen, "EsteMate: a tool for automated power and area estimation in analog top-down design and synthesis", Custom Integrated Circuits Conference, pp.139-142, May 1997

[14] G.Gielen et. Al., "Use of symbolic analysis in analog circuit synthesis", Int. Symp. On Circuits and Systems, Vol.3, pp.2205-2208, 1995

[15] E.Lauwers and G.Gielen, "ACTIF: A high-level power estimation tool for analog continuous-time filters",Proc. ICCAD, San-Jose, CA, pp.193-196, November 2000

[16] R.Castello, I.Bietti, F.Svelto, "High-frequency analog filters in deep-submicron CMOS technology", Proc. ISSCC, 1999, MP4.5

[17] C.Yoo, S-W.Lee, W.Kim, "A +/- 1.5V, 4MHz CMOS continuous-time filter with single-integrator based tuning", JSSC, Vol.33, Jan.1998, pp.18-27

6 MODELS AND ANALYSIS TECHNIQUES FOR SYSTEMATIC DESIGN AND VERIFICATION OF FREQUENCY SYNTHESIZERS

Bart De Smedt, Georges Gielen

ESAT-MICAS

Katholieke Universiteit Leuven, Belgium

E-mail: [Bart.Desmedt, Georges.Gielen]@esat.kuleuven.ac.be

The telecommunications market has shown an explosive growth in the last decade, and has shifted to a consumer commodity, even for wireless applications. To satisfy the customers' need for more mobility and higher data rates, new standards are continuously being developed, such as GSM, GPRS, UMTS, Bluetooth, Hiperlan–2, etc. The shortening time–to–market constraints for these products call for new design techniques to design even better and ever more complex systems in a shorter time.

Re–usability is a key issue in the design methodology to shorten the design cycle. Re–usability in a strict meaning implies a "copy–and–paste" action of an existing component into the new design (hard IP). Especially for analog circuits, however, re–usability in a broader sense is intended. Starting from existing design frameworks and sets of necessary equations, the decision trajectory of previous designs is followed as much as possible. This allows to tune certain design parameters, quickly study its influence on the system performance and tailor the design to different applications, while still exploiting the existing design knowledge.

To enable such a systematic design methodology including re–usability, a top–down design and bottom–up verification trajectory is to be followed [8]. Therefore different abstraction levels are identified first. At each level the behavior of the overall circuit is evaluated using models of the composing blocks. Each model calculates its output variables given the input variables. The format of these variables depends on the abstraction level . Top–down implementation then implies that lower–level model parameters are assigned a value such that the higher–level model meets its specifi-

P. Wambacq et al. (eds.),
Low-Power Design Techniques and CAD Tools for Analog and RF Integrated Circuits, 107–129.

cations at a minimum cost (e.g. minimum power consumption or chip area). After having assigned all variables a value and having designed and laid out every block, a verification of each implementation is performed to take lower–level second–order effects into account in the performance at the higher level.

Such a systematic design methodology is based on two prerequisites. First of all a library of models should be available which enables to describe the behavior of different building blocks at each abstraction level. A specific modeling language (such as VHDL–AMS) allows to implement these models and to describe the interconnectivity of different blocks at a higher level. Secondly, dedicated algorithms are needed to extract and evaluate the high–level specifications during the verification. Typically these cannot be evaluated using standard simulation techniques.

This chapter describes both the necessary models and the simulation algorithm for the design and verification of frequency synthesizers used for telecommunication applications. The chapter is organized as follows. Section 6.1 introduces the different aspects that have to be taken into account when developing a model library. Section 6.2 concentrates on the application domain of frequency synthesizers. The basic terminology is pointed out and an overview of the designer's trade–off is given. Next, an algorithm is presented in section 6.3 to evaluate phase noise in frequency synthesizers. The method is illustrated using the design of a 1.8 GHz CMOS frequency synthesizer. The different models that are needed in the systematic design methodology are then presented in the two following sections. Section 6.4 describes models that can be used for phase noise specification and for settling time evaluation in the top–down design trajectory. The bottom–up verification trajectory is discussed in section 6.5 which presents a model to accurately evaluate the phase noise specification of a complete frequency synthesizer. The chapter concludes with a summary in section 6.6

6.1 ASPECTS OF MODELING

6.1.1 Why modeling ?

Setting up a library with models for various building blocks at the different abstraction levels is a strenuous effort. Before deciding to implement a complete library, a trade–off must be made between the implementation and support effort on the one hand, and the prospective leverage when applying these models afterwards on the other hand.

- Developing behavioral models of a certain block requires a good understanding of the functionality of this block and the different parameters that have an impact on this functionality. Simple models might evaluate quickly, but they can only be applied in a limited set of applications.

- On the other hand, robust models which include many details, are often slow in evaluation. The gain with respect to transistor–level models is often rather small.

This is the well known trade–off between model accuracy and evaluation time. Fortunately the availability of an appropriate model library has a number of advantages to justify all this work.

- The introduction of hierarchy and abstraction in the digital domain allows the designer to handle larger degrees of complexity. The complexity of the analog part also tends to increase (complete transceiver chips are coming). Employing different types of models at different levels can then elevate the analog design abstraction level to manage this increasing complexity.
- Once analog building block models are developed, they can be instantiated to perform system design. When implementing a lower level of the design, the higher–level description is used as a specification and reference.
- Logging the design trajectory is much easier when the behavior of all intermediate stages in this trajectory can be evaluated. Keeping track of the different design decisions is done by adding comments to the pervasive simulation results. Once such a design trajectory is set up, the redesign of a certain component towards other specifications or in another technology can highly be accelerated.

For these reasons, investing in a model library is merely a decision with long–term benefits.

6.1.2 Modeling requirements

When developing models a number of requirements have to be fulfilled.

- The functionality should be modeled in a generic, parameterized way. The model can then cover a wide range of design spaces of underlying lower–level implementations.
- When one wants to evaluate a certain specification, all circuit aspects that influence this specification should be included in the model. In general a trade–off can be made between the accuracy of the model and the necessary evaluation time. In the first stages of a top–down implementation less accurate models can be allowed to get a rough impression of the design space. However, for lower–level implementations and verification later on, a higher degree of accuracy is needed.
- Finally, all these models should be implemented in a standardized manner. The VSI Alliance is putting much effort in the standardization of model interfaces, both for digital designs and for analog components [12]. This will ease the exchange and re–use of models. Not only the interface but also the modeling languages should be standardized. In digital designs the use of VHDL or VERILOG is widespread, both for simulation and synthesis. In the analog domain the VHDL–AMS [10] and VERILOG–AMS [1] standards are available and commercial simulators offering these languages have started to appear.

6.2 INTRODUCTION TO FREQUENCY SYNTHESIZERS

Frequency synthesizers are well known building blocks in telecommunication front–ends [7, 6]. In this chapter a brief introduction to frequency synthesizers is given to get acquainted with the basic terminology and with some design aspects.

First of all the generic synthesizer topology is illustrated. Then the set of principal specifications is presented and the most important design trade–offs are explained.

6.2.1 A typical frequency synthesizer topology

In general a frequency synthesizer for telecommunication applications consists of a control loop which steers the phase and/or frequency of the oscillating output wave (also known as *phase–locked loop* or *PLL*). In some cases more than one control loop is present. Fig. 6.1 shows a typical frequency synthesizer topology.

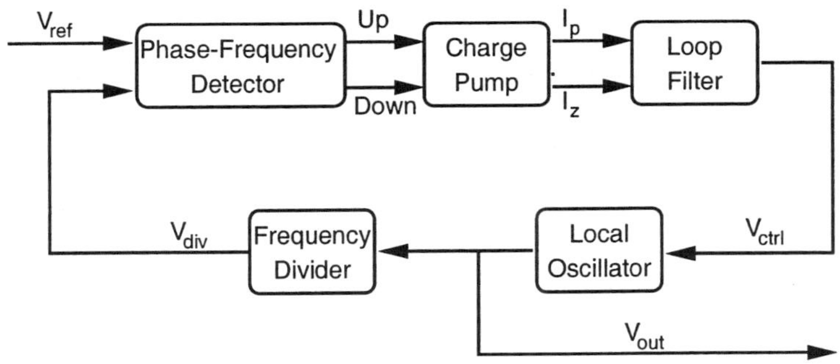

Figure 6.1 Generic frequency synthesizer topology

The local oscillator generates a waveform whose fundamental oscillation frequency is controlled by its input signal. This is either a voltage for a *voltage–controlled oscillator (VCO)* or a current for a *current–controlled oscillator (CCO)*.

In most synthesizers, a frequency divider is present which divides the oscillator's frequency by an integer number *N*. Several different types of frequency dividers exist : both *asynchronous* dividers (also called *prescalers*) and *synchronous* dividers (which exhibit a shorter delay, but draw a lot more current). Some frequency dividers have only one division ratio (i.e. *N*), whereas others can be set up to divide by several integer numbers (so called *dual–* or *multiple–modulus* dividers).

The phase of the waveform coming out of the frequency divider is compared with the phase of a reference source (usually a quartz oscillator), using a *phase detector*. Typically a crystal oscillator is used as reference source. The very first phase detectors consisted of a simple analog multiplier. This topology however is very sensitive to *cycle slips* , i.e. the frequency detector is not able to track the phase of the reference source, but rather lags behind. More recent detectors consist of a digital network containing flip–flops and a delay element. These detectors not only track phase misalignments but also the frequency errors. Hence they are called *phase/frequency detectors*. To improve their linearity around the locking region (when they work in phase–mode), a delay element is often included (to prevent the so–called *dead–zone phenomenon*).

This type of phase/frequency detector outputs two signals : an *up*–signal and a *down*–signal. They steer the switches of the *charge pump* which sources or sinks current into the *loop filter*. Active–up and inactive–down signals at the detector's output indicate that the oscillators' steering signal has to rise and vice versa. When both signals are inactive, no current is fed to or drawn from the filter. Both signals active is not supposed to occur and would only drain current from the power supply without any net signal being fed to the loop filter.

In the loop filter, which is a low–pass filter, high–frequency fluctuations of the charge pump's output are attenuated. The filter's output is therefore a stable signal that is fed to the oscillator.

6.2.2 The design trade–off

In telecommunication applications, the two major specifications for frequency synthesizer design are *phase noise* and *settling time*. These specifications are now elaborated in more detail.

The spectrum of an ideal frequency synthesizer is a single Dirac–impulse at the oscillation frequency. In practice, the power spectral density at the output contains unwanted contributions at other frequencies, including both harmonics and noise skirts around every tone. A typical power spectrum density is shown in Fig. 6.2

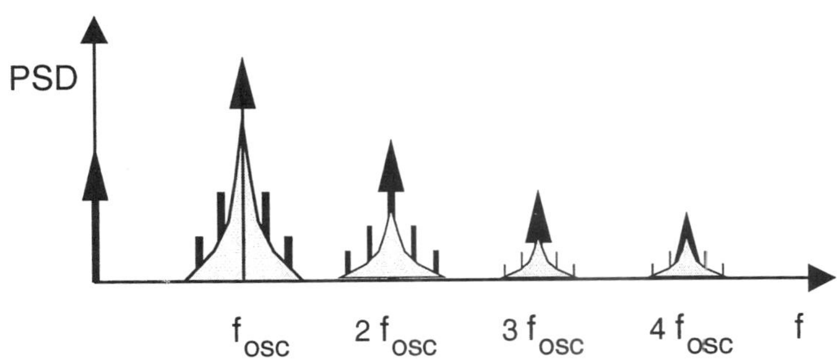

Figure 6.2 Oscillator's power spectral density

Due to the oscillator's signal power at frequencies close to the carrier (i.e. *phase noise, spurious components*), neighboring channels in the antenna signal will fold into the wanted signal band, thereby deteriorating the performance of the receiver. A similar phenomenon occurs in the transmitter chain. For this reason, the power spectral density of the oscillator is strictly specified from the start of the design, depending on the application.

A typical phase noise specification for the GSM protocol is given by :

$$\mathcal{L}\,(\Delta f = 600\text{ kHz}) = -121\text{ dBc/Hz} \tag{6.1}$$

The second principal design specification for a frequency synthesizer is its settling time. In most applications the synthesizer switches between transmit and receive frequencies. The time necessary for the oscillation frequency to settle within a specified level of accuracy is limited. Moreover the settling specification is becoming more and more stringent in modern applications. For DCS–1800 applications, typically 700 μs is allowed for the oscillator to settle within a relative accuracy of 10^{-6}.

From a designer's perspective, it is the loop filter which mainly determines both the phase noise level and the settling performance. Given the local oscillator's performance (which is usually optimized for minimum phase noise), one can exchange good phase noise performance of the overall system for short settling time by tuning the loop filter characteristics.

6.3 PHASE NOISE EVALUATION

As depicted in the previous section, the spectral purity at the oscillator's output is one of the main specifications for synthesizer design. Unfortunately, evaluation of the spectral content of an oscillator is not a trivial operation for several reasons.

- An oscillator operates essentially in a nonlinear and time–varying fashion, which renders traditional small–signal analyses incorrect.
- Most of the time there is a discrepancy between the oscillation frequency on the one hand (e.g. 1.8 GHz) and the frequency offset at which phase noise is specified on the other hand (e.g. 600 kHz). Therefore transient analysis *off–the–shelf* will lead to unacceptably long simulation times to obtain accurate results.
- Finally, due to the large signal swings at the internal oscillator nodes, some of the noise sources are not stationary, which makes classical small–signal noise analyses unsuited for phase noise evaluation.

In [4] a method is described to evaluate the phase noise spectrum at the oscillator's output. This section gives a summary of that method.

The phase noise evaluation method consists of the following three steps :

1. identification and quantification of the noise sources
2. propagation from each noise source to the output node
3. combination of all contributions to obtain the overall phase noise spectrum

These steps are now described in somewhat more detail.

6.3.1 Identification and quantification of noise sources

Apart from the well–known noise equations for integrated devices [11], two specific aspects have to be taken care of.

1. Many integrated LC–tank oscillators make use of one or more on–chip coils. To that end metal structures are laid out in a spiral configuration. Often all available metal layers are laid out on top of each other. These structures however suffer from parasitic resistances that inject noise into the oscillator's signal path. To quantify this influence, an equivalent electrical model of the coil is set

up. Therefore, either a set of first–order approximations is used (using geometrical calculations) or a complete finite–element simulation is performed (to take second–order effects into account such as e.g. skin effect [2]) and fitted to an electrical model. Fig. 6.3 shows the resulting electrical model with quantified parasitic elements for a fairly simple coil.

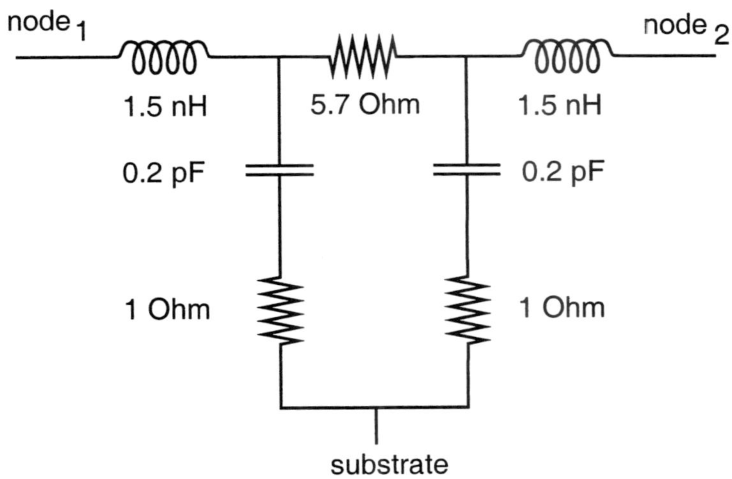

Figure 6.3 Electrical model for a simple integrated coil

2. Secondly some of the noise sources will not operate in a stationary regime. This is in particular the case for active device noise contributions, such as flicker noise and thermal noise in bipolar or MOS transistors. The *cyclostationary noise* injection is taken into account during the evaluation by modulating the noise power spectral density by the momentary large–signal device state variables (e.g. cyclostationary white noise in a MOS transistor is modulated by the momentary transconductance).

6.3.2 Propagation from the noise source to the output node

The next step in the phase noise evaluation is the calculation from each specific noise source to the output. More specifically the spectral contribution at the output node at a specified frequency offset from the carrier has to be calculated. Due to the inherent nonlinear and periodically time–varying nature of an oscillator[1], distinctive frequencies in the noise source spectrum will result in contributions at this specified frequency. The particular values for these frequencies can easily be expressed as the

[1] In general an oscillator operates as a positive feedback system with gain larger than unity. So far everything could be described in terms of linear equations. However there is clearly some sort of an amplitude clipping mechanism necessary to limit the output amplitude. It is this system which introduces the nonlinearities and the harmonics. Noise upconversion however arises from the time–varying nature of the circuit.

sum or the difference of (harmonics of) the carrier frequency and the evaluation frequency (see eq. 6.2 further). A graphical representation is shown in Fig. 6.4.

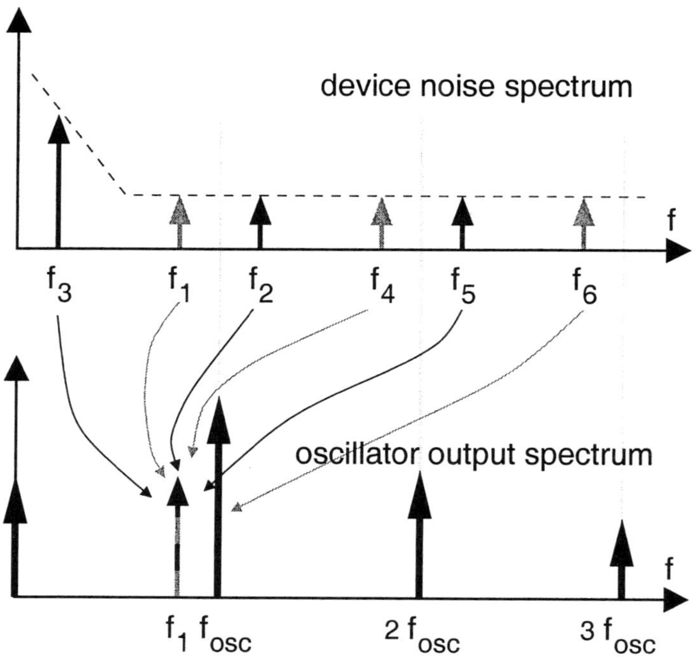

Figure 6.4 Due to the nonlinear and time–varying operation of an oscillator, noise from different frequencies of the noise power spectral density of a noise source (top) contributes to the output spectrum (bottom) at the specified evaluation frequency f_1.

The propagation factor from the noise source to the output spectrum is obtained using simulation techniques. To that end, the device noise spectrum is sampled at the specified frequencies. Current sources are then placed in parallel with the noise–generating device, injecting a sinusoidal current proportional to the square root of the momentary noise power. The magnitude of the current sources is therefore modulated by the proper circuit parameter (e.g. by the momentary diode current in the case of shot noise in a PN–junction). After a large transient simulation containing all these sinusoidal noise currents, the magnitude of the total phase noise at the output is estimated using discrete Fourier transformation techniques.

6.3.3 Combination of all contributions

The previously described technique allows to calculate the propagation η_k to the output from every spectral noise sample. All these individual contributions from different frequencies and from all different noise sources in the circuit are to be summed quadratically :

$$\begin{aligned}\mathcal{L}(\Delta f) &= \mathcal{L}(f_{osc} - f) \\ &= \sum_{\substack{noise \\ sources \\ k}} \left[\sum_{m=1}^{m=\infty} \eta_k^2 \left(\frac{f}{m}\right) + \eta_k^2 \left(m.f_{osc} \pm f\right) \right] \end{aligned} \tag{6.2}$$

However, during the large transient simulation, all contributions are summed linearly at every evaluation frequency[2], making it impossible to distinguish each individual component after simulation and Fourier transformation. This would imply that no two frequency samples can be applied during the same simulation that would result in a contribution at the same evaluation frequency. One can nevertheless evaluate the total phase noise content by launching a separate simulation for each contribution, thereby blowing the number of simulation runs.

To overcome this problem and to calculate the partial contribution of every noise source in only one single simulation, the noise spectra of sources are sampled at closely–spaced frequencies, one for every device noise spectrum, in each frequency region of interest, such that none of the applied frequency samples results in contributions overlapping at the same frequency. An updated frequency selection scheme is shown in Fig. 6.5.

Evaluation and quadratic summation of these partial noise source spectra in a post–simulation extraction step then yields the overall phase noise spectrum. Fig. 6.6 gives an illustration of the global phase noise evaluation for a 1.8 GHz CMOS voltage–controlled oscillator.

6.3.4 Numerical aspects

To achieve accurate results within a reasonable evaluation time, special attention has to be paid to all sources of inaccuracies during the simulations. The dominating simulator inaccuracy contribution originates from aliasing effects [3]. Nevertheless these aliasing contributions during the simulation can be made insignificant by selecting the DFT–frequency grid carefully [4].

To illustrate the ruining effect of aliasing contributions during circuit simulation, two separate simulations are performed. In both cases, only the nominal oscillator was simulated, without instantiating any current source (used to sample the device noise spectrum). The only difference in both set–ups is the timestep between two successive simulated values. After transient simulation, discrete Fourier transformation was performed. Fig. 6.7 shows the two resulting spectra.

[2]Thinking in terms of Kirchhoff's current law, several noise contributions (originating either from different noise sources, or from different current sources of the same noise source) enter a certain node. The simulator accumulates these different portions of noise linearly. If some of these contributions have spectral components at the same frequency, they cannot be isolated from each other using spectral evaluation techniques.

[3]Integration and truncation inaccuracies can be set to a value that is small enough in practical set–ups (i.e. smaller than -140 dBc/Hz)

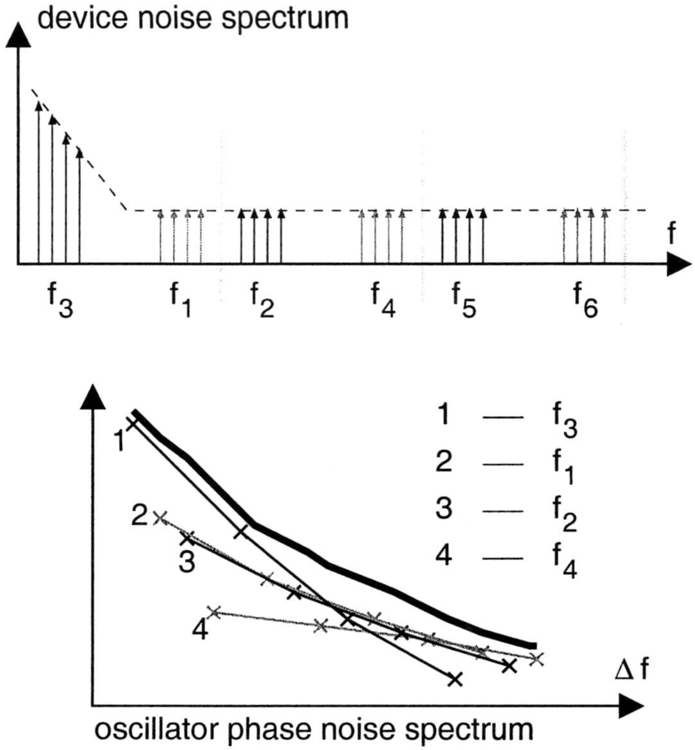

Figure 6.5 Frequency selection scheme to evaluate the global phase noise spectrum in one single simulation run

Comparing both figures clearly shows that aliasing components, folding back into the frequency region around the fundamental carrier, completely ruin the spectral purity that is needed for accurate evaluation of the contributions originating from the applied noise current sources.

One should note that the relative difference between both timesteps in this simulation example (hence the relative frequency misalignment) is only as small as 10^{-4}. This implies that the simulation timestep has to be selected very carefully [4].

6.4 BEHAVIORAL MODELS FOR TOP–DOWN DESIGN

As explained in section 6.1 the design cycle is shortened by re–using previous design knowledge to a large extent. Behavioral models and simulation techniques both in the design and in the verification trajectory provide for the necessary tool set to support such a design methodology.

This section describes two examples of top–down design models. First of all a *functional–level* model is presented which describes the phase noise performance of

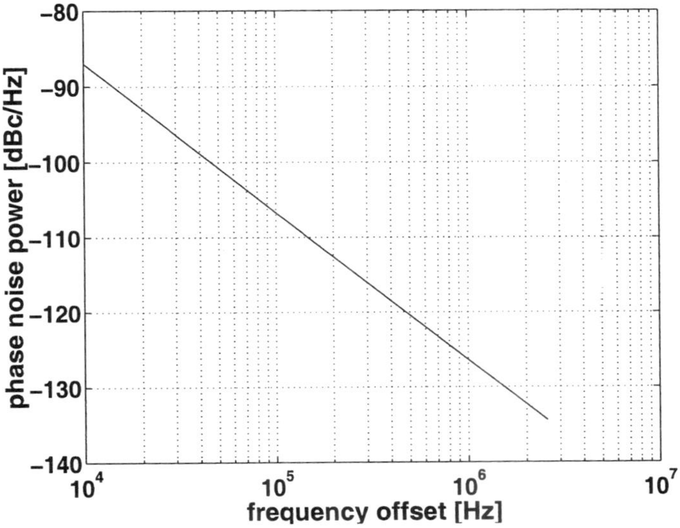

Figure 6.6 Illustration of the global phase noise evaluation for a 1.8 GHz CMOS VCO

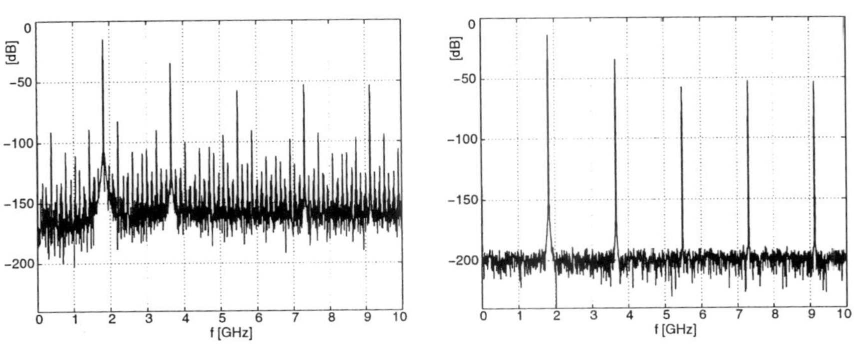

(a) PSD with aliasing

(b) PSD without aliasing

Figure 6.7 Simulated power spectral densities (PSD), showing the importance of an appropriate simulation timestep selection

a complete frequency synthesizer. Secondly a macromodel is presented to evaluate a synthesizer's settling time at an early stage in the design cycle.

6.4.1 Synthesizer models for transceiver functional–level design

In [3] an environment is proposed, called ORCA, for system–level receiver design. The different building blocks of a receiver, and the signals on every node are described in a dedicated manner. All composing blocks (low–noise amplifier, mixer, local oscillator, filter, A/D converter) of a receiver topology are described as behavioral models in the frequency domain. Polynomial functions as well as Dirac–impulses are used to quantify the signal spectra on each node.

Each block in the receiver chain performs wanted operations on the signal (e.g. amplification, frequency translation), but also unwanted operations (e.g. noise addition, harmonic distortion and intermodulation, ...). Once the model parameters of all blocks are determined, the ORCA simulator allows to calculate both the wanted and the unwanted signal transformations. In this way the overall signal propagation and output spectrum is analyzed. To differentiate between alternative implementations, a cost function is defined (e.g. the cost to implement a certain block is proportional to the power consumption and the necessary chip area). The design space can then be searched for the feasible solution with the lowest cost.

To exemplify such a functional model used in this top–down design stage, the frequency synthesizer is considered. In general the power spectral density at the output of such a synthesizer can be described by eqs. :

$$P_{FreqSynth}(f) = P_{Harmonics}(f) + P_{Spurs}(f) + P_{PhaseNoise}(f) \tag{6.3}$$

Typically only the frequency region around the fundamental carrier is of interest. Therefore :

$$P_{Harmonics}(f) = \sum_{k=0}^{k=N_{Harms}} A_k^2.\delta(f - k.f_{LO}) \tag{6.4}$$

$$P_{Spurs}(f) = \sum_{k=0}^{k=N_{Spurs}} \left[S_k^2.\delta(f_{LO} - k.f_{REF}) + S_k^2.\delta(f_{LO} + k.f_{REF})\right] \tag{6.5}$$

$$\begin{aligned} P_{PhaseNoise}(f) = & \left[N_0 + N_0\frac{(\Delta f_1)^2}{(\Delta f)^2} + N_0\frac{\Delta f_2(\Delta f_1)^2}{(\Delta f)^3}\right]u(\Delta f - \Delta f_3) \\ & + \left[N_3 + N_3\frac{\Delta f_4}{\Delta f}\right]u(\Delta f_3 - \Delta f) \end{aligned} \tag{6.6}$$

where :

$$\begin{aligned} \Delta f &= |f - f_{LO}| \\ u(f) &= 1, \qquad f > 0 \\ &= 0, \qquad f < 0 \end{aligned}$$

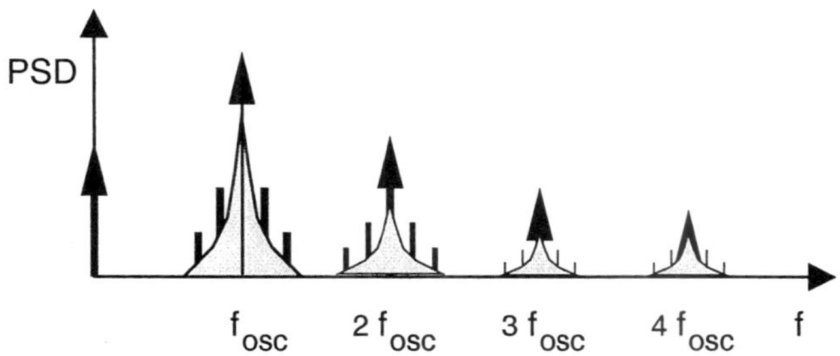

(a) power spectral density of a frequency synthesizer

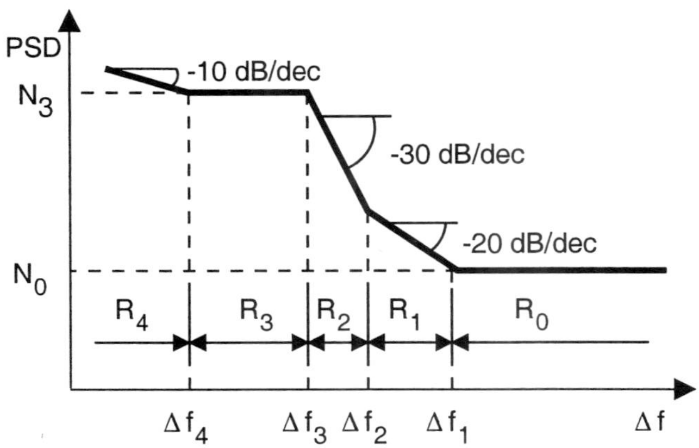

(b) phase noise relation for frequencies relative to the fundamental

Figure 6.8 Power spectral density as a functional model for a frequency synthesizer

and $\delta(f)$ represents the Dirac–function.

An asymptotic simplification of these equations is plotted in Fig. 6.8. Different regions can be identified in the phase noise relation. For frequencies larger than Δf_3, the local oscillator's phase noise is dominant (with successive $1/f^3$, $1/f^2$ and a flat region). For smaller frequency offsets, the noise contributions in the other frequency synthesizer blocks dominate (both white noise and flicker noise contributions).

6.4.2 Top–down behavioral–level models for synthesizer design

Once the specifications for the frequency synthesizer are determined during the functional–level design phase, a lower layer of models has to be available to start the synthesizer's design. At this point in the top–down design the oscillator's operation is linearized. This allows to use the local oscillator's phase as state variable [7, 6]. With these linear models, symbolic expressions are derived to obtain the wanted design equations. The combination of linearization and symbolic expressions results in a drastic reduction of the necessary evaluation time.

Eqs. (6.7)–(6.10) give an overview of the linear models for each frequency synthesizer building block :

$$\text{VCO}: \quad \Phi_{vco} = F_{vco}.\frac{1}{s} = \left(F_0 + K_{vco}.V_{lpf}\right).\frac{1}{s} \tag{6.7}$$

$$\text{DIVN}: \quad \Phi_{divn} = \frac{\Phi_{vco}}{N} \tag{6.8}$$

$$\text{PFD}: \quad V_{pfd} = \Phi_{ref} - \Phi_{divn} \tag{6.9}$$

$$\text{LPF}: \quad V_{lpf} = V_{pfd}.H_{lpf}(s) \tag{6.10}$$

When the topology of the frequency synthesizer is known (consisting of either one or sometimes more control loops), the composing blocks are replaced by their respective models. A SPICE–like netlist is generated which can be fed to the symbolic simulator ISAAC [9]. Using ISAAC the loop transfer function is derived. This yields the design equation necessary to examine loop stability.

Further, noise sources present in the circuit at the block level are also added to the netlist (e.g. loop filter noise, local oscillator's phase noise, ...). The transfer functions from all these noise sources to the oscillator's output are calculated symbolically. From the resulting transfer functions the contribution of each noise source to the output is computed. When these contributions are combined appropriately, the overall closed–loop phase noise results. The resulting equations now indicate which contribution dominates the phase noise in each frequency region. Moreover, since these equations are fully symbolic, they show which design variable to adjust to tune the phase noise towards the specified value.

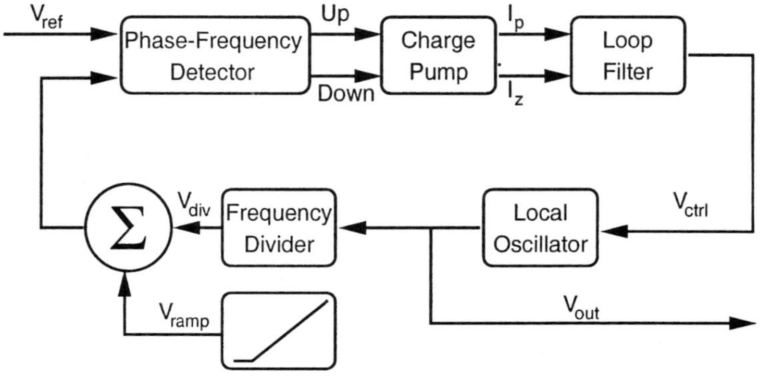

Figure 6.9 Signal source and simulation setup used for settling time evaluations

Finally, to calculate the settling time of the oscillation frequency, the prescaler's division factor is changed abruptly. This is modeled by adding an extra source of disturbance at the output of the divider (Fig. 6.9). This extra signal source generates a ramping voltage, corresponding with a linearly growing phase error. The settling time is then derived in three steps. First the transfer function from this source to the output is calculated. It is then split in partial fractions. Eq. (6.11) shows the general partial fractions formulation :

$$H(s) = \sum_{i=1}^{i=N_1} \frac{A_i}{1+\tau_i.s} + \sum_{j=1}^{j=N_2} B_j.\frac{s+b_j}{s^2+\zeta_j.\omega_j.s+\omega_j^2} \tag{6.11}$$

The first summation sums over all real poles, whereas the second summation enumerates all complex conjugate poles. Remark that the obtained transfer function (6.11) is no longer symbolic in general. Finally, after the transfer function is split, it is transformed into the time domain, using the inverse Laplace transform. This yields the equations necessary to evaluate the settling time of the control loop.

6.4.3 Illustration : settling time evaluation during top–down design

In this section an illustration is given of the settling time evaluation method described in the previous section. To that end, the frequency synthesizer topology of Fig. 6.1 is examined. It contains a simple first–order low–pass loop filter. All necessary equations are derived to calculate the settling time when the frequency division factor is changed. The corresponding netlist is passed to ISAAC. Eq. (6.12) gives the resulting transfer function, from the prescaler disturbance output (V_{ramp}) to the input node of the voltage controlled oscillator (see Fig. 6.9). The latter voltage is linearly proportional to the momentary oscillation frequency, and therefore indicates the oscillation frequency fluctuations as a function of time.

$$H(s) = \frac{-K_{vco}A_{lpf}\left(1+sC_{lpf}R_{lpf_2}\right)}{\frac{K_{vco}A_{lpf}}{N_{div}} + s\left(1+\frac{K_{vco}A_{lpf}C_{lpf}R_{lpf_2}}{N_{div}}\right) + s^2C_{lpf}\left(A_{lpf}R_{lpf1}+R_{lpf_2}\right)} \tag{6.12}$$

The calculated transfer function is multiplied with the Laplace transform of a ramp function. Then the inverse Laplace transform is derived to obtain the response at the output. Fig. 6.10(b) shows the frequency error as a function of time when a ramp function is applied as disturbance after the prescaler. The slope represents an ongoing frequency misalignment , hence modeling the change in frequency division factor (Fig. 6.10(a)). From this plot the settling time can be estimated. For a 10^{-6} relative accuracy (f_{osc} = 1.8 GHz), the loop needs approximately 120 μs to settle for the values used in this analysis.

At this stage all necessary design equations are derived. With these equations the parameters of the synthesizer composing blocks can be determined such that the global

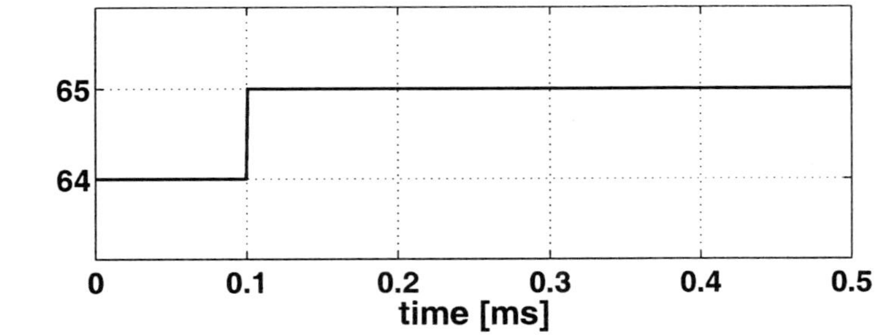

(a) prescaler's division rate

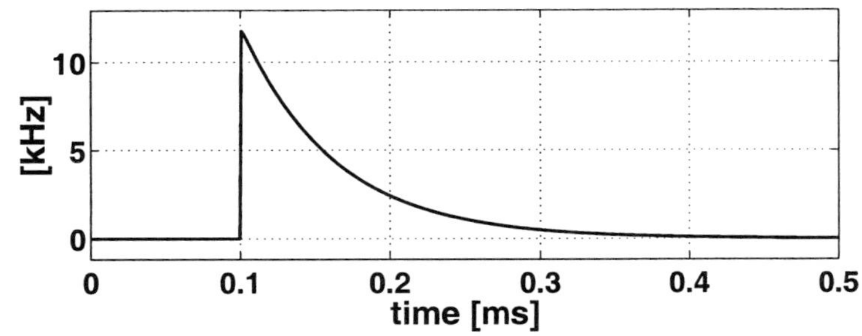

(b) output frequency error

Figure 6.10 Frequency error due to synthesizer settling

specifications are met. Then the implementation of the underlying subblocks is started. Once all subblocks are implemented, the global behavior should be verified. For this purpose other, more accurate models are needed that take into account second–order effects. This verification process is discussed in the following section.

6.5 BEHAVIORAL MODELS FOR BOTTOM–UP VERIFICATION

After implementing the frequency synthesizer down to the lowest level, a final verification of the overall behavior has to be performed. At this stage all nonlinearities are taken into account as accurately as possible. For that reason node voltages are used as state variables rather than the phase of the oscillator wave.

The local oscillator itself introduces the largest nonlinearities in the frequency synthesizer. These nonlinearities, together with the time–varying behavior of the local oscillator circuit, have a dominating influence on the phase noise performance at small frequency offsets. To track the nonlinear behavior, a model is developed of the open loop oscillator (see also [5]). Later on in this section, this model is illustrated with the evaluation of the phase noise spectrum of the overall frequency synthesizer.

6.5.1 Accurate nonlinear model of the local oscillator

The local oscillator model described in this section, exhibits both the static and dynamic performance of a transistor–level oscillator. Depending on the wanted degree of accuracy, the behavioral model is computationally one to two orders of magnitude faster than the corresponding transistor–level model.

First the model composition is explained. Then a verification is performed to compare the model's accuracy with transistor–level results.

Starting from an oscillator implementation, the static transfer function from the input voltage or input current to the fundamental oscillator frequency is extracted from consecutive simulations. Simultaneously the amplitude of the fundamental frequency component and its harmonics is tracked as a function of the input variable. This yields the vectors $f_{stat}\left(v_{in}\right)$, $A_k\left(v_{in}\right)$ and $\phi_k\left(v_{in}\right)$. When calculating the static oscillation frequency during the model implementation, a higher–order interpolation algorithm is applied to the f_{stat} samples surrounding the momentary input value. Fig. 6.11 shows such a typical static frequency relation for a 1.8 GHz voltage controlled oscillator. Note that the curve is not linear.

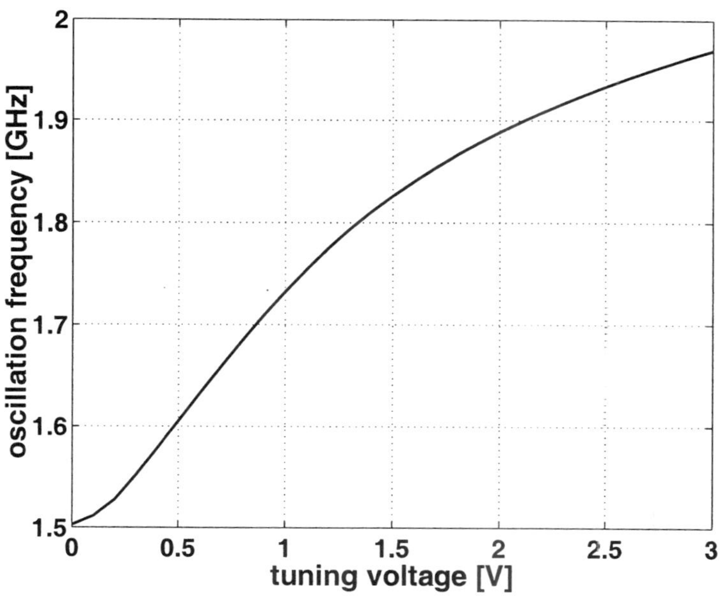

Figure 6.11 Static transfer between tuning voltage and fundamental oscillation frequency of an oscillator

A model including only the static behavior is not accurate enough to evaluate phase noise contributions. Hence the dynamic behavior is superimposed on the static behavior. To estimate this dynamic behavior, a step function is applied to the input variable (Fig. 6.12(a)). The fundamental oscillation frequency (Fig. 6.12(d)) cannot follow this abrupt change immediately. The time constants that correspond with this lagging behavior are incorporated in the transfer function $H_{stat2dyn}(s)$.

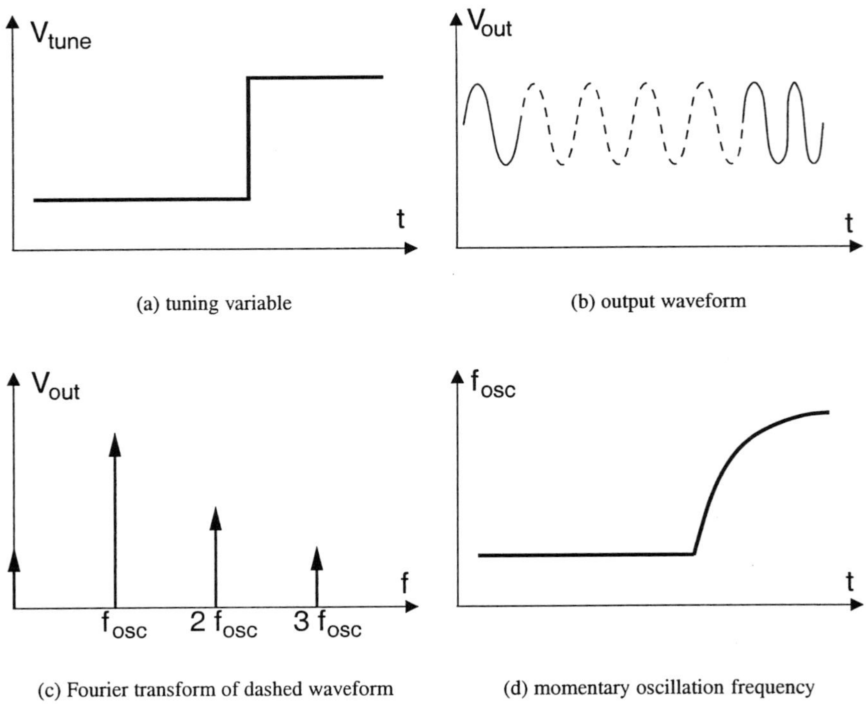

Figure 6.12 Extraction method for the dynamic model parameters of an oscillator

Its coefficients are determined in the following manner. For a given time–varying input variable (Fig. 6.12(a)), the output waveform is simulated. Then a time frame is selected containing a small number of periods of the fundamental waveform (dashed lines in Fig. 6.12(b)). From this time frame the fundamental oscillation frequency is calculated using the discrete Fourier transformation (Fig. 6.12(c)). The time frame is then shifted. For each time shift the instantaneous oscillation frequency is determined. This yields a tracked version of the frequency behavior as a function of time (Fig. 6.12(d)). Note that a wide time frame acts as a low–pass filtering operation on the oscillator signal, thereby hiding some of the high–frequency behavior. In practice, a time frame of four periods wide is large enough to obtain accurate results without corrupting the dynamics of the system.

The parameters of the transfer function $H_{stat2dyn}(s)$ are calculated by mapping the momentary oscillation waveform on the *momentary static frequency.* Therefore the tuning variable is first converted to the static oscillation frequency using the nonlinear relationship derived earlier (i.e. f_{stat} and Fig. 6.11). This static oscillation frequency is used as input for an identification process. The coefficients of $H_{stat2dyn}$ are tuned in such a way that the model's output matches with the calculated momentary frequency behavior. During this identification process, noise influence (e.g. quantization noise during the frequency measurement) is largely attenuated. Robust results are obtained when the oscillator input waveform contains many frequency tones, and thereby activates the model's functionality over a wide frequency range.

So far both the static and the dynamic behavior of the local oscillator have been derived. The output waveform is then calculated as :

$$v_{out}(t) = A_0(v_{in}(t)) + \sum_{k=1}^{k=N} A_k(v_{in}(t))\, sin(\Phi_k(t)) \tag{6.13}$$

$$\Phi_k(t) = \phi_k(v_{in}(t)) + 2\pi \int_{t_0}^{t} k.f_{dyn}(\tau)\, d\tau \tag{6.14}$$

$$f_{dyn}(t) = h_{stat2dyn}(\tau) \otimes f_{stat}(v_{in}(\tau)) \tag{6.15}$$

The output waveform is constructed from a Fourier–series expansion. The harmonics' amplitude, calculated during the static extraction, is modulated by the input variable. The phase of each harmonic sine wave is calculated as the sum of a phase offset $\phi_k\,(v_{in}(t))$ (which is obtained from the static extraction) and the integral with respect to time of the dynamic oscillation frequency. This frequency is written as a convolution of the impulse response $h_{stat2dyn}(\tau)$ (which is the inverse Laplace transform of $H_{stat2dyn}(s)$) with the *momentary static oscillation frequency*, corresponding with the momentary tuning variable.

This oscillator model is used when extracting the phase noise for the closed–loop synthesizer, given the noise generated by all the subblocks . To obtain accurate results, the spectral behavior of the local oscillator model should match very well with the spectral behavior of the transistor–level oscillator for different time–varying input signals. The accuracy of the model is verified by applying a sine wave superimposed on a DC–value on the input node, and comparing the spectra of both the extracted oscillator model and the implemented device–level oscillator description. Fig. 6.13 shows the two resulting spectra. One can identify the fundamental frequency component (straight arrow), the applied sine wave component (dashed arrow) and the newly generated lobe on the other side of the carrier (dash–dotted arrow), both for the transistor–level (straight curve) and the behavioral model (dashed curve) oscillator. In this example, the error on the side lobes between the transistor–level and behavioral model oscillator is below 0.25 dB while the simulation speed–up is larger than 30.

6.5.2 Illustration : phase noise evaluation of a complete frequency synthesizer at the verification stage

In this section the phase noise spectrum of a 1.8 GHz CMOS frequency synthesizer [2] is verified after its design.

Phase noise evaluation of a complete frequency synthesizer is tackled in a hierarchical manner, and can therefore be split up into the following four major steps.

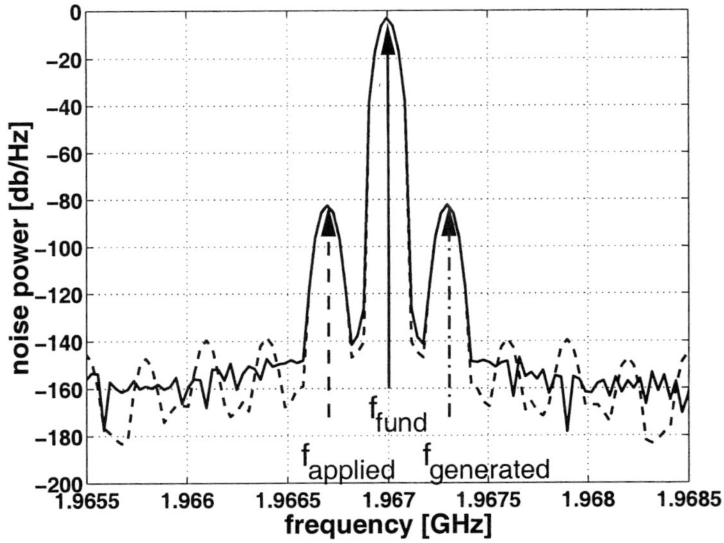

Figure 6.13 Verification of the oscillator model when a sinewave is superimposed on the DC–tuning variable : the straight–line spectrum shows the transistor–level simulations, the dashed–line spectrum shows the behavioral model simulations

1. Find the *steady–state* regime for the overall synthesizer and its composing building blocks
2. Around this working point, quantify the noise contributions of the different building blocks
3. Calculate the noise propagation from each of the building blocks to the synthesizer's output
4. Combine all contributions to obtain the global phase noise spectrum of the synthesizer

Actually the phase noise evaluation method is very similar to the one described in section 6.3 The major difference is the higher position in the design hierarchy at this stage. At the circuit level, a circuit is split up into different elementary components. Here, however, at the behavioral level, the overall system is split up into several building blocks.

This higher level of abstraction has two consequences.

1. First of all no standard models are available for the different building blocks. At the circuit level, all elementary components (transistors, nonlinear capacitors, . . .) have a detailed model that is available in most circuit simulators. The lack of models implies that dedicated models have to be developed. For one of the blocks, the voltage–controlled oscillator, an accurate model was already

developed in section 6.5.1 The other components in the frequency synthesizer are described using structural models.

2. Secondly, there are no noise power spectral densities available for the different building blocks. To solve this problem, the noise contributions of all individual building blocks first have to be calculated. For most blocks, a simple small–signal noise analysis suffices to this end. For the voltage–controlled oscillator however, the phase noise extraction method described in section 6.3 has to be applied, at the correct operating point (i.e. at the correct value for the tuning variable).

Once the models and the noise spectra of the different building blocks are available, the phase noise evaluation method of section 6.3 can be applied at a higher level in the design hierarchy. This is illustrated now.

First of all the parameters of the model defined in section 6.5 (eqs. 6.13–6.15) are extracted from the design. Then the noise spectra of the composing blocks are calculated. Two noise sources are considered here : the loop filter contains dominant noise sources, as well as the voltage–controlled oscillator. The phase noise spectrum of the open–loop VCO is simulated as mentioned in section 6.3 Noise contributions from the loop filter are evaluated using classical small–signal noise analysis techniques.

Sinusoidal current sources are then placed in parallel with the noise–generating subblocks. These current sources sample the noise spectra at carefully selected frequencies. The transfer from these sources to the output is then evaluated. In this way the partial spectra of those contributions is calculated. This yields the global output phase noise spectrum after combining the partial spectra.

Fig. 6.14 shows the resulting phase noise spectrum of the synthesizer. For large frequency offsets the oscillator's phase noise is dominating, whereas for small frequency offsets the filter noise contributions are dominating. Note that these simulation results correspond well with the measured phase noise spectrum for large frequency offsets. However for small offsets, the simulations underestimate the measured phase noise due to some unmodeled noise contributions in the filter.

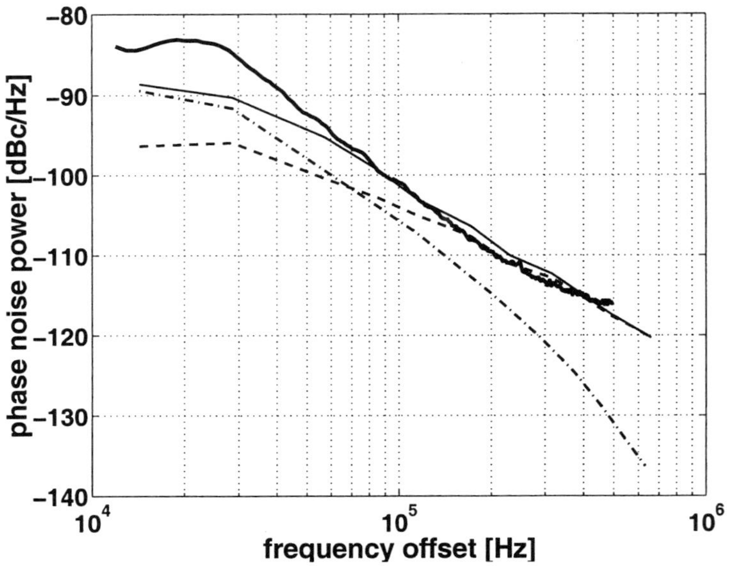

Figure 6.14 Partial and global phase noise spectrum : VCO–spectrum (dashed lines), loop filter spectrum (dash–dotted lines), the global simulated phase noise spectrum (thin straight line) and the measured spectrum (thick straight line)

6.6 SUMMARY

After a brief introduction to the top–down design and bottom–up verification methodology, this chapter has presented the necessary models for the systematic design of a frequency synthesizer used in telecommunication applications. The models are tuned towards the evaluation of the trade–off between the loop settling time and phase noise performance at the output node. Both models for top–down design and bottom–up verification were presented. The validity of these models was illustrated using a 1.8 GHz CMOS frequency synthesizer.

References

[1] Accellera (2000). *VERILOG Analog/Mixed–Signal : language reference manual.* Accellera.

[2] Craninckx, J. (1998). *Wireless CMOS Frequency Synthesizer Design.* Kluwer Academic Publisher, Boston.

[3] Crols, J., Donnay, S., Steyaert, M., and Gielen, G. (1995). A high–level design and optimization tool for analog RF receiver front–ends. In *International Conference on Computer–Aided Design (ICCAD)*, pages 550–553.

[4] De Smedt, B. and Gielen, G. (1997). Accurate simulation of phase noise in oscillators. In *European Solid State Circuits Conference (ESSCIRC)*, pages 204–208, Southampton, UK.

[5] De Smedt, B. and Gielen, G. (1998). Nonlinear behavioral modeling and phase noise evaluation in phase locked loops. In *Custom Integrated Circuits Conference (CICC)*, pages 53–56, Santa Clara, USA.

[6] Egan, W. (1981). *Frequency synthesis by phase lock.* J. Wiley and Sons.

[7] Gardner, F. (1979). *Phaselock techniques.* J. Wiley and Sons.

[8] Gielen, G. (1998). *Top–down design of mixed–mode systems : challenges and solutions*, chapter II.6. Kluwer Academic Publisher. edited by J. Huijsing, R. van de Plassche and W. Sansen.

[9] Gielen, G., Walscharts, H., and Sansen, W. (1989). ISAAC : a symbolic simulator for analog integrated circuits. *IEEE Journal of Solid State Circuits*, 24(6):1587–1597.

[10] IEEE 1076.1 working group (1997). *IEEE standard VHDL 1076.1 language reference manual - analog and mixed–signal extensions to VHDL 1076.* IEEE 1076.1 working group.

[11] Laker, K. and Sansen, W. (1994). *Design of analog integrated circuits and systems.* McGraw-Hill.

[12] VSI Alliance (1998). System chip letter : managing the design reuse challenge. http://www.vsi.org.

7 NONLINEAR SYMBOLIC NETWORK ANALYSIS: ALGORITHMS AND APPLICATIONS TO RF CIRCUITS

Wim Verhaegen, Georges Gielen

ESAT-MICAS
Katholieke Universiteit Leuven, Belgium
E-mail: [Wim.Verhaegen, Georges.Gielen]@esat.kuleuven.ac.be

In the analysis of analog integrated circuits, distortion and intermodulation are important factors. Either they are unwanted, as is the case in linear building blocks like opamps or filters, or they are explicitly wanted to obtain a signal shifted in frequency, as is the case with mixers. Distortion and intermodulation need to be assessed accurately in both cases, which requires time–consuming simulations using the classical numerical approaches. Moreover, the numerical errors accumulate with increasing simulation times, limiting the attainable numerical accuracy.

These limitations can be overcome by employing alternative simulation techniques like harmonic balance (e.g. [15]), multitime analysis (e.g. [21]) and the use of describing functions in circuits with feedback (e.g. [9]). These methods exploit some characteristics of the signal waveforms present in the class of circuits we are interested in: the periodic behavior with a limited set of discrete signal frequencies in the case of harmonic balance, the fact that the signals can be decomposed into contributions at separate frequencies in the case of multitime analysis, and the fact that a finite set of circuit equations suffices for approximating the harmonics when using describing functions for circuits with feedback.

A similar strategy is taken by the method described in this chapter, and which is based on the approximation technique presented in [10, 32, 34]. This technique is a simplified version of the Volterra–series–based approaches presented in [16, 2, 3]. Under some assumptions, of which the fact that the circuit is behaving weakly nonlinearly is the most important one, the Kirchhoff equations for the harmonics can approximately be solved and the harmonic components of the circuit outputs are derived in a fully symbolic way. The advantage of this technique is that a closed–form symbolic

P. Wambacq et al. (eds.),
Low-Power Design Techniques and CAD Tools for Analog and RF Integrated Circuits, 131–154.

formulation is found for the harmonic distortion and intermodulation components in terms of the small–signal parameters and the nonlinearities of the transistors in the circuit, at the expense of a small loss of accuracy. This formulation has some advantages compared to the numerical analysis approaches: the influence of each nonlinear circuit component can be identified, and the obtained formula can be reused at multiple operating points of the circuit. It is thus possible to use the obtained formulas as design equations in a high–level system design environment.

Before going into the details of the algorithm, it is to be noted that similar approaches have been followed in the past to obtain symbolic expressions for the distortion in specific classes of circuits. E.g. the distortion in sampling mixers is analyzed in [36], and a method for analyzing the distortion of analog building blocks in [23]. All symbolic approaches are intrinsically limited someway, and the above publications are no exceptions.

The scope of the algorithm presented here is limited to weakly nonlinear circuits. This means that the circuit characteristics are nonlinear in a smooth way, implying that higher–order contributions are always smaller than lower–order ones, and that the applied signals are small. These restrictions form no problem for the application area studied in this chapter: RF circuits in standard analog IC technology. The algorithm for symbolic distortion analysis of weakly nonlinear analog integrated circuits is presented in section 7.1, its implementation issues are discussed in section 7.2, and it is illustrated with example applications in section 7.3. Finally, conclusions are drawn in section 7.4.

7.1 ALGORITHM

The algorithm used for analyzing weakly nonlinear circuits is explained in [10] and [32]. In addition the nonlinear properties of bipolar and MOS transistors are discussed in extenso in [34]. The analysis algorithm is briefly repeated here for the sake of completeness, after which its application to practical RF circuits will be demonstrated.

7.1.1 Terminology

The following terms are used:

- A *two–dimensional conductance* is a conductance with 2 controlling branch voltages, in this chapter denoted as v_i and v_j.

- A *three–dimensional conductance* is a conductance with 3 controlling branch voltages, in this chapter denoted as v_i, v_j and v_k.

- A nonlinear current is described using its DC component and the derivatives to the controlling branch voltages. The derivatives up to order 3 are described by *nonlinearity coefficients*, which are linked to the derivatives of the large–signal characteristic of the device with the equation (here given for a nonlinearity coefficient of order q for a three–dimensional conductance):

$$K_{q_{mg_1\&ng_2\&(q-m-n)g_3}} = \frac{m!n!(q-m-n!)\partial i\,(v_i,v_j,v_k)}{\partial^m v_i \partial^n v_j \partial^{(q-m-n)} v_k} \qquad (7.1)$$

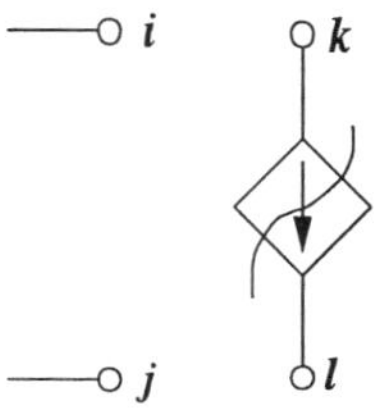

Figure 7.1 Part of a network with a nonlinear transconductance

where g_1, g_2, g_3 are the (trans)conductances controlled by the voltages v_i, v_j, v_k.

- A *phasor* is an imaginary number representing the amplitude and phase of the frequency content of a signal at a given frequency. For the purpose of weakly nonlinear analysis, it suffices to express all frequency components at a linear combination of the input frequencies. Furthermore, we will always deal with 1 or 2 different input frequencies. Given two input frequencies f_1 and f_2, the phasor for the voltage at node x at frequency $mf_1 + nf_2$ is denoted as $V_{x,m,n}$, and the current through an element R is denoted as $I_{R,m,n}$. These phasors thus correspond to Fourier coefficients.

7.1.2 Description of the algorithm fundamentals

Distortion analysis requires a single sinusoidal input, while intermodulation analysis requires two sinusoidal inputs, of which the input frequencies are in general unrelated. At most 2 input frequencies are thus needed for the analysis of distortion and intermodulation. To keep the explained method general, we will model each node voltage and branch current as a sum of phasors for each linear combination of input frequencies f_1 and f_2. E.g. for a node voltage v_x this results in

$$\begin{aligned} v_x(t) &= \sum_{m=0}^{+\infty} \sum_{n=0}^{+\infty} \mathrm{Re}\left(V_{x,m,n} e^{j(m\omega_1 + n\omega_2)t}\right) \\ &= \frac{1}{2} \sum_{m=-\infty}^{+\infty} \sum_{n=-\infty}^{+\infty} V_{x,m,n} e^{j(m\omega_1 + n\omega_2)t} \end{aligned} \tag{7.2}$$

with

$$\omega_1 = 2\pi f_1 \tag{7.3a}$$

$$\omega_2 = 2\pi f_2 \tag{7.3b}$$

The Kirchhoff equations are written down for the circuit under analysis using equations of the type (7.2). Suppose a nonlinear transconductance is present in the circuit between nodes k and l, which is controlled by the voltage between nodes i and j (Fig. 7.1). The branch current through the transconductance is given by

$$\begin{aligned} i(t) &= g_m(v_i(t) - v_j(t)) + K_{2g_m}(v_i(t) - v_j(t))^2 \\ &\quad + K_{3g_m}(v_i(t) - v_j(t))^3 + \cdots \end{aligned} \tag{7.4}$$

In order to get manageable results, it is assumed that expression (7.4) can be truncated after the 3rd–order term without too much loss of accuracy in the final analysis results. This *weak–nonlinearity assumption* is made for each nonlinear element in the circuit. Note that in principle expression (7.4) could be truncated after any term, but in practice we are interested in 2nd– and 3rd–order distortion, which are mainly determined by the 2nd– and 3rd–order nonlinear coefficients. It is also convenient to truncate after the 3rd–order terms, as it proves to be hard to get accurate values for higher–order nonlinearity coefficients. So, using the weak nonlinearity assumption, the branch current is modeled as

$$\begin{aligned} i_{\text{approx}}(t) &= g_m(v_i(t) - v_j(t)) + K_{2g_m}(v_i(t) - v_j(t))^2 \\ &\quad + K_{3g_m}(v_i(t) - v_j(t))^3 \end{aligned} \tag{7.5}$$

Applying Kirchhoff's current law to the partial network shown in figure 7.1 and substituting the node voltages according to (7.2) then yields for the nodes k and l respectively:

$$\begin{aligned} &\cdots + \frac{1}{2} g_m \left(\sum_{m=-\infty}^{+\infty} \sum_{n=-\infty}^{+\infty} V_{i,m,n} e^{j(m\omega_1 + n\omega_2)t} - \sum_{m=-\infty}^{+\infty} \sum_{n=-\infty}^{+\infty} V_{j,m,n} e^{j(m\omega_1 + n\omega_2)t} \right) \\ &+ \frac{1}{4} K_{2g_m} \left(\sum_{m=-\infty}^{+\infty} \sum_{n=-\infty}^{+\infty} V_{i,m,n} e^{j(m\omega_1 + n\omega_2)t} - \sum_{m=-\infty}^{+\infty} \sum_{n=-\infty}^{+\infty} V_{j,m,n} e^{j(m\omega_1 + n\omega_2)t} \right)^2 \\ &+ \frac{1}{8} K_{3g_m} \left(\sum_{m=-\infty}^{+\infty} \sum_{n=-\infty}^{+\infty} V_{i,m,n} e^{j(m\omega_1 + n\omega_2)t} - \sum_{m=-\infty}^{+\infty} \sum_{n=-\infty}^{+\infty} V_{j,m,n} e^{j(m\omega_1 + n\omega_2)t} \right)^3 \\ &+ \cdots = 0 \end{aligned} \tag{7.6a}$$

$$\begin{aligned} &\cdots - \frac{1}{2} g_m \left(\sum_{m=-\infty}^{+\infty} \sum_{n=-\infty}^{+\infty} V_{i,m,n} e^{j(m\omega_1 + n\omega_2)t} - \sum_{m=-\infty}^{+\infty} \sum_{n=-\infty}^{+\infty} V_{j,m,n} e^{j(m\omega_1 + n\omega_2)t} \right) \\ &- \frac{1}{4} K_{2g_m} \left(\sum_{m=-\infty}^{+\infty} \sum_{n=-\infty}^{+\infty} V_{i,m,n} e^{j(m\omega_1 + n\omega_2)t} - \sum_{m=-\infty}^{+\infty} \sum_{n=-\infty}^{+\infty} V_{j,m,n} e^{j(m\omega_1 + n\omega_2)t} \right)^2 \\ &- \frac{1}{8} K_{3g_m} \left(\sum_{m=-\infty}^{+\infty} \sum_{n=-\infty}^{+\infty} V_{i,m,n} e^{j(m\omega_1 + n\omega_2)t} - \sum_{m=-\infty}^{+\infty} \sum_{n=-\infty}^{+\infty} V_{j,m,n} e^{j(m\omega_1 + n\omega_2)t} \right)^3 \\ &+ \cdots = 0 \end{aligned} \tag{7.6b}$$

The powers in equations (7.6a) and (7.6b) are expanded. For equation (7.6a) for example this yields:

$$
\begin{aligned}
\cdots &+ \frac{1}{2} g_m \left(\sum_{m=-\infty}^{+\infty} \sum_{n=-\infty}^{+\infty} \left(V_{i,m,n} - V_{j,m,n} \right) e^{j(m\omega_1 + n\omega_2)t} \right) \\
&+ \frac{1}{4} K_{2g_m} \left(\sum_{m_a=-\infty}^{+\infty} \sum_{n_a=-\infty}^{+\infty} \sum_{m_b=-\infty}^{+\infty} \sum_{n_b=-\infty}^{+\infty} \left(V_{i,m_a,n_a} V_{i,m_b,n_b} \right.\right. \\
&\qquad \left.\left. - 2V_{i,m_a,n_a} V_{j,m_b,n_b} + V_{j,m_a,n_a} V_{j,m_b,n_b} \right) e^{j((m_a+m_b)\omega_1 - (n_a+n_b)\omega_2)t} \right) \\
&+ \frac{1}{8} K_{3g_m} \left(\sum_{m_a=-\infty}^{+\infty} \sum_{n_a=-\infty}^{+\infty} \sum_{m_b=-\infty}^{+\infty} \sum_{n_b=-\infty}^{+\infty} \sum_{m_c=-\infty}^{+\infty} \sum_{n_c=-\infty}^{+\infty} \left(V_{i,m_a,n_a} V_{i,m_b,n_b} V_{i,m_c,n_c} \right.\right. \\
&\qquad - 3V_{i,m_a,n_a} V_{i,m_b,n_b} V_{j,m_c,n_c} + 3V_{i,m_a,n_a} V_{j,m_b,n_b} V_{j,m_c\,n_c} \\
&\qquad \left.\left. - V_{j,m_a,n_a} V_{j,m_b,n_b} V_{j,m_c,n_c} \right) e^{j((m_a+m_b+m_c)\omega_1 + (n_a+n_b+n_c)\omega_2)t} \right) \\
&+ \cdots = 0
\end{aligned}
\tag{7.7}
$$

In order to get the response of the circuit at a frequency $m\omega_1 + n\omega_2$, the contributions at that frequency at the left– and right–hand side of the Kirchoff equations in the form (7.7) are to be identified and the resulting set of equations has to be solved. Starting with the first–order response of the circuit at ω_1, the terms in $e^{j\omega_1 t}$ of equation (7.7) are given by:

$$
\begin{aligned}
\cdots &+ \frac{1}{2} g_m \left(V_{i,1,0} - V_{j,1,0} \right) \\
&+ \frac{1}{4} K_{2g_m} \left(\sum_{m=-\infty}^{+\infty} \left(V_{i,1,0} V_{i,1-m,0} + V_{j,m,0} V_{j,1-m,0} - 2V_{i,m,0} V_{j,1-m,0} \right) \right) \\
&+ \frac{1}{8} K_{3g_m} \left(\sum_{m_a=-\infty}^{+\infty} \sum_{m_b=-\infty}^{+\infty} \left(V_{i,m_a,0} V_{i,m_b,0} V_{i,1-m_a-m_b,0} \right.\right. \\
&\qquad - 3V_{i,m_a,0} V_{i,m_b,0} V_{j,1-m_a-m_b,0} + 3V_{i,m_a,0} V_{j,m_b,0} V_{j,1-m_a-m_b,0} \\
&\qquad \left.\left. - V_{j,m_a,0} V_{j,m_b,0} V_{j,1-m_a-m_b,0} \right) \right) \\
&+ \cdots = 0
\end{aligned}
\tag{7.8}
$$

In order to extract a symbolic formulation from a set of Kirchhoff equations like (7.8), the higher–order phasors need to be pruned from the expressions. This is acceptable for two reasons:

- For most networks the magnitude of the phasors decreases rapidly with increasing order, so they have very little impact.
- The higher–order phasors occur in pairs or triplets in equation (7.8), and since for our applications $|V_{i,m,n}| < 1$, the product of 2 phasors is negligible when compared to a single phasor.

After pruning the higher–order phasors, equation (7.8) is simplified to:

$$\cdots + \frac{1}{2} g_m \left(V_{i,1,0} - V_{j,1,0} \right) + \cdots \quad = \quad 0 \tag{7.9}$$

i.e. the Kirchhoff equation one would write down for node k for the linearized network. This method can be applied to each Kirchhoff equation. We can thus conclude that the 1st–order response of a weakly nonlinear circuit is found by solving the linearized circuit.

This same approximation method is used for calculating higher–order responses. E.g. the output at $\omega_1 + \omega_2$ is found by identifying all terms in $e^{j(\omega_1+\omega_2)t}$ in the Kirchhoff equations. For equation (7.7) this results in

$$\begin{aligned} &\cdots + \frac{1}{2} g_m \left(V_{i,1,1} - V_{j,1,1} \right) \\ &+ \frac{1}{4} K_{2g_m} \left(\sum_{m=-\infty}^{+\infty} \sum_{n=-\infty}^{+\infty} \left(V_{i,m,n} V_{i,1-m,1-n} \right. \right. \\ &\qquad \left. \left. - 2 V_{i,m,n} V_{j,1-m,1-n} + V_{j,m,n} V_{j,1-m,1-n} \right) \right) \\ &+ \frac{1}{8} K_{3g_m} \, (\text{higher–order terms}) + \cdots = 0 \end{aligned} \tag{7.10}$$

The terms related to K_{3g_m} occur again in triplets and are negligible. The sum of terms related to K_{2g_m} is also simplified: the contributions of higher frequencies are negligible with respect to the contributions at ω_1 and ω_2. Taking all this into consideration, equation (7.10) is simplified to

$$\begin{aligned} &\cdots + \frac{1}{2} g_m \left(V_{i,1,1} - V_{j,1,1} \right) \\ &+ \frac{1}{4} K_{2g_m} \left(2V_{i,1,0} V_{i,0,1} - 2V_{i,1,0} V_{j,0,1} - 2V_{i,0,1} V_{j,1,0} + 2V_{j,1,0} V_{j,0,1} \right) \\ &+ \cdots = 0 \end{aligned} \tag{7.11}$$

Introducing the auxiliary variable

$$i_{NL2g_m} \quad = \quad K_{2g_m} \left(V_{i,1,0} - V_{j,1,0} \right) \left(V_{i,0,1} - V_{j,0,1} \right) \tag{7.12}$$

and multiplying the left–hand side and right–hand side of equation (7.11) by 2 then yields

$$\cdots + g_m\left(V_{i,1,1} - V_{j,1,1}\right) + \cdots \quad = \quad -i_{NL2g_m} \tag{7.13}$$

In comparison with equation (7.9), this is the same linearized network which is being analyzed. The only differences are the nature of the unknown variables, which are now phasors instead of node voltages, and the addition of a fictitious current source i_{NL2g_m} which represents the impact of the nonlinearity, while the normal circuit input is zero in this case.

The extraction of a symbolic formula for the higher–order response at a network output is thus performed through *linear analysis* of a *modified network*. The modifications are in the form of adding fictitious current sources which are determined by localized nonlinearities only. Those modifications are described by fixed nonlinearity stamps, which are listed for all 2nd– and 3rd–order nonlinearities in section 7.1.3. The fictitious current sources act as frequency converters in the network, the controlling phasors are propagated from elsewhere in the network through linear transfer functions, and the fictitious current is further propagated to the circuit output of interest through linear transfer functions. So the harmonic in a node voltage v_x is given by

$$V_{x,m,n} \quad = \quad \sum_{i=1}^{n_{NL}} i_{NLq_i} \cdot TF_{q_i \mapsto v_x}(m\omega_1 + n\omega_2) \tag{7.14}$$

where n_{NL} is the number of nonlinearities in the circuit, and $TF_{q_i \mapsto v_x}$ the network function mapping the fictitious current generated by the nonlinearity q_i on the node voltage v_x.

Note that nonlinearities can be nested, as the controlling phasor of a nonlinearity might need to be generated by some other nonlinearity in the network. The order of these recursive phasors, however, decreases with each step back, so the recursion stops at some point, and a closed–form symbolic expression is obtained.

7.1.3 Nonlinearity stamps for 2nd– and 3rd–order analysis

The elaboration of the Kirchhoff equations and the subsequent simplification is easily carried out in an analogous way for other output frequencies and nonlinear elements. For the purpose of the analysis of analog integrated circuits, the possible nonlinear elements are resistors (modeled as nonlinear conductances), transistors (modeled as one-, two- and three-dimensional nonlinear conductances and transconductances), and capacitors. The nonlinearity stamps used for building up fictitious current sources like the one in equation (7.12) are listed in table 7.1 for an output at $|\omega_1 \pm \omega_2|$, table 7.2 for $2\omega_1$, table 7.3 for $|2\omega_1 \pm \omega_2|$, and table 7.4 for $3\omega_1$.

Note that all phasors $V_{i,x,y}$, $V_{j,x,y}$, $V_{k,x,y}$ in tables 7.1 through 7.4 denote phasors related to the controlling *branch* voltages v_i, v_j, v_k for the nonlinearities. The indices to the g's and C's correspond to the controlling voltages: 1 for i, 2 for j and 3 for k.

Furthermore, the nonlinearity stamps for the two– and three–dimensional conductances denote only the cross–terms resulting from the interaction from the two (or three) controlling voltages of the two–dimensional (or three–dimensional) conduc-

Table 7.1 Nonlinearity stamps for response at $|\omega_1 \pm \omega_2|$

Nonlinear element	*Nonlinearity stamp for response at* $|\omega_1 \pm \omega_2|$
(trans)conductance	$K2_{g_1} V_{i,1,0} V_{i,0,\pm1}$
capacitance	$j(\omega_1 \pm \omega_2) K2_{C_1} V_{i,1,0} V_{i,0,\pm1}$
2D conductance	$\frac{1}{2} K2_{g_1 \& g_2} (V_{i,1,0} V_{j,0,\pm1} + V_{i,0,\pm1} V_{j,1,0})$
3D conductance	0

Table 7.2 Nonlinearity stamps for response at $2\omega_1$

Nonlinear element	*Nonlinearity stamp for response at* $2\omega_1$
(trans)conductance	$\frac{1}{2} K2_{g_1} V_{i,1,0}^2$
capacitance	$j\omega_1 K2_{C_1} V_{i,1,0}^2$
2D conductance	$\frac{1}{2} K2_{g_1 \& g_2} V_{i,1,0} V_{j,1,0}$
3D conductance	0

Table 7.3 Nonlinearity stamps for response at $|2\omega_1 \pm \omega_2|$

Nonlinear element	*Nonlinearity stamp for response at* $|2\omega_1 \pm \omega_2|$
(trans)conductance	$K2_{g_1} (V_{i,1,0} V_{i,1,\pm1} + V_{i,0,\pm1} V_{i,2,0}) + \frac{3}{4} K3_{g_1} V_{i,1,0}^2 V_{i,0,\pm1}$
capacitance	$(2j\omega_1 \pm \omega_2) \left[K2_{C_1} (V_{i,1,0} V_{i,1,\pm1} + V_{i,0,\pm1} V_{i,2,0}) + \frac{3}{4} K3_{C_1} V_{i,1,0}^2 V_{i,0,\pm1}\right]$
2D conductance	$\frac{1}{2} K2_{g_1} [V_{i,0,\pm1} V_{j,2,0} + V_{i,1,0} V_{j,1,\pm1} + V_{i,1,\pm1} V_{j,1,0} + V_{,2,0} V_{j,0,\pm1}] + \frac{1}{4} K3_{2g_1 \& g_2} \left[2V_{i,0,\pm1} V_{i,1,0} V_{j,1,0} + V_{i,1,0}^2 V_{j,0,\pm1}\right] + \frac{1}{4} K3_{g_1 \& 2g_2} \left[2V_{i,1,0} V_{j,0,\pm1} V_{j,1,0} + V_{i,0,\pm1} V_{j,1,0}^2\right]$
3D conductance	$\frac{1}{4} K3_{g_1 \& g_2 \& g_3} [V_{i,0,\pm1} V_{j,1,0} V_{k,1,0} + V_{i,1,0} V_{j,0,\pm1} V_{k,1,0} + V_{i,1,0} V_{j,1,0} V_{k,0,\pm1}]$

tance. E.g. for a two–dimensional conductance the fictitious current source is a combination of these cross–terms and the two one–dimensional contributions from both controlling voltages. The complete fictitious current source for an output response at

Table 7.4 Nonlinearity stamps for response at $3\omega_1$

Nonlinear element	*Nonlinearity stamp for response at* $3\omega_1$
(trans)conductance	$K_{2_{g_1}} V_{i,1,0} V_{i,2,0} + \frac{1}{4} K_{3_{g_1}} V_{i,1,0}^3$
capacitance	$3j\omega_1 \left[K_{2_{C_1}} V_{i,1,0} V_{i,2,0} + \frac{1}{4} K_{3_{C_1}} V_{i,1,0}^3\right]$
2D conductance	$\frac{1}{2} K_{2_{g_1 \& g_2}} \left[V_{i,1,0} V_{j,2,0} + V_{i,2,0} V_{j,1,0}\right]$ $+\frac{1}{4} K_{3_{2g_1 \& g_2}} V_{i,1,0}^2 V_{j,1,0} + \frac{1}{4} K_{3_{g_1 \& 2g_2}} V_{i,1,0} V_{j,1,0}^2$
3D conductance	$\frac{1}{4} K_{3_{g_1 \& g_2 \& g_3}} V_{i,1,0} V_{j,1,0} V_{k,1,0}$

$2\omega_1$ thus equals (see table 7.2)

$$i_{NL2_{g_1 \& g_2}} = \frac{1}{2} K_{2_{g_1}} V_{i,1,0}^2 + \frac{1}{2} K_{2_{g_2}} V_{j,1,0}^2 + \frac{1}{2} K_{2_{g_1 \& g_2}} V_{i,1,0} V_{j,1,0} \qquad (7.15)$$

7.2 IMPLEMENTATION OF THE ALGORITHM

In order to obtain practical results from the algorithm fundamentals described above, some additional functionality is needed. Firstly the symbolic (and numerical, as will be explained below) representation of many linear transfer functions need to be generated. It would be tedious to do this by hand, and even infeasible for networks of medium to large complexity. Luckily this task has been automated already by symbolic analysis tools, as is described in many publications [1, 22, 28, 17, 24, 25, 11, 10, 18, 33, 29, 13, 35, 8, 14, 31, 19, 26, 6, 12, 5, 27, 20, 30]. The generation of a numerical transfer function as described in [7] and the symbolic approximation algorithm described in [6] have been used for the application examples shown in this chapter. Both algorithms have been implemented in the Symba linear analysis environment. All linear circuit characteristics of the example circuits in section 7.3 are derived using Symba.

Secondly, after introducing all fictitious current sources in the circuit, the closed–form formula for the response at the output is very long for all but the extremely trivial networks. The interpretation of these formulas is not easy, but in most cases they can be drastically simplified by pruning all terms with a minimal contribution to the total response. In order to achieve this *numerical screening*, a numerical evaluation of all contributions must be performed over the frequency range of interest. This requires that all nonlinearity coefficients and linear transfer functions are evaluated numerically, which in turn requires numerical values for all small–signal parameters. These parameters are obtained from numerical simulation results in some design point. The slight disadvantage of this is that the network can no longer be parameterized when screening is applied, as the validity of the result is only guaranteed for the chosen parameter combination of the design point. Note that this step corresponds to the pruning of terms in expression (7.14).

Thirdly, after the numerical screening is finished, a full symbolic formulation of the result is obtained by substituting all transfer functions by their symbolic representation using Symba. These symbolic representations are in turn lengthy for networks of medium to high complexity, so for the sake of interpretability they need to be simplified as well. To this purpose one of the two simplification schemes implemented in Symba can be used:

- a shorter symbolic formulation can be generated by allowing an approximation error on magnitude and phase over a frequency interval of interest;
- or the number of terms generated per coefficient in the numerator and denominator can be limited to a given maximum number.

In the latter case the accuracy of the final result of course needs to be verified against the exact result. Note that one might not want to obtain a fully symbolic expression anyway, as the result with all transfer functions as plain unexpanded symbols already yields quantitative insight into the relative importance of the nonlinear elements. The required level of granularity needs to be considered on a case–to–case basis.

In summary, all the functionality needed for reading and manipulating a network, calculating the nonlinearity coefficients, constructing the network response and manipulating the final result has been implemented in an interactive nonlinear analysis engine which cooperates with Symba that generates the linear expressions. This interaction is illustrated in figure 7.2.

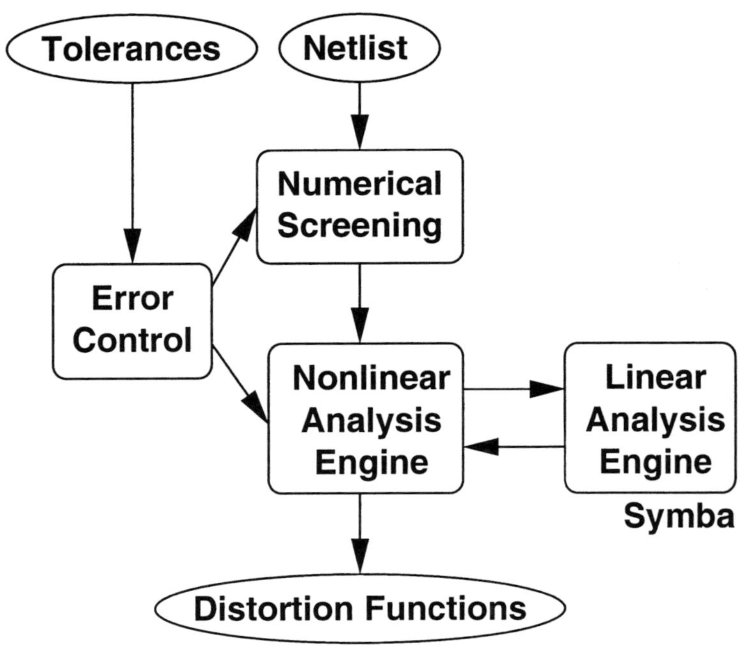

Figure 7.2 The nonlinear analysis engine and its environment

7.2.1 Error control algorithm

One block in figure 7.2, the error control, deserves some more attention. It is logical that the trade–off between the length and the accuracy of the result depends largely on the error control methods that are used. A large range of possibilities is available here, but we will restrict ourselves to the following sequence of error control methods:

- An exact value of the harmonic component under analysis is calculated numerically. The result is a frequency–dependent function that will be used in the subsequent approximation steps as a numerical reference. It is thus supposed that this reference is accurate. Note, however, that this is only true as long as the weak–nonlinearity assumption holds.

- First a number of terms of the harmonic component will be removed by numerical screening. Due to the nature of the typical expression for a harmonic component (see equation 7.14), this process is split into two steps:

  - Pairs of (almost) cancelling terms are identified and replaced by their difference as an explicit representation. This provides a better overall picture on all the terms, as the impact of each individual term might be high, while the net combined impact of both terms is low. The difference of two such terms will also be referred to as a term below.

  - All terms are tested for pruning in order of ascending weight. The pruning of a term is accepted if the updated approximation of the harmonic component stays within a predefined magnitude tolerance $\mathcal{T}_{mag,NS}$ and phase tolerance $\mathcal{T}_{pha,NS}$ from the numerical reference.

- Secondly, all linear circuit functions which are still present in the harmonic component after screening, are approximated by Symba using a magnitude tolerance $\mathcal{T}_{mag,TF}$ and phase tolerance $\mathcal{T}_{pha,TF}$ for each transfer function. Very permissive values are chosen for the error tolerances $\mathcal{T}_{mag,TF}$ and $\mathcal{T}_{pha,TF}$ at the start, but after the approximation of all transfer functions, the numerical evaluation of the harmonic component using those approximations is compared to the numerical reference using tolerances $\mathcal{T}_{mag,SA}$ and $\mathcal{T}_{pha,SA}$ for the symbolic approximation. If the approximation is not valid, the tolerances $\mathcal{T}_{mag,TF}$ and $\mathcal{T}_{pha,TF}$ are made stricter and a new iteration is started. This process is repeated until the error tolerances $\mathcal{T}_{mag,SA}$ and $\mathcal{T}_{pha,SA}$ are met. The choice of the tolerances $\mathcal{T}_{mag,SA}$ and $\mathcal{T}_{pha,SA}$, as well as $\mathcal{T}_{mag,NS}$ and $\mathcal{T}_{pha,NS}$, will be discussed below in section 7.2.2.

  Note that a high number of iterations does not at all deteriorate the time efficiency of this algorithm. The terms of the coefficients of the circuit functions are always generated in the same order of decreasing weight; so a regeneration with stricter tolerances $\mathcal{T}_{mag,TF}$ and $\mathcal{T}_{pha,TF}$ can simply reuse previous results and only has to select some extra terms to reach the tighter error tolerances.

- Finally, parameters characterizing the nonlinear behaviour of the circuit — such as IP3, IM3, ... — can be derived from one or more harmonic components and linear circuit functions. This derivation is strictly speaking a post–processing step to the nonlinear analysis engine, and not part of it. When calculating these parameters based on approximations, similar steps as the ones described above are to be undertaken to control the error on these derived characteristics.

7.2.2 Error tolerances

A typical error tolerance $\mathcal{T}_x$, where x can be either magnitude or phase, is shown in figure 7.3. It consists of a maximum limit for positive deviations (Δx^+) and negative deviations (Δx^-) from the numerical reference. Δx^+ and Δx^- are piecewise linear functions of the frequency f. In the shown example, the tolerance is checked for all frequencies between f_0 and f_3. Also note that the tolerance is symmetrical between f_0 and f_1, but asymmetrical thereafter.

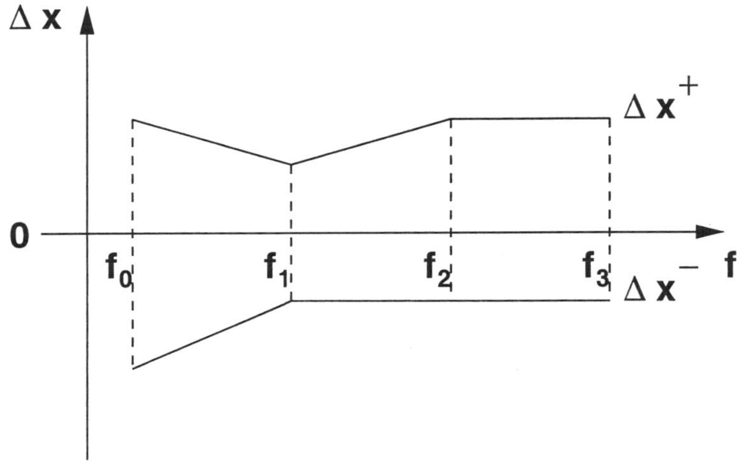

Figure 7.3 Graphical representation of a generic error tolerance $\mathcal{T}_x$

As explained in section 7.2.1, the accuracy of the end result of the nonlinear analysis engine (i.e. an approximation of the harmonic component), is controlled by the error tolerances $\mathcal{T}_{mag,NS}$ and $\mathcal{T}_{pha,NS}$ for numerical screening, and $\mathcal{T}_{mag,SA}$ and $\mathcal{T}_{pha,SA}$ for symbolic approximation. However, the user only needs to specify tolerances $\mathcal{T}_{mag,user}$ and $\mathcal{T}_{pha,user}$ on the end result. There is thus a degree in freedom left in picking the error tolerances for numerical screening and symbolic approximation.

Note that the *errors* made during these steps are incremental: the total error on the complex response will be $\epsilon_{NS}(f)$ after numerical screening, and $\epsilon_{NS}(f) + \epsilon_{SA}(f)$ after a subsequent symbolic approximation. The final error on the result is thus, $\epsilon_{tot}(f) = \epsilon_{NS}(f) + \epsilon_{SA}(f)$, and this needs to satisfy the error tolerances $\mathcal{T}_{mag,user}$ and $\mathcal{T}_{pha,user}$. It is thus only evident to choose the error tolerances $\mathcal{T}_{mag,SA} = \mathcal{T}_{mag,user}$ and $\mathcal{T}_{pha,SA} = \mathcal{T}_{pha,user}$.

This leaves still a degree of freedom in the choice of the tolerances $\mathcal{T}_{mag,NS}$ and $\mathcal{T}_{pha,NS}$. On the one hand, it is evident that a result will always be found as long as those tolerances are stricter than $\mathcal{T}_{mag,user}$ resp. $\mathcal{T}_{pha,user}$, as this ensures that $|\epsilon_{NS}(f)| < |\epsilon_{tot}(f)|$ for all frequencies f. On the other hand, the length of the resulting expression varies with the degree of strictness that is applied during numerical screening. In order to remove this ambiguity, we assume that any arbitrary set of error tolerances $(\mathcal{T}_{mag,*}, \mathcal{T}_{pha,*})$ cannot be exactly covered alone by pruning terms from expression 7.14. It is thus possible to choose $\mathcal{T}_{mag,NS} = \mathcal{T}_{mag,user}$ and $\mathcal{T}_{pha,NS} = \mathcal{T}_{pha,user}$ and use the margin left during numerical screening for symbolic approximation. Practical results confirm this conclusion, so it is applied to all circuits under analysis.

7.3 APPLICATIONS OF NONLINEAR SYMBOLIC ANALYSIS

7.3.1 Two–stage miller–compensated opamp

For the purpose of illustrating the use of the nonlinear analysis engine, the distortion in a relatively simple opamp is studied first. The Miller opamp shown in figure 7.4 is chosen as illustratory circuit. The opamp has a differential input at the matched source–coupled pair M1A and M1B, and a single–ended output. The unity feedback obtained with the resistor Rfb configures this opamp as a voltage follower.

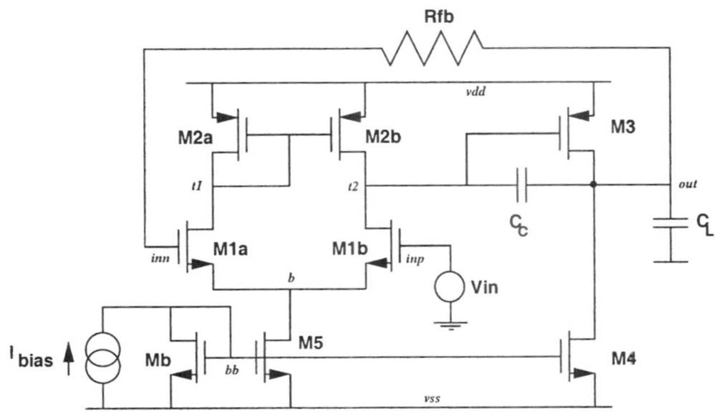

Figure 7.4 Schematic of the Miller opamp with feedback

As a reference, the exact linear response of the voltage follower is shown in figure 7.5. Note that the response starts falling off at a unity gain frequency of around 10MHz. The small peaking around 1MHz and the low phase margin are characteristics of this specific circuit design, and are not relevant for our purpose of studying the distortion.

This circuit has been analyzed by the nonlinear analysis engine with an error tolerance of 5dB on the magnitude and 10° on the phase of the result, for all values of the input frequency $f_{in} \in [1E3, 1E9]Hz$. The application of all the fictitious current sources results in a second–order response at the output $V_{(out,0),2}$ which is the product

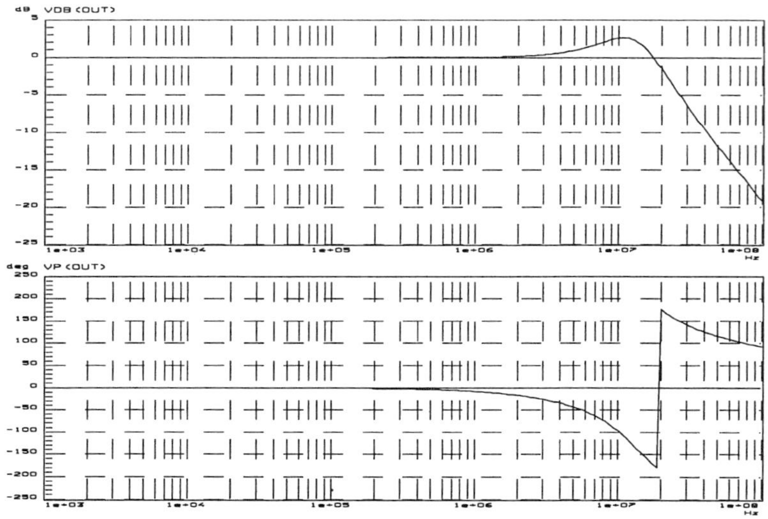

Figure 7.5 Bodeplot of the output voltage of the Miller opamp with feedback

of $\frac{v_{in}^2}{2}$ and the sum of the following contributions, which are graphically shown in figure 7.6:

$$K_{2_{gm\&go},M1B} H_{v_{inp,0}\mapsto v_{t2,b}}(j\omega) H_{v_{inp,0}\mapsto v_{inp,b}}(j\omega) H_{i_{DS,M1B}\mapsto v_{out,0}}(2j\omega) \tag{7.16a}$$

$$K_{2_{gm\&go},M1A} H_{v_{inp,0}\mapsto v_{t1,b}}(j\omega) H_{v_{inp,0}\mapsto v_{inn,b}}(j\omega) H_{i_{DS,M1A}\mapsto v_{out,0}}(2j\omega) \tag{7.16b}$$

$$K_{22_{gm},M1B} H_{v_{inp,0}\mapsto v_{inp,b}}(j\omega) H_{v_{inp,0}\mapsto v_{inp,b}}(j\omega) H_{i_{DS,M1B}\mapsto v_{out,0}}(2j\omega) \tag{7.16c}$$

$$K_{22_{gm},M1A} H_{v_{inp,0}\mapsto v_{inn,b}}(j\omega) H_{v_{inp,0}\mapsto v_{inn,b}}(j\omega) H_{i_{DS,M1A}\mapsto v_{out,0}}(2j\omega) \tag{7.16d}$$

$$\begin{gathered}(K_{22_{go},M2A} + K_{2_{gm\&go},M2A} + K_{22_{gm},M2A})* \\ H_{v_{inp,0}\mapsto v_{t1,vdd}}(j\omega) H_{v_{inp,0}\mapsto v_{t1,vdd}}(j\omega) H_{i_{DS,M2A}\mapsto v_{out,0}}(2j\omega)\end{gathered} \tag{7.16e}$$

$$K_{22_{gm},M2B} H_{v_{inp,0}\mapsto v_{t1,vdd}}(j\omega) H_{v_{inp,0}\mapsto v_{t1,vdd}}(j\omega) H_{i_{DS,M2B}\mapsto v_{out,0}}(2j\omega) \tag{7.16f}$$

$$K_{22_{gm},M3} H_{v_{inp,0}\mapsto v_{t2,vdd}}(j\omega) H_{v_{inp,0}\mapsto v_{t2,vdd}}(j\omega) H_{i_{DS,M3}\mapsto v_{out,0}}(2j\omega) \tag{7.16g}$$

$$K_{22_{go},M1B} H_{v_{inp,0}\mapsto v_{t2,b}}(j\omega) H_{v_{inp,0}\mapsto v_{t2,b}}(j\omega) H_{i_{DS,M1B}\mapsto v_{out,0}}(2j\omega) \tag{7.16h}$$

$$K_{22_{go},M1A} H_{v_{inp,0}\mapsto v_{t1,b}}(j\omega) H_{v_{inp,0}\mapsto v_{t1,b}}(j\omega) H_{i_{DS,M1A}\mapsto v_{out,0}}(2j\omega) \tag{7.16i}$$

$$K_{22_{go},M3} H_{v_{inp,0}\mapsto v_{out,vdd}}(j\omega) H_{v_{inp,0}\mapsto v_{out,vdd}}(j\omega) H_{i_{DS,M3}\mapsto v_{out,0}}(2j\omega) \tag{7.16j}$$

$$K_{22_{go},M4} H_{v_{inp,0}\mapsto v_{out,vss}}(j\omega) H_{v_{inp,0}\mapsto v_{out,vss}}(j\omega) H_{i_{DS,M4}\mapsto v_{out,0}}(2j\omega) \tag{7.16k}$$

$$K_{2_{gm\&go},M3} H_{v_{inp,0}\mapsto v_{out,vdd}}(j\omega) H_{v_{inp,0}\mapsto v_{t2,vdd}}(j\omega) H_{i_{DS,M3}\mapsto v_{out,0}}(2j\omega) \tag{7.16l}$$

$$K_{22_{go},M5} H_{v_{inp,0}\mapsto v_{b,vss}}(j\omega) H_{v_{inp,0}\mapsto v_{b,vss}}(j\omega) H_{i_{DS,M5}\mapsto v_{out,0}}(2j\omega) \tag{7.16m}$$

$$K_{2_{gm\&go},M2B} H_{v_{inp,0}\mapsto v_{t2,vdd}}(j\omega) H_{v_{inp,0}\mapsto v_{t1,vdd}}(j\omega) H_{i_{DS,M2B}\mapsto v_{out,0}}(2j\omega) \tag{7.16n}$$

$$K_{22_{go},M2B} H_{v_{inp,0}\mapsto v_{t2,vdd}}(j\omega) H_{v_{inp,0}\mapsto v_{t2,vdd}}(j\omega) H_{i_{DS,M2B}\mapsto v_{out,0}}(2j\omega) \tag{7.16o}$$

Note that this is a simplified view already, as the zero contributions have been omitted. Also note that the nonlinear coefficients of transistor M2A have an identical

Table 7.5 Second–order nonlinearity coefficients for the transistors in the Miller opamp

Transistor	$K_{2_{2g_o}}$	$K_{2_{g_m\&g_o}}$	$K_{2_{2g_m}}$
M1A	-1.9967e-08	2.19228e-05	0.000927819
M1B	-2.90793e-08	2.27662e-05	0.000919779
M4	-2.10673e-06	6.60911e-05	0.00185052
M5	-8.68249e-07	3.90932e-05	0.000860076
MB	-1.97237e-06	4.684e-05	0.000830105
M2A	6.86074e-07	-8.7469e-06	-0.000469537
M2B	6.50754e-07	-8.09528e-06	-0.000471967
M3	-2.62388e-06	-1.52991e-05	-0.000968071

impact in contribution (7.16e) because M2A has been configured as a diode (see figure 7.4).

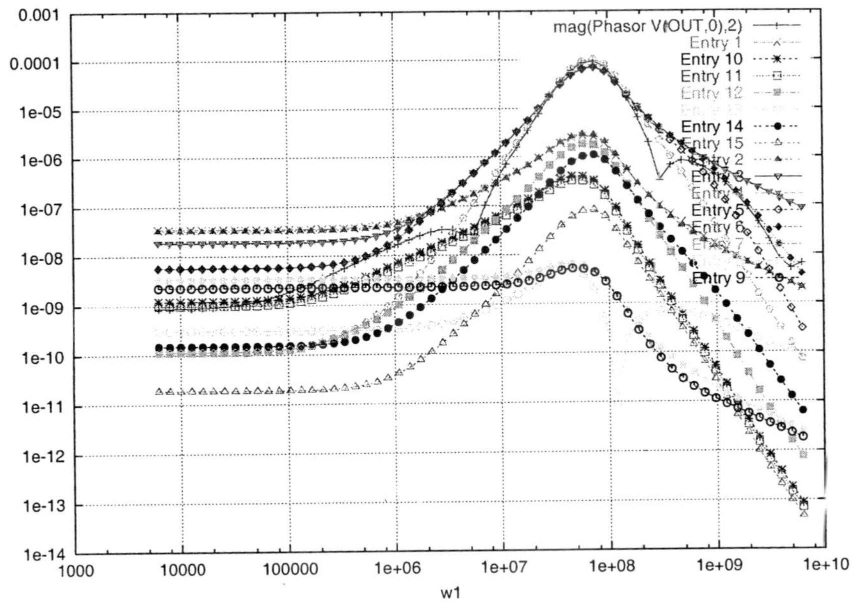

Figure 7.6 Frequency–dependent magnitudes of all contributions to, and the total phasor $V_{(out,0),2}$ for the Miller opamp

The nonlinear coefficients are calculated as the derivatives of a transistor characteristic fitted through points obtained using a numerical simulator, and are given in table 7.5.

Note that the nonlinear coefficients for M1A and M1B, resp. M2A and M2B are approximately equal. This is a consequence of the matching of those transistors. The small deviations result from the asymmetrical biasing conditions caused by the single–ended output.

In order to interpret the expression for $V_{(out,0),2}$ and obtain an expression for, say, the 2nd order output–referred intercept point $OIP2$, the contributions (7.16a) to (7.16o) are further simplified. This is achieved in the following steps:

1. First, the entries which contribute only a little to the value of $V_{(out,0),2}$ are eliminated by numerical screening. The frequency–dependent weights of all entries (7.16a) to (7.16o) prior to numerical screening are calculated with the formula

$$w_i(f) = \frac{|e_i(f)|}{\left|V_{(out,0),2}(f)\right|} \tag{7.17}$$

and are shown in figure 7.7. Note that some weights exceed the value of 1.0, which is possible because those entries are cancelled out by others, an effect which is hidden by the magnitude based weight formula (7.17). Those pairs are identified during the numerical screening, though, and are eliminated if possible.

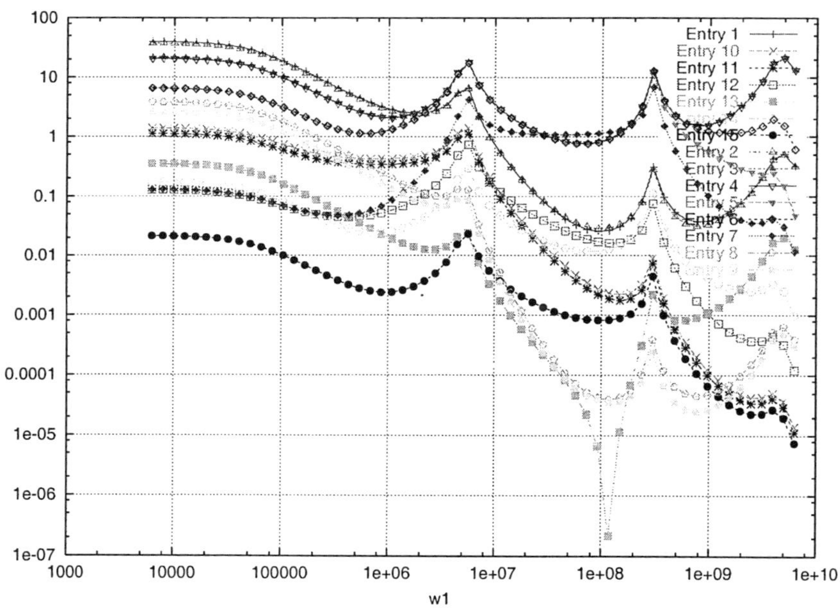

Figure 7.7 Frequency dependent relative weights of all contributions to $V_{(out,0),2}$ for the Miller opamp

The result of the numerical screening for this specific case, is that entries (7.16a through 7.16d) and (7.16h) through (7.16o) can be omitted.

2. Secondly, the transfer functions found in the remaining entries (7.16e), (7.16f) and (7.16g) are approximated to obtain simple, interpretable expressions. The expressions are generated automatically using Symba, and are relatively simple due to the unity–feedback that is applied to the circuit:

$$H_{v_{inp,0}\mapsto v_{t1,vdd}} \approx -1 \tag{7.18a}$$

$$H_{i_{DS,M2A}\mapsto v_{out,0}} \approx \frac{-1}{g_{m_{M1B}}} \tag{7.18b}$$

$$H_{i_{DS,M2B}\mapsto v_{out,0}} \approx \frac{1}{g_{m_{M1B}}} \tag{7.18c}$$

$$H_{v_{inp,0}\mapsto v_{t2,vdd}} \approx -1 \tag{7.18d}$$

$$H_{i_{DS,M3}\mapsto v_{out,0}} \approx \frac{-1}{g_{m_{M3}}} \tag{7.18e}$$

The transfer functions (7.18a) to (7.18e) are substituted into the entries (7.16e) to (7.16g) of $V_{(out,0),2}$. Taking into account the matching of the transistor pairs M1A and M1B, the following small–signal elements have approximately identical values:

$$g_{m_{M1A}} \approx g_{m_{M1B}} \tag{7.19}$$

These small signal elements are therefore replaced by the following approximating variables:

$$g_{m_{M1}} = \frac{g_{m_{M1A}} + g_{m_{M1B}}}{2} \tag{7.20}$$

It is also to be noted from table 7.5 that

$$K_{2_{2g_m},M2A} \gg K_{2_{2g_o},M2A} + K_{2_{g_m\&g_o},M2A} \tag{7.21}$$

It is thus allowed to discard the right–hand side of the inequality (7.21) in the entry (7.16e) of $V_{(out,0),2}$.

All these substitutions and eliminations result in

$$V_{(out,0),2} = \frac{v_{in}^2}{2}\left(-\frac{K_{2_{2g_m},M3}}{g_{m_{M3}}} + \frac{K_{2_{2g_m},M2B} - K_{2_{2g_m},M2A}}{g_{m_{M1}}}\right) \tag{7.22}$$

3. Finally the input–referred second–order intercept point $IIP2$ is calculated, which is defined as the input amplitude for which $|V_{out}| = |V_{out,2}|$, i.e.

$$V_{in} = \frac{V_{in}^2}{2}\frac{\left|-g_{m_{M1}}K_{2_{2g_m},M3} + g_{m_{M3}}\left(K_{2_{2g_m},M2B} - K_{2_{2g_m},M2A}\right)\right|}{g_{m_{M1}}g_{m_{M3}}} \tag{7.23}$$

So $IIP2$ is given by

$$IIP2 = \frac{2g_{m_{M1}}g_{m_{M3}}}{\left|-g_{m_{M1}}K_{2_{2g_m},M3} + g_{m_{M3}}\left(K_{2_{2g_m},M2B} - K_{2_{2g_m},M2A}\right)\right|} \tag{7.24}$$

For the design point chosen for the Miller opamp, the terms in the numerator and denominator in equation (7.24) have the following values:

$$2g_{m_{M1}}g_{m_{M3}} = 824.5 \times 10^{-9} \quad (7.25a)$$

$$-g_{m_{M1}}K_{2_{2g_m},M3} = 424.11 \times 10^{-9} \quad (7.25b)$$

$$g_{m_{M3}}\left(K_{2_{2g_m},M2B} - K_{2_{2g_m},M2A}\right) = -7.5656 \times 10^{-9} \quad (7.25c)$$

As a result the contribution of $K_{2_{2g_m},M3}$ in the denominator of (7.24) is dominant, and the expression can be simplified to:

$$IIP2 = \left|\frac{2g_{m_{M3}}}{K_{2_{2g_m},M3}}\right| \quad (7.26)$$

which equals 1.944V for the chosen design point.

Note that the output–referred second–order intercept point $OIP2$ equals $IIP2$ due to the unity feedback configuration of the opamp.

The obtained result illustrates the application of symbolic analysis on an example circuit. However, for the analysis of practical circuits some additional constraints need to be taken into account. One such constraint is the intrinsic mismatch between transistor parameters due to variations in the production process. These mismatches are also candidate sources of distortion, and need to be taken into account to get realistic results. Note that the mismatch expressed for transistors M2A and M2B in equation (7.24) is a second–order effect resulting from asymmetric biasing. This is not the same as mismatch caused by variations in the production process, which we will refer to as *first–order mismatch* in the remainder of the chapter.

In order to illustrate the impact of first–order mismatch on distortion, the derivation of $OIP2$ has been repeated for the same Miller opamp with a 10% deviation introduced on the widths of transistors M1A and M2A. Note that this deviation is equivalent to a threshold voltage or electron mobility deviation, with parameters depending on the chosen technology. The impact of the mismatch on the first–order behaviour of the circuit is minor, and the approximation of $V_{(out,0),2}$ by the entries (7.16a) to (7.16o) is still valid, as can easily be verified in the same way as it was done for the circuit with no first–order mismatch.

There are however quite a few differences in the numerical values of the symbolic parameters. As a result the small signal parameters of matching devices can no longer be described by a single symbol as in equation (7.20). A symbolic formulation analogous to (7.26) for $OIP2$ must thus be expressed in terms of the different values for $g_{m_{M1A}}$ and $g_{m_{M1B}}$. Numerical screening of the entries (7.16a) to (7.16o) reveals that again the entries (7.16e), (7.16f) and (7.16g) are dominant. This results in the following expression for $OIP2$ for the Miller opamp with mismatch:

$$IIP2_{mm} = \left|\frac{2g_{m_{M1b}}g_{m_{M3}}}{g_{m_{M3}}\left(K_{2_{2g_m},M2B} - K_{2_{2g_m},M2A}\right) - g_{m_{M1B}}K_{2_{2g_m},M3}}\right| \quad (7.27)$$

A numerical evaluation of (7.27) yields $IIP2_{mm} = 1.817$V, which is slightly lower than the equivalent value with no first–order mismatch.

7.3.2 Downconverting mixer

As a second application example, the gain factor in a downconverter is studied. The circuit topology which was proposed in [4] is shown in figure 7.8. A detailed topology of the low–frequency opamp used in this mixer is shown in figure 7.9.

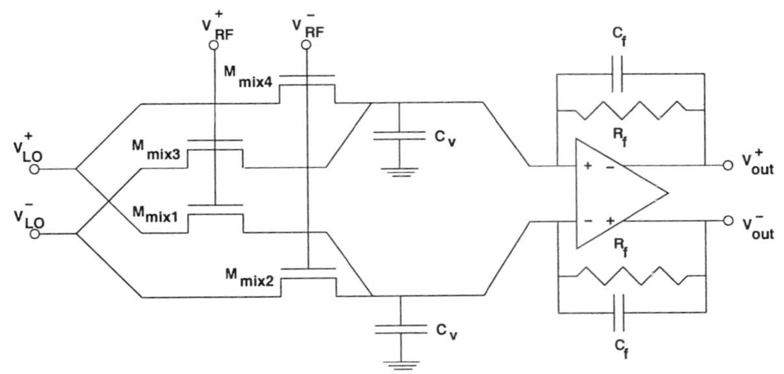

Figure 7.8 Topology of the downconverter mixer

The downconverter topology consists of two main parts. The first stage consists of four intercoupled passtransistors Mmix1 to Mmix4, that convert the differential RF and LO signals to a differential current. This current contains the product of the input signals, but also higher harmonics. Since the circuit is a downconverter, we are only interested in the LF behavior, and this is filtered out by the LF opamp in the second stage. The capacitors C_V are added between the stages to keep the loads of the passtransistors at a virtual ground at all times. Note that the mixer must be designed carefully to keep transistors Mmix1 to Mmix4 in the linear region at all times. For a full explanation on this circuit, the reader is referred to [4].

We are interested in the conversion gain of the mixer. This gain factor is the amplitude of the output signal at frequency $|f_{RF} - f_{LO}|$, and is thus analyzed as the second–order intermodulation product IM2 of the input signals at frequencies f_{RF} and f_{LO}. As a result, two input sources V_{RF} and V_{LO} are applied at the circuit given to the nonlinear analysis engine, and the inputs shown in figure 7.8 are derived from them using voltage–controlled voltage sources so that $V_{RF}^{+} = V_{RF}$, $V_{RF}^{-} = -V_{RF}$, $V_{LO}^{+} = V_{LO}$ and $V_{LO}^{-} = -V_{LO}$. A mixer frequency of $f_{LO} = 1.3GHz$ is chosen, and an RF input at $f_{RF} = 1.3005GHz$ is applied. The amplitudes of the input signals are also limited by the design, and are set to $A_{LO} = 445mV$ and $A_{RF} = 5\mu V$ for the presented results.

Similarly to the example of the Miller opamp in section 7.3.1, the results will be valid for a user–specified range of input frequencies and error margins. In the present example, we have two input frequencies f_{LO} and f_{RF} that can be varied, for which the nonlinear analysis engine provides the following choices:

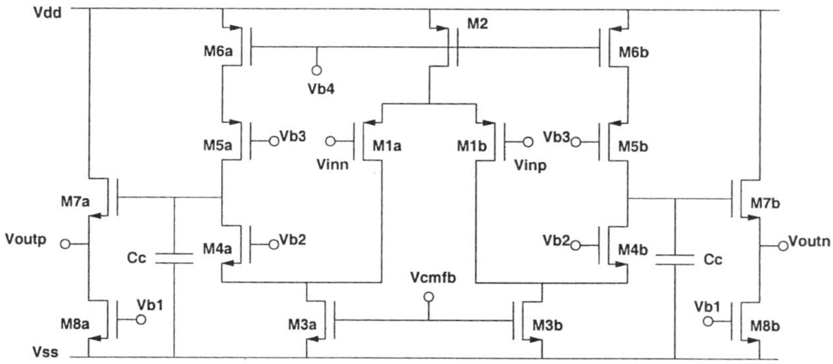

Figure 7.9 Topology of the LF-opamp used in the downconverter mixer

- f_{LO} is varied in the interval $[f_{LO_{min}}, f_{LO_{max}}]$, while f_{RF} is kept at a constant value;
- f_{LO} is varied in the interval $[f_{LO_{min}}, f_{LO_{max}}]$, and $f_{RF} = \alpha f_{LO}$;
- f_{LO} is varied in the interval $[f_{LO_{min}}, f_{LO_{max}}]$, and f_{RF} is varied in the interval $[f_{RF_{min}}, f_{RF_{max}}]$;
- f_{LO} is varied in the interval $[f_{LO_{min}}, f_{LO_{max}}]$, and f_{RF} is varied in the interval $\left[\frac{f_{LO}}{\delta}, \delta f_{LO}\right]$.

The second option is chosen for analyzing the downconverter mixer, as this assures that the frequency of the downconverted signal is in the bandwidth of the LF–opamp. The error tolerances for this circuit were set to a 2dB deviation on the magnitude and a 10° deviation on the phase of the response, for $f_{LO} \in [1.3E6, 1.3E12]\,Hz$.

The nonlinear analysis engine computes a symbolic result containing 76 entries for this problem. However, only the following 8 entries are non–zero and are thus an accurate representation of the frequency component of v_{out} at the frequency $f_2 - f_1$ (with $f_2 = f_{RF}$ and $f_1 = f_{LO}$):

$$\frac{A_{LO}A_{RF}}{2} K_{2_{gm\&go}, Mmix1} H_{v_{vlo,0} \mapsto v_{mixn,vlop}}(j\omega_2)\, H_{Mmix1.ids \mapsto v_{out,0}}(j\omega_2 - j\omega_1) \quad (7.28a)$$

$$\frac{A_{LO}A_{RF}}{2} K_{2_{gm\&go}, Mmix2} H_{v_{vlo,0} \mapsto v_{mixn,vlon}}(j\omega_2)\, H_{Mmix2.ids \mapsto v_{out,0}}(j\omega_2 - j\omega_1) \quad (7.28b)$$

$$\frac{A_{LO}A_{RF}}{2} K_{2_{gm\&go}, Mmix4} H_{v_{vlo,0} \mapsto v_{mixp,vlop}}(j\omega_2)\, H_{Mmix4.ids \mapsto v_{out,0}}(j\omega_2 - j\omega_1) \quad (7.28c)$$

$$\frac{A_{LO}A_{RF}}{2} K_{2_{gm\&go}, Mmix3} H_{v_{vlo,0} \mapsto v_{mixp,vlon}}(j\omega_2)\, H_{Mmix3.ids \mapsto v_{out,0}}(j\omega_2 - j\omega_1) \quad (7.28d)$$

$$A_{LO}A_{RF} K_{22_{gm}, Mmix2} H_{v_{vlo,0} \mapsto v_{vrfn,vlon}}(j\omega_2)\, H_{Mmix2.ids \mapsto v_{out,0}}(j\omega_2 - j\omega_1) \quad (7.28e)$$

$$A_{LO}A_{RF} K_{22_{gm}, Mmix1} H_{v_{vlo,0} \mapsto v_{vrfp,vlop}}(j\omega_2)\, H_{Mmix1.ids \mapsto v_{out,0}}(j\omega_2 - j\omega_1) \quad (7.28f)$$

$$A_{LO}A_{RF} K_{22_{gm}, Mmix3} H_{v_{vlo,0} \mapsto v_{vrfp,vlon}}(j\omega_2)\, H_{Mmix3.ids \mapsto v_{out,0}}(j\omega_2 - j\omega_1) \quad (7.28g)$$

$$A_{LO}A_{RF} K_{22_{gm}, Mmix4} H_{v_{vlo,0} \mapsto v_{vrfn,vlop}}(j\omega_2)\, H_{Mmix4.ids \mapsto v_{out,0}}(j\omega_2 - j\omega_1) \quad (7.28h)$$

The weights of contributions (7.28a) to (7.28d) are 25%, while the weights of contributions (7.28e) to (7.28h) are negligible. This implies that the mixer works as expected, i.e. the nonlinear coefficient $K_{2_{gm\&go}}$ gives rise to a harmonic component at frequency $f_{RF} - f_{LO}$ in the mixer current, which is then converted to a voltage using the RF–opamp. No additional intermodulation terms are generated in the LF–opamp, as there are no points where signals at frequencies f_{RF} resp. f_{LO} are present. Hence the large number of zero–contributions in the symbolic output of the nonlinear analysis engine.

7.4 CONCLUSIONS

Based on a approximation algorithm for the Kirchhoff equations in a weakly nonlinear circuit, an algorithm for symbolically calculating harmonic components has been presented. An implementation of this algorithm in the linear–analysis environment Symba has been demonstrated on some applications, illustrating the tight interaction with linear analysis and the multitude of simplification techniques that is needed to get interpretable yet accurate symbolic results. A number of prerequisites need to be satisfied in order to achieve this:

- the weakly nonlinear assumption must hold;
- the nonlinear coefficients must be approximated accurately in order to get a reliable numerical screening and good final result.

The latter condition can be difficult to meet for circuits with complex transistor models, especially in the case of third–order nonlinear coefficients. The nonlinear analysis engine offers the options of obtaining the coefficients numerically through interpolation, or calculating them analytically from a model in an extensible library. The latter approach is definitely preferred for practical applications, but requires some work which can become quite tedious for complex models.

The use of the nonlinear analysis engine in the design process is twofold. The numerical screening facility allows an early identification of the possible design problems, making it a verification tool, and the symbolic results can be used as design equations in the design process, making it a design tool.

References

[1] Alderson, G. E. and Lin, P. M. (1973). Computer generation of symbolic network functions: A new theory and implementation. *IEEE Transactions on Circuit Theory*, 20(1):48–56.

[2] Bussgang, J., Ehrman, L., and Graham, J. (1974). Analysis of nonlinear systems with multiple inputs. *Proceedings IEEE*, 62:1088–1118.

[3] Chua, L. O. and Ng, C.-Y. (1979). Frequency-domain analysis of nonlinear systems: Formulation of transfer functions. *IEE Journal of Electron. Circuits. Syst.*, 3(6):257–269.

[4] Crols, J. and Steyaert, M. (1994). A full CMOS 1.5 GHz highly linear broadband downconversion mixer. In *Proceedings European Solid–State Circuits Conference*, pages 248–251.

[5] Daems, W., Gielen, G. G. E., and Sansen, W. (1999). Circuit complexity reduction for symbolic analysis of analog integrated circuits. In *Proceedings Design Automation Conference*, pages 958–963.

[6] Dobrovolny, P. (1998). Design tool for symbolic analysis based on the matroid intersection theory. In *Proceedings IEEE International Symposium on Circuits and Systems*.

[7] Fernández Fernandéz, F. V., Guerra, O., Rodríguez-García, J. D., and Rodríguez-Vázquez, A. (1999). Corrections to symbolic analysis of large analog integrated circuits: The numerical reference generation problem. *IEEE Transactions on Circuits and Systems-II: Analog and Digital Signal Processing*, 46(4):485–486.

[8] Fernández Fernandéz, F. V., Rodríguez-Vázquez, A., Martin, J. D., and Huertas, J. L. (1993). Approximating nested format symbolic expressions. *Alta Frequenza – Rivista Di Elettronica*, 5(6):326–335.

[9] Floberg, H. and Mattisson, S. (1995). Symbolic distortion analysis of nonlinear elements in feedback amplifiers using describing functions. *Journal of Circuit Theory and Applications*, 23(4):345–356.

[10] Gielen, G. G. E. and Sansen, W. (1991). *Symbolic Analysis for Automated Design of Analog Integrated Circuits*. Kluwer Academic Publishers.

[11] Gielen, G. G. E., Walscharts, H., and Sansen, W. (1989). ISAAC: a symbolic simulator for analog integrated circuits. *IEEE Journal of Solid-State Circuits*, 24(6):1587–1597.

[12] Guerra, O., Rodríguez-García, J. D., Roca, E., Fernández Fernandéz, F. V., and Rodríguez-Vázquez, A. (1998). A simplification before and during generation methodology for symbolic large–circuit analysis. In *Proceedings International Conference on Electronic Circuits and Systems*, volume 3, pages 81–84.

[13] Hassoun, M. M. and McCarville, K. S. (1993). Symbolic analysis of large-scale networks using a hierarchical signal flowgraph approach. *Journal of Analog Integrated Circuits and Signal Processing*, 3:31–42.

[14] Hsu, J. and Sechen, C. (1994). Fully symbolic analysis of large analog integrated circuits. In *Proceedings Custom Integrated Circuits Conference*, pages 21.4.1–21.4.4.

[15] Kundert, K., White, J., and Sangiovanni-Vincentelli, A. L. (1990). *Steady–state methods for simulating analog and microwave circuits*. Kluwer Academic Publishers.

[16] Kuo, Y. (1973). Distortion analysis of bipolar transistor circuits. *IEEE Transactions on Circuit Theory*, CT–20:709–716.

[17] Liberatore, A. and Manetti, S. (1988). SAPEC: A personal computer program for the symbolic analysis of electric circuits. In *Proceedings IEEE International Symposium on Circuits and Systems*, pages 897–900.

[18] Lin, P. M. (1991). *Symbolic Network Analysis*. Elsevier.

[19] Nebel, G., Kleine, U., and Pfleiderer, H. (1995). Symbolic pole/zero calculation using SANTAFE. *IEEE Journal of Solid-State Circuits*, 30(7):752–761.

[20] Pi, T. and Shi, C.-J. R. (2000). Multi-terminal determinant decision diagrams: A new approach to semi-symbolic analysis of analog integrated circuits. In *Proceedings Design Automation Conference*, pages 19–22.

[21] Rowchowdhury, J. (1998). MPDE methods for efficient analysis of wireless systems. In *Proceedings Custom Integrated Circuits Conference*, pages 451–454.

[22] Sannuti, P. and Puri, N. N. (1980). Symbolic network analysis: an algebraic formulation. *IEEE Transactions on Circuits and Systems*, 27(8):679–687.

[23] Sansen, W. (1999). Distortion in elementary transistor circuits. *IEEE Transactions on Circuits and Systems-II: Analog and Digital Signal Processing*, 46(3):315–325.

[24] Seda, S. J., Degrauwe, G. R., and Fichtner, W. (1988a). Lazy-expansion symbolic expression approximation in SYNAP. In *Proceedings IEEE/ACM International Conference on Computer Aided Design*, pages 310–317.

[25] Seda, S. J., Degrauwe, G. R., and Fichtner, W. (1988b). A symbolic analysis tool for analog circuit design automation. In *Proceedings IEEE/ACM International Conference on Computer Aided Design*, pages 488–491.

[26] Shi, C.-J. R. and Tan, X.-D. (1997). Symbolic analysis of large analog circuits with determinant decision diagrams. In *Proceedings IEEE/ACM International Conference on Computer Aided Design*.

[27] Shi, C.-J. R. and Tan, X.-D. (2000). Canonical symbolic analysis of large analog circuits with determinant decision diagrams. *IEEE Transactions on Computer-Aided Design of Integrated Circuits and Systems*, 19(1):1–18.

[28] Singhal, K. and Vlach, J. (1981). Symbolic circuit analysis. *Proceedings IEE*, 128(Part G):81–86.

[29] Sommer, R., Hennig, E., Dröge, G., and Horneber, E.-H. (1993). Equation-based symbolic approximation by matrix reduction with quantitative error prediction. *Alta Frequenza – Rivista Di Elettronica*, 5(6):317–325.

[30] Verhaegen, W. and Gielen, G. G. E. (2001). Efficient DDD–based symbolic analysis of large linear analog circuits. In *Proceedings Design Automation Conference*.

[31] Wambacq, P., Fernández Fernandéz, F. V., Gielen, G. G. E., Sansen, W., and Rodríguez-Vázquez, A. (1995). Efficient symbolic computation of approximated small-signal characteristics. *IEEE Journal of Solid-State Circuits*, 30(3):327–330.

[32] Wambacq, P., Gielen, G. G. E., Kinget, P. R., and Sansen, W. (1999). High-frequency distortion analysis of analog integrated circuits. *IEEE Transactions on Circuits and Systems-II: Analog and Digital Signal Processing*, 46(3):335–345.

[33] Wambacq, P., Gielen, G. G. E., and Sansen, W. (1992). A cancellation–free algorithm for the symbolic analysis of large analog circuits. In *Proceedings IEEE International Symposium on Circuits and Systems*, pages 1157–1160.

[34] Wambacq, P. and Sansen, W. (1998). *Distortion Analysis of Analog Integrated Circuits*. Kluwer Academic Publishers.

[35] Yu, Q. and Sechen, C. (1994). Approximate symbolic analysis of large analog integrated circuits. In *Proceedings IEEE/ACM International Conference on Computer Aided Design*, pages 664–671.

[36] Yu, W. and Leung, B. H. (1999). Distortion analysis of MOS track–and–hold sampling mixers using time–varying volterra series. *IEEE Transactions on Circuits and Systems-II: Analog and Digital Signal Processing*, 46(2):101–113.

8 APPROACHES TO FORMAL VERIFICATION OF ANALOG CIRCUITS

Lars Hedrich, Walter Hartong

Institute of Microelectronic Circuits and Systems,
University of Hannover, Germany

E-mail: [hedrich, hartong]@ims.uni-hannover.de

8.1 FORMAL VERIFICATION: THE ALTERNATIVE APPROACH

In the digital world formal verification is an attractive alternative to simulation. Currently, this technique is evolving to a widely used verification method, indicated by a growing number of commercial vendors (see e.g. [1, 2, 3]). For analog circuits no comparable techniques are known yet. However, the same problems driving the development of such tools in the digital world can be found for analog circuits. These are shorter design cycles, more complex designs in terms of transistor counts and increased need for first time correct designs. Formal verification has some important advantages over traditional validation methods like circuit simulation. It gives a strong mathematical correctness proof of the entire circuit behavior, increases the quality of the design and prevents redesigns. Formal verification is a push-button solution without the need for developing test stimuli saving much time in the design cycle.

Traditionally, formal verification methods are divided into model checking and equivalence checking. The former proves that given statements (theorems) are fulfilled by the designed circuit. The statements are derived from the handwritten specification. In this chapter we restrict ourselves to equivalence checking methods, proving that two given system descriptions are equivalent. For analog systems a general definition can be defined as follows:

> "Formal verification for analog systems proves/disproves that the input/output behavior of a target system is equal to that of a specifying system. The proof is valid for all possible input stimuli and for the entire system behavior."

The general procedure for formal verification of analog systems is shown in *Figure 8.1*. It takes two system descriptions as input, e.g. behavioral models, transistor netlists, or extracted netlists. In general they are given on different levels of abstraction.

P. Wambacq et al. (eds.),
Low-Power Design Techniques and CAD Tools for Analog and RF Integrated Circuits, 155–191.

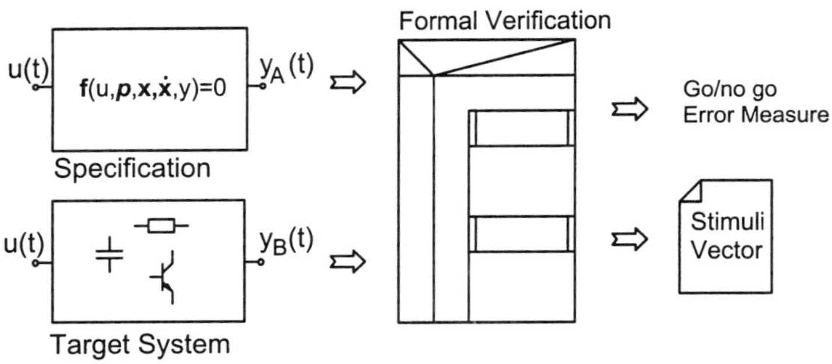

Figure 8.1 Formal verification procedure. The output is a go/no go flag and, if necessary, a stimuli vector for error state excitation.

The two system descriptions are referred to as specification and target system, respectively. This distinction is necessary for certain verification tasks where the meaning of the specification description is different from the target system description. In other verification tasks the two descriptions have equivalent meanings and therefore are not distinguished. The output of the verification algorithm is a go/no go flag or an error indicating the deviation between the two systems. Additionally, stimuli vectors can be generated for traditional DC, AC, or transient analysis. These vectors will drive the systems into states where the errors are visible using traditional simulations. This will help the designer to observe an error in a traditional simulation environment. Using this technique, it is possible to smoothly integrate a formal verification tool into a standard analog design flow.

8.1.1 Design Flow

In a traditional design flow for analog circuits, simulation is the main design validation method. Based on simulation, some approaches exist to make the validation process more efficient, e.g. automatic data sheet extraction [4] and worst-case analysis or yield estimation [5, 6]. Some of these methods are able to take parameter and design tolerances into account.

If formal verification is included within an analog design flow (see *Figure 8.2*), most of the validation tasks can be replaced or supported by the formal verification tool. For equivalence checking, two machine-readable system descriptions are needed. Therefore, the first task in such a design flow is to generate a behavioral model from the specification. This can be regarded as an executable specification, which is used as the reference model for simulation and formal verification tasks. With this model, simulations have to be carried out to check its correctness. The following design steps can then be validated by comparing the results to the behavioral model using formal verification.

The main advantages of this procedure in comparison with traditional simulation based design flows are:

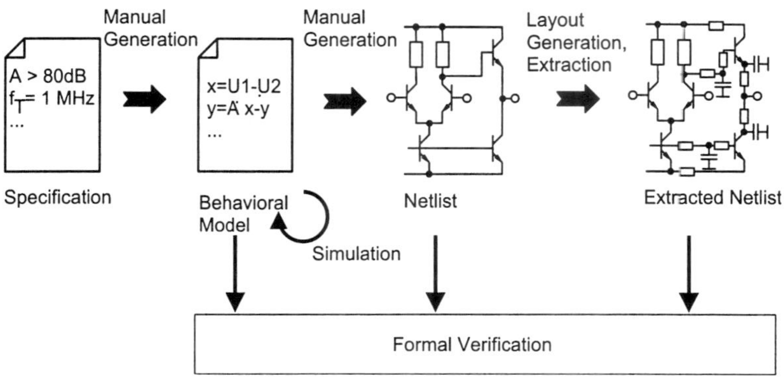

Figure 8.2 Design flow using formal verification

- No stimuli needed: Formal verification works independently from input stimuli. Therefore, the generation of stimuli is not required and it is not possible to miss essentially required stimuli.
- Proof: All design flaws will be identified. At first sight this results in a larger effort for designing the circuit and the behavioral model because the verification process will uncover all side effects and known but not modeled properties of the circuit and the specification. However, this effort is regained at least if a redesign is prevented.

In the next sections three approaches to formal verification for different circuit classes are presented.

8.1.2 Circuit Description

In contrast to digital logic, analog systems are continuous in signal values and time. Therefore, it is not reasonable to check exact identity of the two systems, because even a small deviation of a system parameter may lead to a negative verification result. Due to this, formal verification of analog circuits has to handle deviations in the behavior of both systems. This can be done in two ways. First, the parameters can be represented as ranges or more formally as intervals. Second, a deviation measure has to be defined to indicate the error magnitude. If this error is below a predefined threshold, the systems are equal from a practical point of view. In this approach much of the proof character of formal verification is missing. However, in the general case of nonlinear dynamic circuits, the use of intervals or ranges is prohibitive due to the problem complexity.

The problem of proving the equivalence of two circuits with parameter tolerances can be described by showing the equivalence of two non-autonomous systems of nonlinear first-order differential equations having parameters with tolerances. In this chapter we restrict ourselves to single-input single-output (SISO) circuits. One system consists of n nonlinear first order differential equations

$$S : \left\{ \begin{array}{l} f_1(\mathbf{x}(t), \dot{\mathbf{x}}(t), u(t), \boldsymbol{p}) = 0, \\ f_2(\mathbf{x}(t), \dot{\mathbf{x}}(t), u(t), \boldsymbol{p}) = 0, \\ \dots \\ f_n(\mathbf{x}(t), \dot{\mathbf{x}}(t), u(t), \boldsymbol{p}) = 0 \end{array} \right\} \tag{8.1}$$

with the input variable u(t), the vector of n system variables (e.g. currents or voltages)

$$\mathbf{x}(t) = [x(t)_1, x(t)_2, ..., x(t)_n]^T, \tag{8.2}$$

and the vector of m parameters $\boldsymbol{p}$ (e.g. resistance or gain) with their tolerances p_i

$$\boldsymbol{p} = [\, p_1, p_2, ..., p_m]^T. \tag{8.3}$$

Tolerance parameters are printed in italics and tolerance parameter vectors are printed in bold italics. A possible description for tolerance parameters is the use of finite intervals. This gives the possibility to use an exact proof in contrast to statistical modeling like Gaussian normal distribution. Additionally, inequalities from specifications can be modeled easily by choosing a large interval. A tolerance parameter interval is described by its lower and upper bound:

$$p_i = \left[\underline{p}_i .. \overline{p}_i\right] \tag{8.4}$$

In this case, $\boldsymbol{p}$ is an m-dimensional hypercube in the parameter space.

Tolerance parameters can have different meanings as they describe process tolerances or specifications. Especially for being pessimistic, a distinction between the specification and its tolerance parameters and the target circuit and its tolerance parameters has to be made. This will be explained in detail in the following sections. In general, an output variable y(t) has to be identified for each circuit.

Mathematically the formal verification problem means to prove that the solution of the target differential equation system lies within the solution defined by the specification and its tolerances for all input stimuli in a specified range and for all defined parameter variations. However, solving a non-autonomous nonlinear differential equation system with interval techniques is far too complex for realistic circuit sizes. Therefore, the problem class has to be restricted.

8.1.2.1 Circuit Classes. Since analog systems can be divided into different classes, different formulations of the verification problem are possible. Three main classes of analog systems can be distinguished based on the following criteria:

- linear/nonlinear,
- static/dynamic,
- parameter tolerance/nominal values.

The three main classes used in our approaches are: static nonlinear systems with tolerances, dynamic linear systems with tolerances and dynamic nonlinear systems without tolerances.

For each of these classes a special formal verification algorithm will be described in the following sections. A suitable algorithm for the overall class including all other classes - nonlinear dynamic systems with tolerances - has not been found yet. This is mainly caused by the algorithms runtime complexity.

Table 8.1 Different circuit classes and properties of the formal verification algorithms

Circuit classes	nonlinear, static, tolerances	linear, dynamic, tolerances	nonlinear, dynamic
Input stimuli	none	none	none
Proof	strongly mathematical	strongly mathematical	numerical error
Underlying mathematics	interval arithmetic	interval arithmetic	state space theory
Specification inequalities	handled by finite intervals	handled by finite intervals	-
Parameter Tolerances	handled by finite intervals	handled by finite intervals	-
Distinction between specification and target circuit	yes	yes	no

8.1.2.2 Other Approaches. Existing approaches for hybrid systems use known verification methodologies for digital systems verification and adapt them more or less to analog or partially analog systems [7, 8].

Some other approaches for analog systems have been proposed to compute performance characteristics from a target circuit and to compare it with given specifications, like *yield estimation* [9], *worst-case analysis* [10, 11, 12] or *design centering* [13]. Most of these approaches are based on probabilistic methods giving good and reliable results for the mentioned tasks. However, they are not able to prove that a circuit with tolerance parameters fulfills a certain specification for all parameter combinations, because the tolerance parameters are modeled by a probabilistic density function, or the computed regions are non-pessimistic approximations like convex polyhedrons. In general, they are based on simulation experiments and therefore dependent on input stimuli. The algorithms described below have more formal character. The idea behind is to confirm the absence of errors.

8.2 LINEAR DYNAMIC SYSTEMS

The class of linear dynamic circuits can easily be described in the frequency domain using transfer functions H(s). This enables the use of a well established powerful theory.

The main idea is to compare the value sets of the complex transfer functions $\mathrm{H_{spec}}$ and $\mathrm{H_{tar}}$. Related work in computing value sets can be found in [14, 15, 16].

8.2.1 Linear Circuit Description

Symbolic analysis methods [17] are able to calculate transfer functions using transistor netlists or behavioral model descriptions. The resulting transfer function is a parameterized description of the entire system behavior:

$$\mathrm{H}(\mathrm{s},\boldsymbol{p}), \quad \boldsymbol{p} = [p_1, p_2, ..., p_{\mathrm{m}}], \tag{8.5}$$

where $\boldsymbol{p}$ is the parameter interval vector of m tolerance parameters p_i. The tolerance parameters are assumed to be independent from each other and are represented by a finite real interval $p_i = [\underline{\mathrm{p}}_{\mathrm{i}} \,..\, \overline{\mathrm{p}}_{\mathrm{i}}]$.

8.2.2 Basic Algorithm

A safe approximation $\tilde{A}$ of the value sets $A = \{\mathrm{H}(\mathrm{s} = \mathrm{j}\omega_0, \mathbf{p}) \,|\, \mathbf{p} \in \boldsymbol{p}\}$ for each transfer function is computed. A safe approximation for the target transfer function at a fixed frequency $\tilde{A}_{tar}$ is an outer enclosure of the value set A_{tar}:

$$\tilde{A}_{\mathrm{tar}} \supseteq A_{\mathrm{tar}} = \{\mathrm{H_{tar}}(\mathrm{s} = \mathrm{j}\omega_0, \mathbf{p}_{\mathrm{tar}}) \,|\, \mathbf{p}_{\mathrm{tar}} \in \boldsymbol{p}_{\mathrm{tar}}\}. \tag{8.6}$$

A safe approximation at a fixed frequency $\mathrm{s} = \mathrm{j}\omega_0$ for the specifying transfer function $\tilde{A}_{\mathrm{spec}}$ is an inner enclosure of the value set A_{spec}:

$$\tilde{A}_{\mathrm{spec}} \subseteq A_{\mathrm{spec}} = \{\mathrm{H_{spec}}(\mathrm{s} = \mathrm{j}\omega_0, \mathbf{p}_{\mathrm{spec}}) \,|\, \mathbf{p}_{\mathrm{spec}} \in \boldsymbol{p}_{\mathrm{spec}}\}. \tag{8.7}$$

In *Figure 8.3* an example for the outer and inner enclosure $\tilde{A}_{\mathrm{tar}}$ and $\tilde{A}_{\mathrm{spec}}$ is shown.

If $\tilde{A}_{\mathrm{tar}} \subset \tilde{A}_{\mathrm{spec}}$ holds, $\mathrm{H_{tar}}$ fulfills $\mathrm{H_{spec}}$ at a specific frequency ω_0. The extension to a proof for a frequency interval is described in Section 8.2.5.

8.2.3 Outer Enclosure

An outer enclosure of a transfer function value set can be calculated by an interval extension using complex interval arithmetic[18, 19]. Due to the interval arithmetic properties, the outer enclosure can be overestimated. However, dividing the interval vector $\boldsymbol{p}$ into a union of subinterval vectors $\boldsymbol{p}_{\mathrm{k}}$ according to

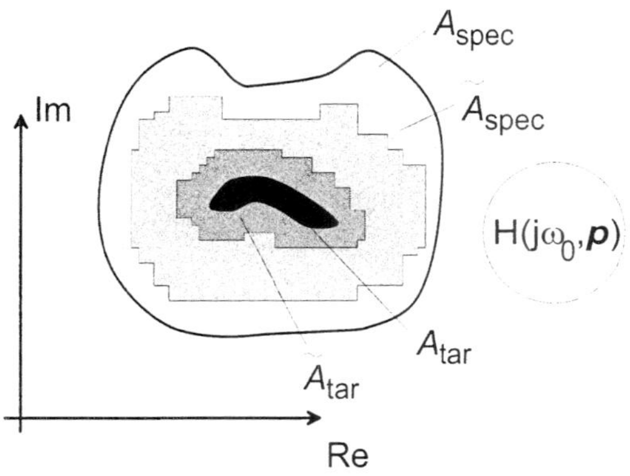

Figure 8.3 Value sets A_{tar}, A_{spec} and enclosures $\tilde{A}_{\mathrm{tar}}$, $\tilde{A}_{\mathrm{spec}}$ for $\mathrm{H}_{\mathrm{tar}}$, $\mathrm{H}_{\mathrm{spec}}$ at a fixed frequency

$$\boldsymbol{p} = \bigcup_{\mathrm{k}=1}^{\mathrm{m}} \boldsymbol{p}_{\mathrm{k}} \ , \tag{8.8}$$

the unified outer enclosure

$$\tilde{A}'(\boldsymbol{p}) = \bigcup_{\mathrm{k}=1}^{\mathrm{m}} \tilde{A}_{\mathrm{k}}(\boldsymbol{p}_{\mathrm{k}}) \ \ \text{for m} \to \infty \tag{8.9}$$

converges to the tight or exact enclosure of the value set $A(\boldsymbol{p})$ (see [20]). Based on this result, the tightest possible outer enclosure for the target transfer function $\mathrm{H}_{\mathrm{tar}}$ is computed by subdividing the intervals recursively. For each subdivision the algorithm chooses a parameter that leads to the smallest overestimation. This technique divides only those parameter intervals that contribute to the overestimation. If a parameter does not have any impact on the overestimation, it will not be divided leading to much less subdivided intervals. The runtime complexity is exponential in terms of the contributing parameters $\mathrm{O}(\gamma^{\mathrm{n}_{\mathrm{contr.param.}}})$, where γ is the number of interval subdivisions per parameter.

After subdivision, the resulting complex intervals are geometrically combined into a single circumscribing polygon. See *Figure 8.4* for an example using the transfer function:

$$H_1 = \frac{a}{(1+b^2)\cdot 0.1 \cdot \mathrm{j} + 0.1 \cdot a + 1} \ , \quad a = [4.5..6] \, , \ b = [-0.4 .. 0.6]. \tag{8.10}$$

The circumscribing polygon $\partial\tilde{A}$ of the enclosure $\tilde{A}$ is used by the proposed algorithms to determine the enclosure area. However, for the ease of understanding, the region itself is used instead of its boundary in the remainder of the paper.

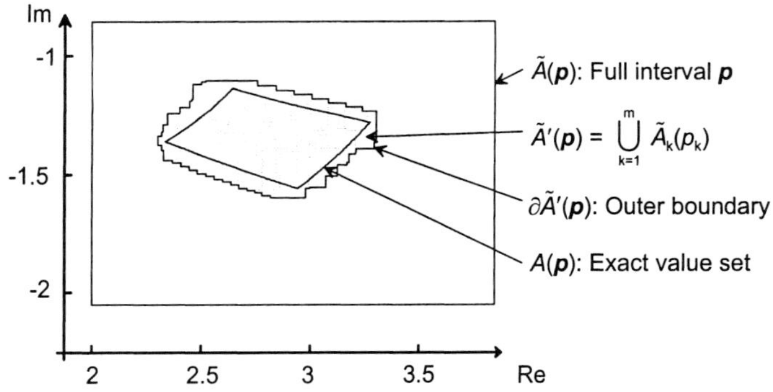

Figure 8.4 Computation of an outer enclosure of the function H_1

For transfer functions with many parameters the overestimation problem is a substantial disadvantage of the algorithm. However, if e.g. the transfer function is given in a linearly dependent form (with the interval parameters p_i as independent coefficients of the numerator and denominator polynoms, [14, 21]),

$$H_1 = \frac{p_{m+n+1}s^m + ... + p_{n+2}s + p_{n+1}}{p_n s^n + p_{n-1}s^{n-1} + ... + p_1 s + p_0}, \ \boldsymbol{p} = [p_0, p_1, ..., p_{m+n+1}] \tag{8.11}$$

a more efficient algorithm for calculating the outer enclosures can be selected. For example, the algorithm explained in Section 8.2.4.2 can be used.

8.2.4 Inner Enclosure

From the viewpoint of interval analysis, the calculation of an inner enclosure is a much larger problem, because the inclusion property of the interval arithmetic [19] only facilitates the safe computation of outer enclosures. However, if the class of functions is restricted to those having a simply connected region for the complex value set, an inscribing polygon can be constructed which is proven to belong to the value set.

8.2.4.1 Safe Path Between Two Points. Assuming that the value set A of a complex function H is a simply connected region, then every closed curve, which lies safely within the value set, encloses a part of the value set. Since the curve containing the whole value set is unknown, an approximation for this optimal enclosure curve is used.

First, a number of points within the value set is calculated. All neighboring points are connected by segments of a circle as shown in *Figure 8.5.* It is possible to determine the segment radius in a way, that each segment lies totally within the value set. Since a segment of a circle is an inconvenient data structure, the segment is replaced by a save polygon approximation.

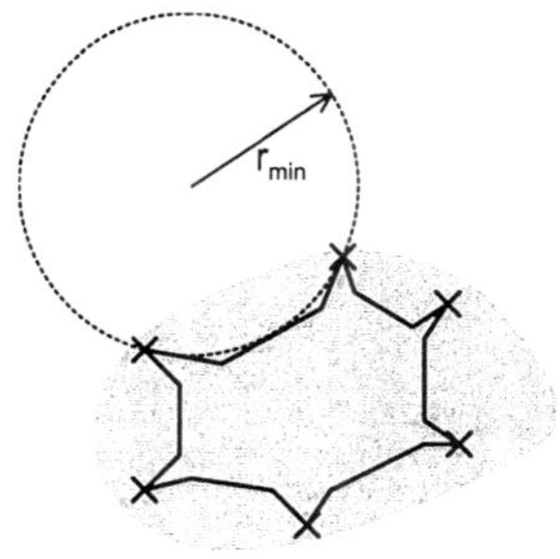

Figure 8.5 Determining a safe path between two points

In *Figure 8.6*a) the (unknown) curve ρ (plotted in bold) with its endpoints A and B is shown. The curve ρ is given by a complex function F(q) of one parameter $q = [\underline{\mathrm{q}} \,..\, \overline{\mathrm{q}}]$, while the endpoints can be calculated according to $A = \{\mathrm{Re}(\mathrm{F}(\underline{\mathrm{q}})), \mathrm{Im}(\mathrm{F}(\underline{\mathrm{q}}))\}$, $B = \{\mathrm{Re}(\mathrm{F}(\overline{\mathrm{q}})), \mathrm{Im}(\mathrm{F}(\overline{\mathrm{q}}))\}$. The radius of curvature of the parameterized curve ρ is defined by

$$\mathrm{r}(\mathrm{q}) = \frac{(\mathrm{R}' + \mathrm{I}')^{\frac{3}{2}}}{\mathrm{R}' \cdot \mathrm{I}'' - \mathrm{I}' \cdot \mathrm{R}''},$$

$$\mathrm{R}' = \frac{\partial \mathrm{Re}(\mathrm{F}(\mathrm{q}))}{\partial \mathrm{q}}, \ \mathrm{R}'' = \frac{\partial \mathrm{R}'(\mathrm{q})}{\partial \mathrm{q}}, \tag{8.12}$$

$$I' = \frac{\partial Im(F(q))}{\partial q}, \ I'' = \frac{\partial I'(q)}{\partial q}.$$

Using interval arithmetic, a lower bound $\underline{\mathrm{r}}$ for the radius of curvature can be computed.

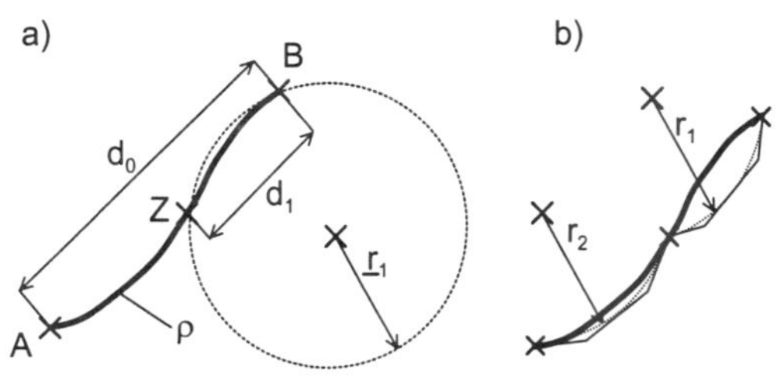

Figure 8.6 a) Maximum curvature and intermediate point Z on a curve b) safe path approximated by a polygon

If the radius is smaller than half the distance between the two endpoints A and B ($1/2 \cdot \mathrm{d}_0 > \underline{\mathrm{r}}_1$ in *Figure 8.6*a), the interval q is divided into two intervals by inserting an additional point $\mathrm{Z} = \left\{\mathrm{Re}(\mathrm{F}(\frac{\underline{\mathrm{q}}+\overline{\mathrm{q}}}{2})), \mathrm{Im}(\mathrm{F}(\frac{\underline{\mathrm{q}}+\overline{\mathrm{q}}}{2}))\right\}$ at half the interval of q. The generated intervals $\left[\underline{\mathrm{q}} \,..\, \frac{\underline{\mathrm{q}}+\overline{\mathrm{q}}}{2}\right]$, $\left[\frac{\underline{\mathrm{q}}+\overline{\mathrm{q}}}{2} \,..\, \overline{\mathrm{q}}\right]$ are then processed in the same way as the original interval. If the radius of curvature is larger than half the distance between the endpoints (i.e. $r_1 > {}^1\!/_2 \cdot d_1$), the parameterized curve ρ can be safely approximated by a polygon as shown in *Figure 8.6*b).

This algorithm has a linear runtime complexity in terms of interval subdivisions γ: $O(\gamma)$. Only one parameter (q) has to be subdivided.

8.2.4.2 Inner Enclosure of a Value Set Using Curvature Examination.

The computation of a path that lies safely within the value set, is divided into two steps. First, a sequence of parameterized curves and their endpoints, which belong to the value set, is computed. In the second step, the enclosures of the curves are determined by means of the algorithm of Section 8.2.4.1. Combining the inner boundaries of the curve enclosures leads to the desired inner enclosure of the value set. The corresponding algorithm reads as given in *Table 8.2*:

Table 8.2 Algorithm for computing an inner enclosure

```
Inner Enclosure (F(p),p)
  Divide parameter hypercube into a set of 2-dimensional faces
  Map the faces into the H(jω)-space
  Compute the enveloping polygon E of the faces in the H(jω)-space:
  E = [e1, e2, ..ek]
  for each segment (ei, ei+1) of the polygon E do
    Compute curvature driven path ((ei, ei+1), F(p),
    p = [pei ..pei+1])
  return inscribing polygon of Ã
```

In *Figure 8.7* a simple example is presented in order to explain the main steps of the algorithm.

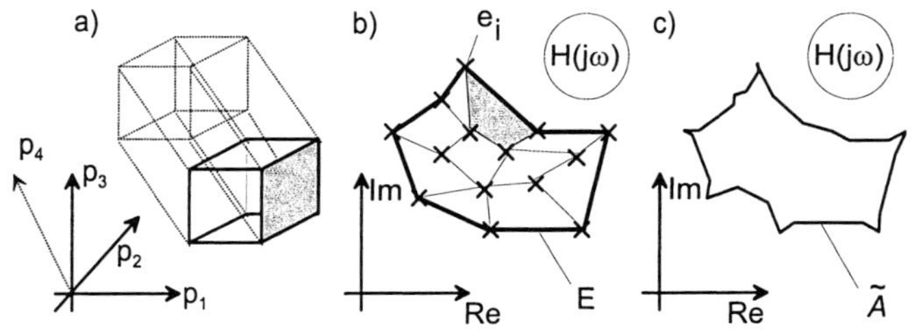

Figure 8.7 a) Faces of the 4-dimensional hypercube in the parameter space, b) faces and enveloping polygon mapped to the H(jω)-space, c) resulting inscribing polygon

Because the enveloping polygon E is determined in a sequence with counterclockwise steps, the safe path can be computed using only an inner polygon approximation of the circle segment in order to get a pessimistic enclosure. The result is a polygon, which describes the safe inner enclosure $\tilde{A}$ by its boundary.

One reason for computing the enveloping polygon from the image of the parameter cube faces and not from the corners using a simple convex hull approach [22] is that along the edges of the faces only one parameter is varying and all others are constant. This results in a much lower overestimation for the interval computation of the radius of curvature and, therefore, reduces computation time and complexity.

Comparing this algorithm with the outer enclosure calculation (Section 8.2.3), it has to be noticed that the runtime of the inner enclosure algorithm is much lower. This results form the nearly constant runtime of the safe path algorithm (Section 8.2.4.1).

8.2.5 Extended Algorithm Including Frequency Interval

With the previously defined algorithms for computing inner and outer enclosures, a proof that the target system fulfills a specifying system, is possible at a single frequency ω_0. In order to extend the proof to a given frequency interval, this frequency interval is introduced as an additional interval parameter.

An approach handling the whole frequency interval in one step is not feasible, because a 3-dimensional calculation and a proof that the target value set V_{tar} lies inside the specifying value set V_{spec} would necessarily be an extreme tough problem. Hence, the frequency interval is divided into several small intervals (see *Figure 8.8*) and for each interval a projection of the frequency interval onto the H(jω) plane at one frequency is carried out. This results in a 2-dimensional pessimistic inclusion check for each frequency interval.

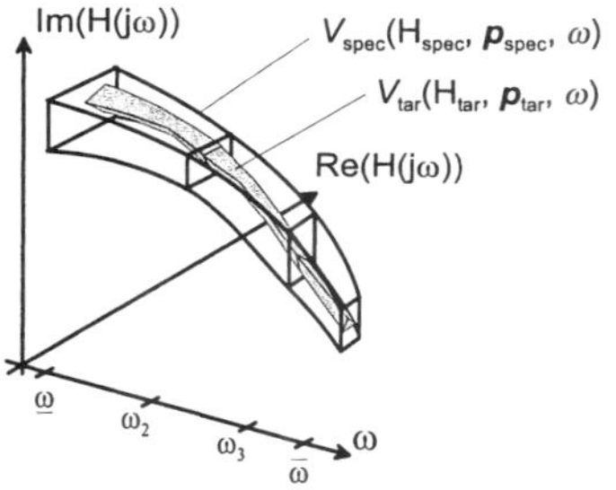

Figure 8.8 Principle of verification for a frequency range

In *Figure 8.9*a) the 3-dimensional value set for a frequency interval $\omega = [\underline{\omega} .. \overline{\omega}]$ is shown. *Figure 8.9*b) displays the corresponding 2-dimensional projection of this structure. The dark shaded volume V_{tar} is mapped to the dark shaded region A_{tar}. The face $A_{spec}(\underline{\omega})$ is hatched in *Figure 8.9*a) and *Figure 8.9*b). The other faces of the 3-dimensional V_{spec} value set parallel to the frequency axes are called the frequency faces $F_{spec,p_i}(\omega, p_i, \mathrm{p})$. They have only one parameter p_i and ω given as interval variables. In addition to the condition that $A_{tar}(\omega)$ lies inside $A_{spec}(\underline{\omega})$, all frequency faces must not touch $A_{tar}(\omega)$. If both conditions hold, it is proven that V_{tar} lies completely inside V_{spec} (see *Figure 8.9*c)

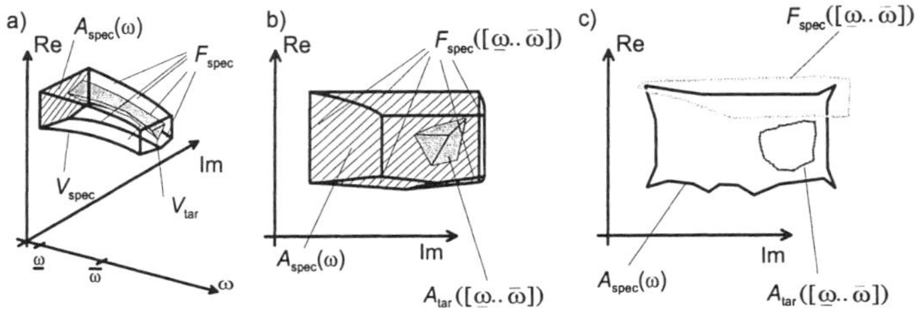

Figure 8.9 a) 3-dimensional value sets, b) 2-dimensional projection, c) inner and outer enclosures of faces

In *Table 8.3* the corresponding algorithm is presented. Note, that it tries to calculate the outer enclosure for the faces using curvature examination (algorithm of Section 8.2.4.2), because this method computes a tighter outer enclosure much faster than the direct interval method (algorithm of Section 8.2.3). In order to use this algorithm for determining outer enclosures, it has to be proved that the value set is bounded by its edges. This is checked for the faces $F_{\mathrm{spec},p_i}(\omega, p_i, \mathrm{p})$ having two interval parameters ω, $p_i$by investigating the perpendicular area vector being strictly positive or negative using interval arithmetic.

8.2.6 Example: gmC-Filter

The circuit to be verified is a CMOS gmC biquad filter [23, 24]. The cutoff frequency is about 70 kHz.

8.2.6.1 Circuit Description. In *Figure 8.10* a schematic of the filter is shown.

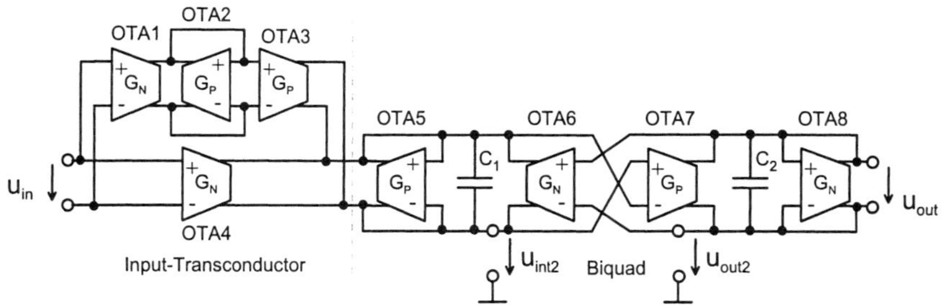

Figure 8.10 Basic configuration of the biquad filter

The ideal transfer function can be determined as

$$\mathrm{H(s)} = \frac{\mathrm{G_M^2 C_1 C_2}}{\mathrm{2s^2 + s(G_M C_1 + G_M C_2) + G_M^2 C_1 C_2}}, \quad \mathrm{G_M = G_N = G_P}. \tag{8.13}$$

Table 8.3 Formal verification algorithm for linear dynamic systems

---

Formal Verification of linear dynamic systems
(H_{spec}, H_{tar},$\omega = [\underline{\omega}..\overline{\omega}]$, $\boldsymbol{p}_{spec}$, $\boldsymbol{p}_{tar}$)
  Divide ω into k subintervals [ω_1, ω_2,..., ω_k]
  for each ω_i **do**
    Compute $\tilde{A}_{tar}$ = outer enclosure (H_{tar}, ω_i, $\boldsymbol{p}_{tar}$)
    Compute $\tilde{A}_{spec,\underline{\omega}_i}$ = inner enclosure (H_{spec},$\underline{\omega}_i$, $\boldsymbol{p}_{spec}$)
    Compute $\tilde{A}_{spec,\omega_i}$ = inner enclosure (H_{spec}, $\overline{\omega}_i$, $\boldsymbol{p}_{spec}$)
    for all edges p_j of $\tilde{A}_{spec,\underline{\omega}_i}$ **do**
      if $n \subset \mathbb{R}^+$ or $n \subset \mathbb{R}^-$ **then**
        Compute $\tilde{F}_{spec,p_j}$ = outer enclosure using curvatures (H_{spec}, ω_i, $p_{j,spec}$, $\mathbf{p}_{spec}$)
        (Section 8.2.4.2)
      else
        Compute $\tilde{F}_{spec,p_j}$ = outer enclosure (H_{spec}, ω_i, $p_{j,spec}$, $\mathbf{p}_{spec}$)
        (Section 8.2.3)
      if not ($\tilde{A}_{tar} \subseteq \tilde{A}_{spec,\underline{\omega}_i} \wedge \forall p_j : \tilde{A}_{tar} \cap \tilde{F}_{spec,p_j} = \emptyset$) **then**
        if frequency interval is too small **then**
          return false
        else
          Reduce frequency interval ω_i
  return true

The OTAs labeled with $G_P$in *Figure 8.10* consist of PMOS transistors. G_N represents an NMOS-transistor OTA (see [24]). The input transconductor implements a double G_M stage in order to compensate the 6 dB loss in the biquad section. The circuit consists of about 40 transistors in the filter stage and another 40 transistors for the biasing circuitry.

The specification of the filter reads as follows:

- Gain: $0\text{dB} \pm 0.2\text{dB}$
- Corner frequency: $70 \text{ kHz} \pm 10 \text{ kHz}$

Some relevant parameter tolerances for the technology are given as:

- Global threshold voltage tolerance of NMOS transistors: $\Delta V_{TP} = \pm 18\text{mV}$
- Global threshold voltage tolerance of PMOS transistors: $\Delta V_{TP} = \pm 16\text{mV}$
- Global mobility $\mu_{0P} = \pm 3.3\%$
- Global mobility $\mu_{0N} = \pm 1.8\%$
- Mismatch transconductance of NMOS and PMOS transistors calculated from Area, ΔV_T and $\Delta\mu_{0,N}$: $\Delta G_N \approx \Delta G_P = \pm 0.14\%$

8.2.6.2 Experimental Results. The executable specification of the filter is verified versus the transistor level netlist using the verification algorithm developed for linear circuits. In this case, the nonlinear behavior of the circuit is neglected by linearizing the circuit at the DC operating point. The verification result is true for the whole frequency range from 10 Hz to 1 MHz. This range has been divided into to 297 subintervals. Two of these frequency subintervals are shown in *Figure 8.11*.

8.3 NONLINEAR STATIC SYSTEMS UNDER PARAMETER TOLERANCES

Since the analog input-output behavior is a continuous function, the difference between two output characteristics always differs from zero in practice. In addition, all system parameter variations result in a variation of the output characteristics and there is always a variation of parameter values in reality. To solve these problems, parameter tolerances are taken into account during the formal verification process as we have seen it in Section 8.2.

8.3.1 Algorithm

The algorithm verifies the static input-output behavior of a given target circuit netlist versus a specification system. If we assume a static behavior, then the related equation systems (Equation (8.1)) can be rewritten as

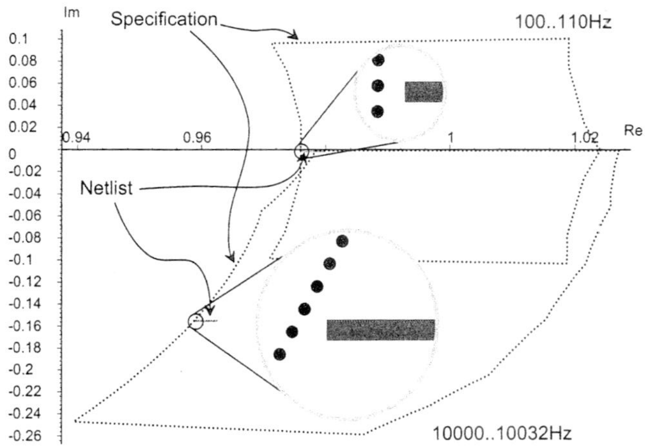

Figure 8.11 Verification result of executable specification versus netlist for a frequency subinterval at 100..110 Hz and 10000..10032 Hz. The magnifications are principle drawings

$$S : \left\{ \begin{array}{l} f_1(\mathbf{x}(t), u(t), \boldsymbol{p}) = 0, \\ f_2(\mathbf{x}(t), u(t), \boldsymbol{p}) = 0, \\ \ldots \\ f_n(\mathbf{x}(t), u(t), \boldsymbol{p}) = 0 \end{array} \right\}. \tag{8.14}$$

Due to the restriction to static systems, it is now an algebraic equation system. The solution of this nonlinear equation system (8.14) with interval parameters can be described as a set of output characteristics in an n-dimensional solution space including all output characteristics with arbitrary parameter sets within the intervals.

There are two different solution spaces: one for the target system and the other for the specification. If the target system's solution space is totally included within the specification solution space, then the specification is satisfied for all parameter variations within the given intervals. That means, the formal verification result is positive (see *Figure 8.12*).

However, in most cases it is not possible to find solution spaces directly, because it is not possible to solve a nonlinear equation system algebraically. Normally, iterative Newton algorithms are used to find a good estimation for the solution. To take intervals into account, we solve the nonlinear equation systems by using interval mathematics [18]. This enables us to get provable enclosures of the solution spaces.

To guarantee an exact verification result it is necessary to work with pessimistic outer and inner enclosures of the solution spaces like in Sections 8.2.3 and 8.2.4.

8.3.2 Outer Enclosure

In general, it is not possible to find the exact solution space of a nonlinear equation system containing interval parameters. Therefore, an arithmetic interval procedure is

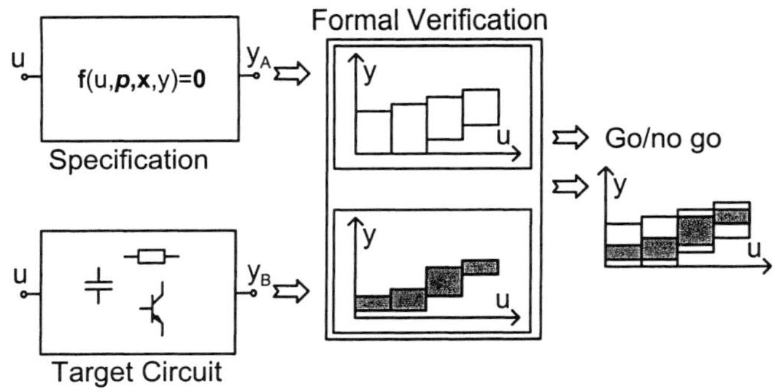

Figure 8.12 Principle of formal verification for nonlinear static systems

used to find an enclosure of the solution. By using a real interval extension, the exact solution is enclosed by a solution interval (*Figure 8.13*).

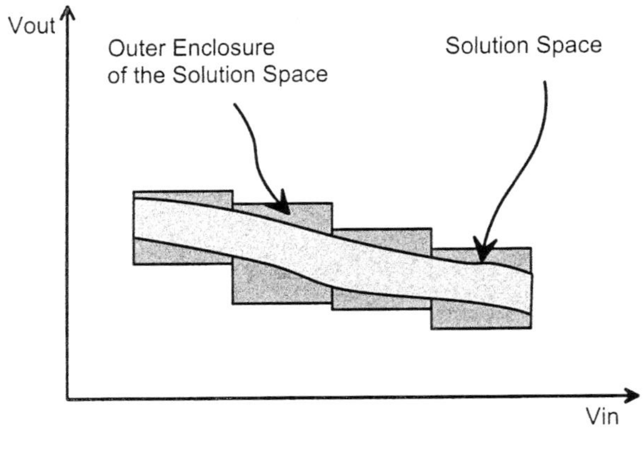

Figure 8.13 Outer enclosures

It is possible to solve the nonlinear interval equation system by the interval Newton algorithm [25]. A given start interval defines the input region for the following iteration. During the Newton iteration this space gets smaller and smaller until only an outer enclosure of all solutions (contained in the start intervals) remains. Solutions that are not enclosed in the start interval are ignored. The enclosure quality depends on the equation system and the number of interval parameters and the interval diameters.

Within the Newton algorithm the inverse Jacobian matrix is used. In case of interval computation, this is an inverse interval matrix. It is not possible to invert an interval matrix directly. For small systems a symbolic matrix inversion can be calculated in an acceptable time. Thereby, the inverse interval matrix can easily be calculated by interval extension of the inverse symbolic matrix. It is also possible to determine the inverse interval matrix by iterative procedures. However, in either way it takes a high computation time to find the inverse interval matrices for large systems [26].

Therefore, we use the Krawczyk algorithm [25, 27] instead of the interval Newton algorithm. Although based on the interval Newton algorithm it does not need an inverse interval Jacobian matrix. The Krawczyk algorithm uses a real inverse Jacobian matrix, which can be calculated numerically by a standard solver.

The standard Krawczyk algorithm solves equation systems with interval variables and real parameters. Our equation system contains interval variables and interval parameters. Thus, we use a straightforward expansion of the Krawczyk algorithm [28].

The expanded Krawczyk algorithm is defined as follows:

$$\begin{aligned} \boldsymbol{x}_{\nu+1} = \Big\{ & \mathbf{x}_\nu - \left(\mathbf{F}'\left(\mathbf{x}_\nu\right)\right)^{-1} \cdot \boldsymbol{f}(\mathbf{x}_\nu, \boldsymbol{p}) + \\ & \left(\mathbf{I} - \left(\mathbf{F}'\left(\mathbf{x}_\nu\right)\right)^{-1} \cdot \boldsymbol{F}'\left(\mathbf{x}_\nu, \boldsymbol{p}\right)\right) \cdot \left(\boldsymbol{x}_\nu - \mathbf{x}_\nu\right) \Big\} \cap \boldsymbol{x}_\nu . \end{aligned} \tag{8.15}$$

In this equation italic variables mark the interval values. Vector $\mathbf{x}_\mathrm{v}$ has arbitrary values within the start interval $\boldsymbol{x}_v$. However, normally it is chosen to be the midpoint of $\boldsymbol{x}_v \cdot \boldsymbol{f}(\mathbf{x}_\nu, \boldsymbol{p})$ describes interval expansion of the nonlinear equation system using the real vector $\mathbf{x}_v$ and the interval vector $\boldsymbol{p}$. $\mathbf{I}$ is the unity matrix, $\boldsymbol{F}'(\mathbf{x}_\nu, \boldsymbol{p})$ is the interval expansion of the Jacobian matrix of $\boldsymbol{f}(\mathbf{x}_\nu, \boldsymbol{p})$ and $(\mathbf{F}'(\mathbf{x}_\nu))^{-1}$ is the numerical inverse of the Jacobian matrix.

8.3.2.1 Start Intervals.

Determination of the start intervals is an important topic. The intervals have to be large enough to contain all system solutions but on the other hand, the intervals should be small to have a good convergence behavior and to minimize the interval intersections. Therefore, the start intervals are generated in two steps. First, the start values are estimated and secondly the start values are verified and – if necessary – expanded (*Table 8.4*). In the first estimation a few Monte Carlo simulations are used to get an idea of the solution space. The borders of this estimation are defined by the extreme values of the simulation results in each dimension. To get an outer enclosure, the start values must be larger than the real solution. Therefore, the estimated solution space is expanded in each direction and dimension by a user defined factor.

Then, the outer enclosure is calculated. Under the assumption that the solution space is a connected area and at least partial contained within the start value (which is proven by the Monte Carlo simulation points), the start value correctness can be determined as follows: if the outer enclosure does not touch the start value borders in each dimension, then the start values (and also the calculated outer enclosure) contain the whole solution space. If this condition is not true, the start values may not contain all solutions. In this case the start values must be expanded and the calculation starts again (*Table 8.4*).

Another reason for the violation of the above condition might be that the Overestimation of the interval mathematics is too large. In that case, the outer enclosure touches the start value border despite the fact that all solutions are contained in the start values. The overestimation can be reduced by interval intersection. However, it is not possible to decide whether the violation of the enclosure criteria is caused by the overestimation being too large or by the start values being too small.

Table 8.4 Outer enclosure algorithm

```
Outer Enclosure
  Estimate start intervals
  do while enclosure not smaller than start intervals
    do while solution gets smaller
      Execute Krawcyk algorithm
    if maximum interval intersections reached then
      Expand start intervals
    else
      Execute interval intersection
  return enclosure
```

8.3.2.2 Solution Space Deviation. Normally, it is not possible to calculate the overestimated solution in one step, because of the overestimating interval mathematics (the interval $[-\infty\ ..\ 10^{28}]$ might be a correct result but is not useful for further calculations) and the convergence behavior of the Krawczyk algorithm.

Both effects can be reduced by an interval intersection. That means, that one of the involved intervals is divided into two parts at the midpoint and the calculation is started for both parts separately. The result is given by the combination of both partial results. It is always smaller or equal to the direct calculation, without any reduction of the exact calculation. By applying the interval intersection recursively, a requested accuracy can be reached (see equations (8.8), (8.9)).

The disadvantage of the interval intersection method is the runtime complexity of the whole algorithm. For every intersection the runtime is doubled. Since every interval variable and every interval parameter must be intersected, algorithm's runtime complexity is exponential in the number of variables and parameters in the worst case.

$$\mathrm{R}_{\mathrm{total}} = O\left(\gamma^{(\mathrm{n}_{\mathrm{Variable}}+\mathrm{n}_{\mathrm{Parameter}})}\right) \tag{8.16}$$

To reduce the regular runtime, a heuristic is used to determine the sequence of interval intersections. Since the first interval intersection is valid for the whole following computation tree, this step is an extremely sensitive factor regarding runtime. The best intersection sequence is estimated by an adaptive algorithm, that takes the success of previous intersections into account. This enables the algorithm to react flexibly to different equation systems and the actual operation point.

Interval intersection can be applied theoretically until all intervals are smaller than the computation accuracy. Further intersections make no sense. However, to avoid infinite or extreme runtimes, it is necessary to interrupt the intersection process earlier.

There is no way to determine the required number of interval intersections because the real solution space is unknown. Therefore, the overestimation cannot be determined either. We use a heuristic to control the intersection process. It interrupts the intersection at the maximum depth of the intersection tree. The maximum depth is adapted during the formal verification process to react to different equation systems and different input voltage regions.

Despite these heuristic improvements, the worst case runtime complexity is still exponential. Therefore, this method is restricted to small system sizes or simple behavioral models for large systems. The number of variables and interval parameters is the critical factor that influences the runtime dramatically.

8.3.3 Inner Enclosure

To get a correct verification result, it is necessary to calculate an inner enclosure for the specifications equation system. With standard interval methods, only outer enclosures can be calculated. Thus, it is not possible to compute the inner enclosure directly. The underestimated solution is re-conducted by two outer enclosures, which can be calculated by the presented Krawczyk algorithm (hatched areas in *Figure 8.14*).

At first, two real parameter sets are chosen from the interval parameter set. By solving the real equation system at one input voltage it is controlled, that the solution point lies under and over (regarding the output voltage) the outer enclosure boundaries respectively. If such parameter sets are found, two outer enclosures are calculated using the Krawczyk algorithm. The actual input voltage range is the only interval value in this calculation. Therefore, overestimation is low.

Under the assumption that the solution space is a connected area without holes, the region between the two outer enclosures – if existing – is an inner enclosure for the solution of the equation system (white areas in *Figure 8.14*).

As mentioned in the beginning of this chapter, the verification condition is fulfilled if the outer enclosure of the target system lies completely inside the inner enclosure of the specification. If this is not true for one input voltage interval, the formal verification has failed in that area. This fault could be a real design error in the target system or a result of the interval mathematics overestimation. This must be checked by the system designer. If the verification result is positive, no further simulation is necessary. The design correctness is proven mathematically.

8.3.4 Example: Differential Pair

To get an impression of the formal verification process we present two verification results for a bipolar differential pair amplifier. The first verification run fails. After some changes in the nominal parameter set, the target circuit can be successfully verified in the second verification process. The specification is the same in both runs.

Figure 8.15 shows the schematic of the differential pair circuit used. The bipolar transistors are modeled by the Gummel Poon model. The input voltage range is Vin = -0,25 V to 0,25 V. This interval is divided into 32 subintervals, which are examined separately. All circuit parameters aside from three are given as nominal values (shown in *Figure 8.15*).

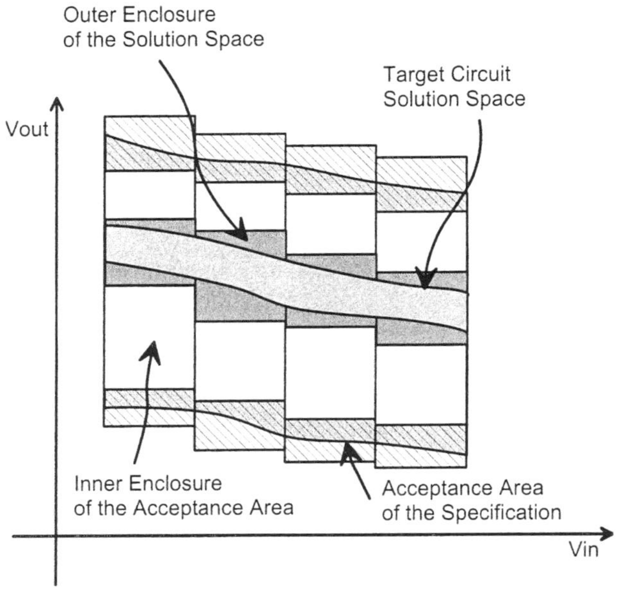

Figure 8.14 Inner and outer enclosures

Three parameters have interval values:

- Vcc = 11.9 V - 12.1 V,
- α_f (q1) = 0.987 - 0.9916,
- Rc1 = 9900 Ω - 10100 Ω.

The specification is modeled by two hyperbolic tangent functions. The functions are moved in horizontal and vertical direction to get an acceptance region for the amplifier characteristics (dotted lines in *Figure 8.17*).

Figure 8.16 shows the verification result of the first run. The outer enclosure is colored gray. It is surrounded by the inner enclosures (white boxes).

The verification result is true in 23 cases of the 32 input intervals at the beginning. The remainder of the input intervals is divided into smaller intervals and examined again. After input interval intersection, the positive input voltages can be verified successfully. The input voltage intersection is easily found at Vin = 0.08 V.

After input interval intersection, the verification result is still negative for some intervals of the negative input voltage part. Even after 80 interval intersections (*Figure 8.16*), some unverified parts remain (indicated by missing white boxes).

At this point the computation was interrupted. The circuit does probably not fulfill the specification. The circuit behavior is outside or very close to the specification border. Therefore, we change some nominal values to improve the output characteristics of the differential pair.

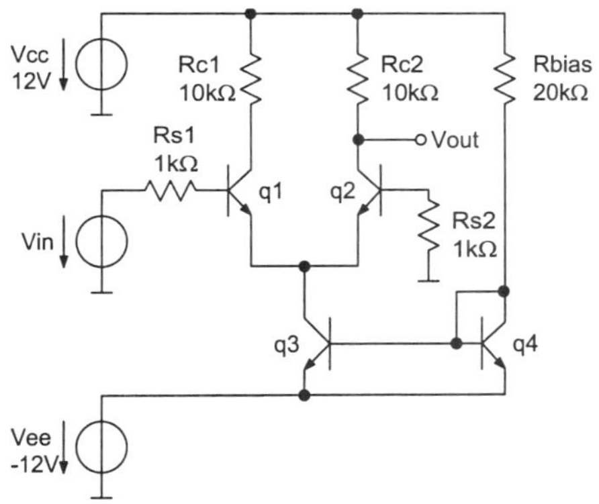

Figure 8.15 Schematic of differential pair

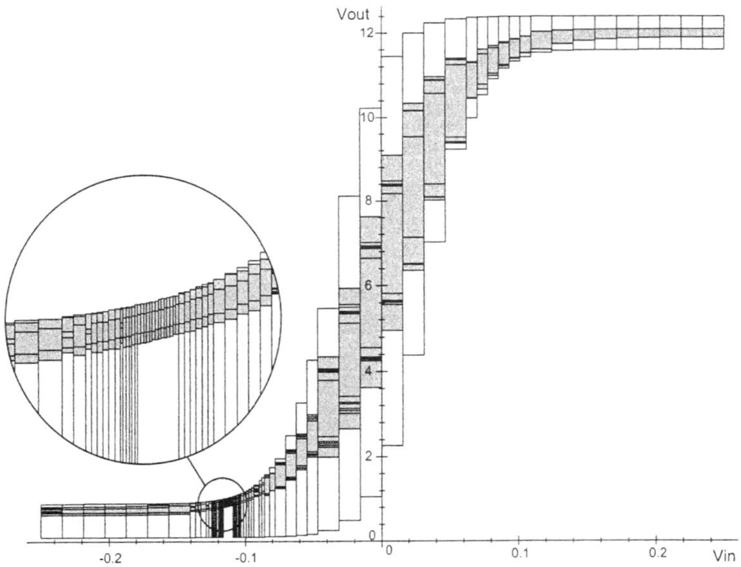

Figure 8.16 Unsuccessful formal verification result

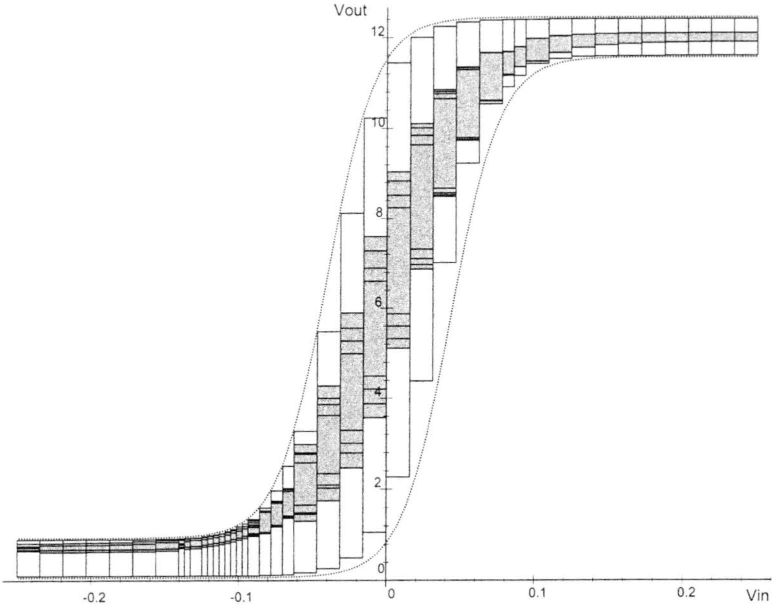

Figure 8.17 Successful formal verification result

A second run with three parameter values changed has been carried out. The specification is the same as used in first run.

- Rc1 = 15900 Ω - 16100 Ω,
- Rc2 = 16 kΩ,
- Rbias = 32 kΩ.

The first verification result is nearly the same as in the first run. Most of the input intervals can be verified. The others are divided and examined again. *Figure 8.17* shows the verification result with 44 intersections of the input voltage interval. At this step all parts can be formally verified. That means, it is formally proved, that the circuit fulfills the specification with respect to the given interval parameters.

8.4 NONLINEAR DYNAMIC SYSTEMS WITH NOMINAL PARAMETERS

The following approach is restricted to nominal parameters. Therefore, it is not possible to perform an inclusion proof as in the foregoing approaches. Furthermore, target systems and specifications need not to be distinguished.

8.4.1 State Space Description

For the case of nonlinear dynamic circuits with nominal parameters, the system of equations (8.1) can be rewritten as

$$\mathrm{S}: \left\{ \begin{array}{l} \mathrm{f}_1(\mathbf{x}(\mathrm{t}), \dot{\mathbf{x}}(\mathrm{t}), \mathrm{u}(\mathrm{t})) = 0, \\ \mathrm{f}_2(\mathbf{x}(\mathrm{t}), \dot{\mathbf{x}}(\mathrm{t}), \mathrm{u}(\mathrm{t})) = 0, \\ \ldots \\ \mathrm{f}_\mathrm{n}(\mathbf{x}(\mathrm{t}), \dot{\mathbf{x}}(\mathrm{t}), \mathrm{u}(\mathrm{t})) = 0 \end{array} \right\} \tag{8.17}$$

where u(t) describes the input signal. One of the circuit variables **x**(t) is the output variable. In control theory, the output variable's equation g is separated from the system of equations **f**. To ensure this, an additional equation g with the additional output variable y(t) can always be added to the equation system resulting in the system

$$\mathrm{S}_{\mathrm{ctl}}: \left\{ \begin{array}{rcl} \mathrm{f}_1(\mathbf{x}(\mathrm{t}), \dot{\mathbf{x}}(\mathrm{t}), \mathrm{u}(\mathrm{t})) & = & 0, \\ \mathrm{f}_2(\mathbf{x}(\mathrm{t}), \dot{\mathbf{x}}(\mathrm{t}), \mathrm{u}(\mathrm{t})) & = & 0, \\ & \ldots & \\ \mathrm{f}_\mathrm{n}(\mathbf{x}(\mathrm{t}), \dot{\mathbf{x}}(\mathrm{t}), \mathrm{u}(\mathrm{t})) & = & 0, \\ \mathrm{g}(\mathrm{y}(\mathrm{t}), \mathbf{x}(\mathrm{t}), \mathrm{u}(\mathrm{t})) & = & 0 \end{array} \right\} \tag{8.18}$$

We use this separation for better understanding only. A subset of the circuit variables is composed of the generalized state space variables $\mathbf{x}^z(\mathrm{t})$ occurring in a derived form within the state equations. The remaining variables are called non-state variables $\mathbf{x}^n(\mathrm{t})$. The z generalized state variables and the input u(t) build up the extended state space.

A SISO system described by state space equations **f** and the output equation g can be represented graphically by a vector field of the derivatives of the generalized state variables $\dot{\mathbf{x}}^z(\mathrm{t})$ in the extended state space and a scalar field of the output variable y(t) (see *Figure 8.18*). This graphical representation describes the entire nonlinear dynamic behavior of the system.

8.4.2 Algorithm

The basic idea of the proposed verification approach for nonlinear dynamic systems is to compare two geometrical descriptions of two systems A and B [29]. The objective is to determine whether the vector fields $\dot{\mathbf{x}}^z_A(\mathbf{x}^z_A, \mathrm{u})$, $\dot{\mathbf{x}}^z_B(\mathbf{x}^z_B, \mathrm{u})$ and the scalar fields $\mathrm{y}_A(\mathbf{x}^z_A, u)$, $\mathrm{y}_B(\mathbf{x}^z_B, u)$ are equal or not. The values of the non-state variables $\mathbf{x}^n$ are determined by the values of the state variables $\mathbf{x}^z$. Therefore, they are not considered explicitly.

8.4.2.1 Nonlinear Mapping of State Space Descriptions. In general, two systems do not have the same internal state variables because they are represented by different implementations on possibly different levels of abstraction. In this case, the simple method described above is not able to identify systems with similar input-output behavior.

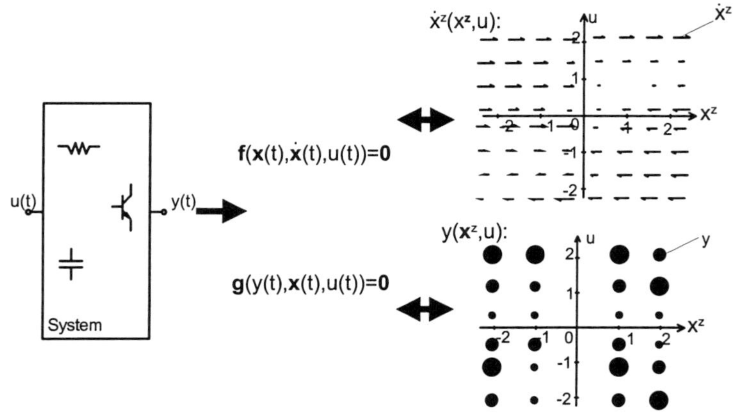

Figure 8.18 Equivalent descriptions of a system as implicit state space equations and as vector and scalar fields in the extended state space

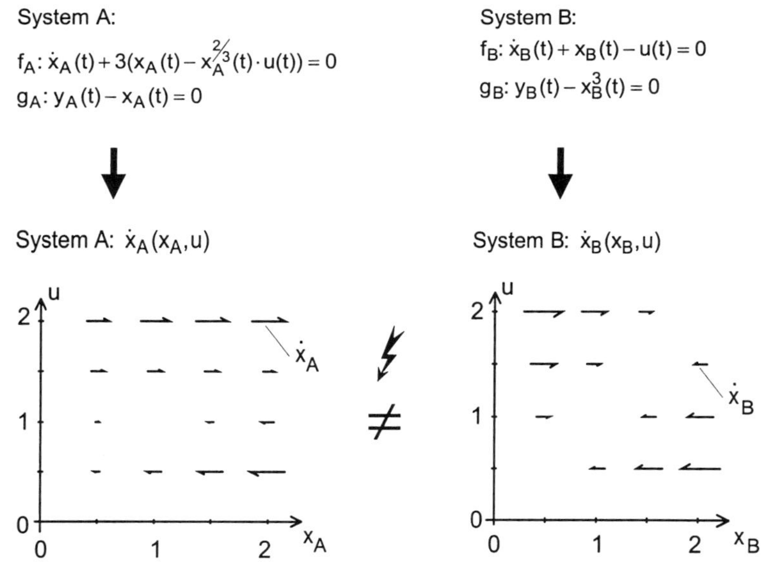

Figure 8.19 Systems with different state encoding but similar input-output behavior

Consider the two systems in *Figure 8.19*. The systems A and B are equal with respect to their input-output behavior: $\mathbf{f}_B$ can be derived from $\mathbf{f}_A$ using $x_A = x_B^3$. However, the differential equations and the vector and scalar fields are different (see *Figure 8.19*). Therefore, mappings $x_A^{z_r} = t(x_A)$ and $x_B^{z_r} = t(x_B)$ have to be found which uniquely map the vectors x_A, x_B onto the state vectors $x_A^{z_r}$, $x_B^{z_r}$. After this mapping, the vector field $\dot{x}_A^{z_r}$ in the virtual extended state space of the transformed system A can be compared to the vector field $\dot{x}_B^{z_r}$. For this example using $x_A^{z_r} = x_A^{\frac{1}{3}}$ and $x_B^{z_r} = x_B$ leads to two identical vector fields $\dot{x}_A^{z_r}$ and $\dot{x}_B^{z_r}$ (see *Figure 8.20*). The same result can be obtained for the scalar fields y_A and y_B.

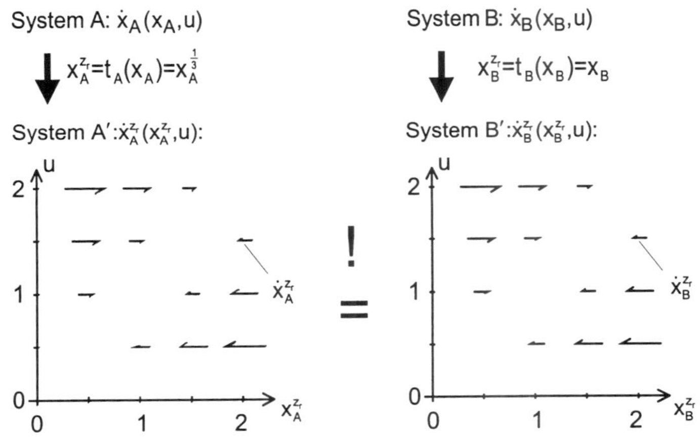

Figure 8.20 Systems' equal vector fields after transformation

Using this result, we can define our verification approach more precisely: If two systems have equal vector fields $\dot{\mathbf{x}}_A^{z_r}$, $\dot{\mathbf{x}}_B^{z_r}$ and scalar fields y_A , y_B

$$\begin{aligned} \dot{\mathbf{x}}_A^{z_r}(\mathbf{x}_A^{z_r}, u) &= \dot{\mathbf{x}}_B^{z_r}(\mathbf{x}_B^{z_r}, u) \;\wedge \\ y_A(\mathbf{x}_A^{z_r}, u) &= y_B(\mathbf{x}_B^{z_r}, u) \;, \end{aligned} \tag{8.19}$$

resulting from mappings t_A and t_B, then the original systems A and B have equal in-/output behavior.

An analytical computation of the nonlinear mapping functions t_A , t_B and the mapped vector and scalar fields is in general not possible. Therefore, a numerical computation has to be used.

8.4.2.2 Sampling the State Space. The extended state space of the two systems is sampled in order to find the mapping functions iteratively (see similar sampling method in system identification [30]). The boundaries of the extended state space are determined by the maximum excitation of the variables and the input. Assuming a finite step size, this leads to a finite set of points at which a comparison has to be carried out. The sampling is stopped if a value of the variables is larger than a predefined bound. Typically, a small multiple of the supply voltage is an appropriate bound for all node voltages of a circuit. The step size of the sampling algorithm is determined by a step size control algorithm, which is explained in Section 8.4.2.5. The basic verification algorithm reads as given in *Table 8.5*:

Table 8.5 Verification algorithm for nonlinear dynamic systems

---

Verification of Nonlinear Dynamic Systems
  for every input value in a predefined range **do**
  Do DC analysis in order to get an initial state vector $\mathbf{x}_{\mathrm{DC_A}}, \mathbf{x}_{\mathrm{DC_B}}$
    for every sample point in the state space in predefined ranges **do**
      Compute solution $\mathbf{x}_{0_A}, \mathbf{x}_{0_B}$ of nonlinear equations system
      Compute linear mapping matrices $\mathbf{T}_{\mathrm{A}}, \mathbf{T}_{\mathrm{B}}$ of
        nonlinear equations system
      Calculate $f_{\dot{x}}$ (error of the state derivatives vector fields) and
      Calculate f_y (error of the output scalar field) and
      Adjust $\mathbf{x}_{\mathrm{A}}$ (procedure explained in Section 8.4.2.4)
      Compute new step sizes $\Delta\mathbf{x}^{z_r}$ and state vectors for
        next sample point:
      $\mathbf{x}_{\mathrm{new_A}} = \mathbf{x}_{0_A} + \mathbf{T}_{\mathrm{A}^{-1}} \cdot \Delta\mathbf{x}^{z_r}, \mathbf{x}_{\mathrm{new_B}} = \mathbf{x}_{0_B} + \mathbf{T}_{\mathrm{B}^{-1}} \cdot \Delta\mathbf{x}^{z_r}$
  if $f_{\dot{x}}, f_y$ for all sample points ¡ error margin **then**
    return true
  else
    return false.

Two loops are necessary to sample the extended state space. In the outer loop, a DC analysis is required because the operating point is used as an initial state vector. The inner loop starts with the computation of the linear mapping matrices $\mathbf{T}_{\mathrm{A}}$ and $\mathbf{T}_{\mathrm{B}}$ for the actual sample point using a linearized state space description, because only for linear systems a canonical representation and a corresponding mapping can be found (see Section 8.4.2.3). After linearization, absolute and relative errors are calculated for the vector fields and the scalar fields, respectively. Afterwards, the states of circuit A are adjusted to meet exactly the states of circuit B. This has to be done to prevent summation of small deviations to larger errors. Finally, a new step size and the resulting new large signal state variables $\mathbf{x}_{\mathrm{new_A}}, \mathbf{x}_{\mathrm{new_B}}$ are computed.

By iterative evaluation of the inner loop, the nonlinear discrete mappings

$$\mathbf{x}_{\mathrm{A}}^{z_r} = \mathbf{t}_{\mathrm{A}}(\mathbf{x}_{\mathrm{A}}), \quad \mathbf{x}_{\mathrm{B}}^{z_r} = \mathbf{t}_{\mathrm{B}}(\mathbf{x}_{\mathrm{B}}) \tag{8.20}$$

are constructed point by point while stepping through the state space. This is shown in *Figure 8.21* for three executions of the inner loop. At symbol ①, the verification run starts with the DC operating point for the input value of $u = 0.5$. The starting points in the original and the virtual extended state space correspond to each other per definition. After computing the mapping matrices $\mathbf{T}_A$ and $\mathbf{T}_B$, the step size $\Delta\mathbf{x}^{z_r}$ is determined. Now the next state variable values in the original extended state space can be computed by evaluating $\mathbf{x}_{new_A} = \mathbf{x}_{0_A} + \mathbf{T}_A^{-1} \cdot \Delta\mathbf{x}^{z_r}$ and $\mathbf{x}_{new_B} = \mathbf{x}_{0_B} + \mathbf{T}_B^{-1} \cdot \Delta\mathbf{x}^{z_r}$. In the virtual extended state space the next points are determined using $\Delta\mathbf{x}^{z_r}$. The resulting points are marked with ②. The second evaluation of the inner loop results in the points labeled with ③.

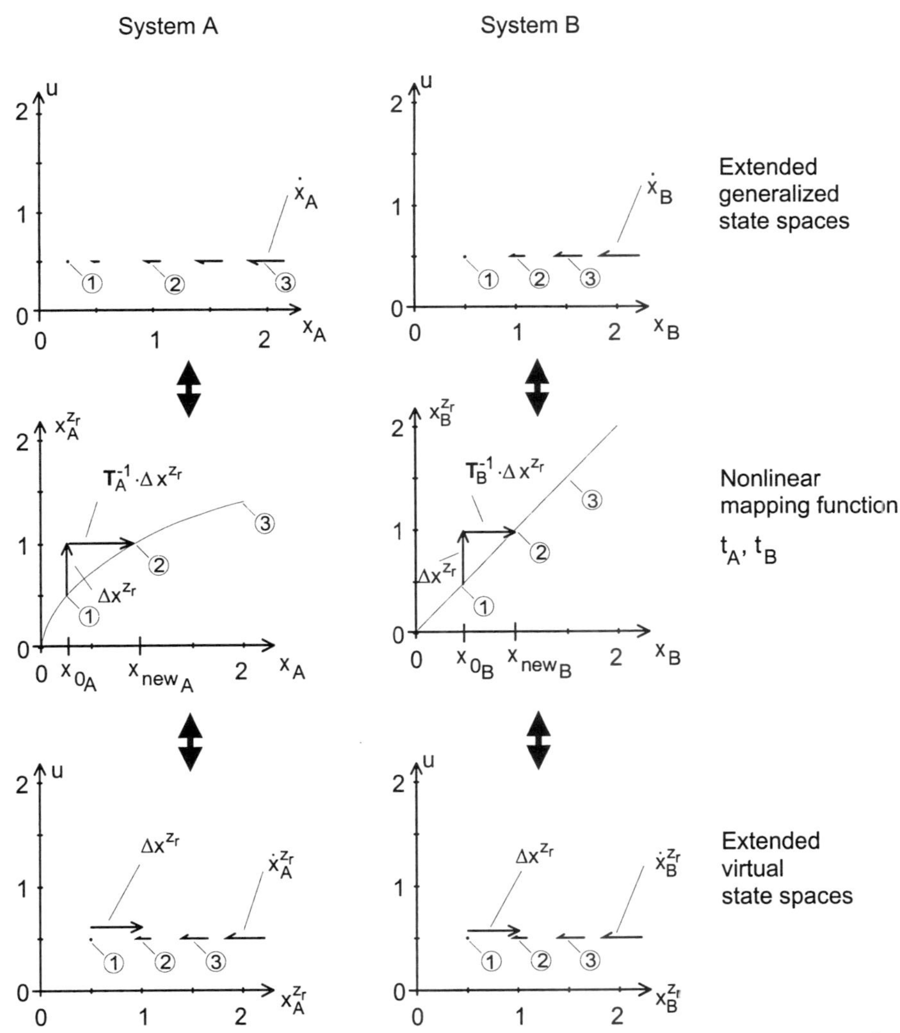

Figure 8.21 Construction of the nonlinear mapping for the example of *Figure 8.19*

8.4.2.3 Obtaining the Linear Mapping Matrices. In general, it is impossible to calculate the nonlinear mappings $\mathbf{x}_{A}^{z_r} = \mathbf{t}_A(\mathbf{x}_A)$ and $\mathbf{x}_{B}^{z_r} = \mathbf{t}_B(\mathbf{x}_B)$ directly from the nonlinear system functions $\mathbf{f}_A$ and $\mathbf{f}_B$. However, a direct mapping at a particular sample point can be calculated based on linearized systems. The three main steps in this task are

- linearizing the system in the operating point,
- computing a canonical form and
- calculating a transformation matrix $\mathbf{T}$

for both systems.

In order to compute a canonical form, the linearized systems have to be transformed into an explicit state space description [31]:

$$\begin{aligned} \dot{\mathbf{x}}^{z_e}(t) &= \mathbf{A} \cdot \mathbf{x}^{z_e}(t) + \mathbf{b} \cdot u(t) \\ y(t) &= \mathbf{c}^T \cdot \mathbf{x}^{z_e}(t) + d \cdot u(t) \\ \mathbf{x}^{z_e} &= T_{sp} \cdot \mathbf{x}, \text{ with } \mathbf{x} \in \mathbb{R}^n,\ \mathbf{x}^{z_e} \in \mathbb{R}^{z_e}, \mathbf{T}_{sp} \in \mathbb{R}^{z_e \times n} \end{aligned} \tag{8.21}$$

The n system variables $\mathbf{x}$ are transformed by a transformation matrix $\mathbf{T}_{sp}$ into the z_e explicit state variables $\mathbf{x}^{z_e}$. In general, the number z_e of explicit state variables is smaller than the number z of time derivatives (see Section 8.4.1) due to linear dependent state variables (e.g. a floating capacitor introduces two state variables in the original equation system $\mathbf{f}$ resulting in one explicit state variable). Afterwards, the system matrices $\mathbf{A}$ of both systems are transformed into a sorted diagonal form. A second transformation matrix $\mathbf{T}_{dia}$ is used to sort the eigenvalues according to their magnitude:

$$\begin{aligned} &\mathbf{A}' = \mathbf{T}_{dia} \cdot \mathbf{A} \cdot \mathbf{T}_{dia}^{-1},\ \mathbf{x}^{z_e\prime} = \mathbf{T}_{dia} \cdot \mathbf{x}^{z_e}, \\ &\text{with } \mathbf{x}^{z_e\prime} \in \mathbb{R}^{z_e}, \mathbf{A}' \in \mathbb{R}^{z_e \times z_e}, \mathbf{T}_{dia} \in \mathbb{R}^{z_e \times z_e} \end{aligned} \tag{8.22}$$

The resulting systems can have different orders $z_{e,A} \neq z_{e,B}$. Then, only the k smallest negative eigenvalues are used in the following calculation. The number of used eigenvalues z_r can be either user defined or the smallest number of eigenvalues in both systems:

$$\begin{aligned} &\mathbf{A}'' = \mathbf{T}_{red} \cdot \mathbf{A}' \cdot \mathbf{T}_{red}^{-1},\ \mathbf{x}^{z_r\prime\prime} = \mathbf{T}_{\mathbf{dia}} \cdot \mathbf{x}^{z_e\prime}, \\ &\text{with } \mathbf{x}^{z_r\prime\prime} \in \mathbb{R}^{z_r}, \mathbf{A}'' \in \mathbb{R}^{z_r \times z_r}, \mathbf{T}_{red} = [\mathbf{I}|\mathbf{0}] \in \mathbb{R}^{z_r \times z_e} \end{aligned} \tag{8.23}$$

At this step, the dimension of the state space is explicitly fixed to z_r for both systems. This technique is comparable to the dominant poles model order reduction. An alternative solution is to further reduce both systems using other model order reduction techniques [32, 33].

Different scales of the state vectors are eliminated by using the vectors $\mathbf{b}'$ (in case of a good controllability) or the vectors $\mathbf{c}'^T$ (in case of a good observability) for an additional scaling condition. The complete algorithm is shown below.

Table 8.6 Algorithm for linear transformation generation at local operating point

---

Determine Linear Mapping Matrices $\mathbf{T}_A$ and $\mathbf{T}_B$
  Linearize systems A and B at the sample point.
  Calculate the state space descriptions and the
    corresponding transformation
    matrices $\mathbf{T}_{sp_A}$ and $\mathbf{T}_{sp_B}$.
  Calculate the sorted diagonal form.

$$\begin{aligned}
\dot{\mathbf{x}}_A^{z_E'}(t) &= \mathbf{A}_A' \cdot \dot{\mathbf{x}}_A^{z_E'} + \mathbf{b}_A' \cdot u(t) \\
y_A(t) &= c_A'^T \cdot \dot{\mathbf{x}}_A^{z_E'} + d_A' \cdot u(t) \\
\dot{\mathbf{x}}_B^{z_E'}(t) &= \mathbf{A}_B' \cdot \dot{\mathbf{x}}_B^{z_E'} + \mathbf{b}_B' \cdot u(t) \\
y_B(t) &= c_B'^T \cdot \dot{\mathbf{x}}_B^{z_E'} + d_B \cdot u(t)
\end{aligned}$$

    and the corresponding transformation matrices $\mathbf{T}_{dia_A}$ and $\mathbf{T}_{dia_B}$.
  if the numbers of eigenvalues of $\mathbf{A}_A'$ and $\mathbf{A}_B'$ are not equal **then**
    reduce the system A | B to the order of
    z_r: $\mathbf{T}_{dia,A|B} = \mathbf{T}_{red} \cdot \mathbf{T}_{dia,A|B}$
  if $||\mathbf{b}_A'|| >> \left\|\mathbf{c}_A'^T\right\| \vee ||\mathbf{b}_B'|| >> \left\|\mathbf{c}_B'^T\right\|$ **then**
    scale transformation matrices according to
    $\mathbf{c}_A'^T \cdot \mathbf{T}_{dia_A} \stackrel{!}{=} \mathbf{c}_B'^T \cdot \mathbf{T}_{diaB}$.
  else
    scale transformation matrices according to
    $\mathbf{T}_{dia_A} \cdot \mathbf{b}_A' \stackrel{!}{=} \mathbf{T}_{dia_B} \cdot \mathbf{b}_B'$.
  return $\mathbf{T}_A = \mathbf{T}_{dia_A} \cdot \mathbf{T}_{sp_A}$ and $\mathbf{T}_B = \mathbf{T}_{dia_B} \cdot \mathbf{T}_{sp_B}$.

8.4.2.4 Adjustment of the Operating Point. During the iterative construction of the nonlinear mapping $\mathbf{x}_A^{z_r} = \mathbf{t}_A(\mathbf{x}_A)$ and $\mathbf{x}_B^{z_r} = \mathbf{t}_B(\mathbf{x}_B)$ small errors in the calculated sample points $\mathbf{x}_A$ and $\mathbf{x}_B$ or in the linear transformation matrices T_A, $\mathbf{T}_B$ can add up to a large error (see the dots in *Figure 8.22*). It can be seen, that new sample points deviate from the equilibrium by forced steps.

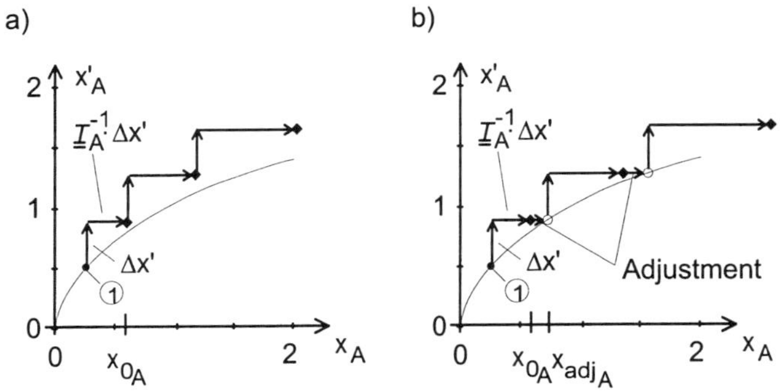

Figure 8.22 Computation of the nonlinear mapping function $\mathbf{x}_A^{z_r} = \mathbf{t}_A(\mathbf{x}_A)$ for the example of *Figure 8.19*: a) without, b) with adjusting the operating point

An iterative adjustment algorithm avoids this problem using the condition $f_{\dot{x}} = \|\dot{\mathbf{x}}_A^{z_r} - \dot{\mathbf{x}}_B^{z_r}\| \overset{!}{=} 0$ to adjust $\mathbf{x}_A$ by a modified Quasi-Newton optimization method [34]. The condition is derived from the assumption that the functions f_A and f_B are equal. Hence, the error $f_{\dot{x}}$ has to be zero. The adjusted operating point $\mathbf{x}_{A,adj}$ meets the exact mapping function (see *Figure 8.22*b).

8.4.2.5 Step Size Control. In order to avoid time consuming equidistant sampling of the state space but maintain high accuracy at regions with nonlinearities, an individual step size to the next sample point is calculated. This is an essential improvement compared to the first version of the algorithm [27].

The steps from the actual calculated sample point to the next are performed each by changing only one state variable x_j at a time. A vector of the circuit variables' second derivatives with respect to the virtual state space variable x_j is computed numerically

$$\mathbf{x}'' = \left[\frac{\partial^2 x_1}{(\partial x_j)^2}, \ldots, \frac{\partial^2 x_i}{(\partial x_j)^2}, \ldots, \frac{\partial^2 x_n}{(\partial x_j)^2}\right]^T . \tag{8.24}$$

This is done for each of the two systems A and B separately. The individual variable step size δx_k for each variable x_k is computed using the maximum of an absolute and a relative error

$$\forall k = 1..n: \quad \delta x_k = \max\left(\sqrt{\frac{2 \cdot \mathrm{abserr}}{|x_k''|}}, \sqrt{\frac{2 \cdot \mathrm{relerr} \cdot |x_k|}{|x_k''|}}\right). \tag{8.25}$$

The smallest step size of these individual variable step sizes is used as a pessimistic step size for the next step of the state space variable x_j

$$\delta x_j = \min_{k=1}^{n} (\delta x_k). \tag{8.26}$$

At the beginning of the sampling, the step size is limited to a predefined step size.

For linear systems or weakly nonlinear regions in an extended state space of a system, the step size becomes large leading to less sample points. This is supported by handling each virtual state space variable separately, which enables different step sizes for each of them.

8.4.3 Transient Error Stimuli Generation

From the presented algorithm result two errors can be measured, one for the vector field of state derivatives and one for the scalar field of the output variables. All errors are determined for each sample point in the state space. Additionally, the operating point is known at every sample point. With this information, an investigation of the design flaw can be carried out. It is easy to find the region in the state space, where the largest error is located. The large signal values of the circuit variables at that location will be an aid to find the flaw. For circuit designers, debugging based on state variables is unusual and not well known. Hence, a simulation, leading to an appropriate error state of the circuits, will be preferred. An approach of computing a transient stimulus, which leads to an error state, will be presented in this section.

The vector of the state derivatives is - geometrically interpreted - a pointer to the next state the circuit will reach after an infinitesimal time-step. Connecting all state points for a certain time slice constructs a trajectory. This is implicitly done during a transient analysis (see *Figure 8.23*).

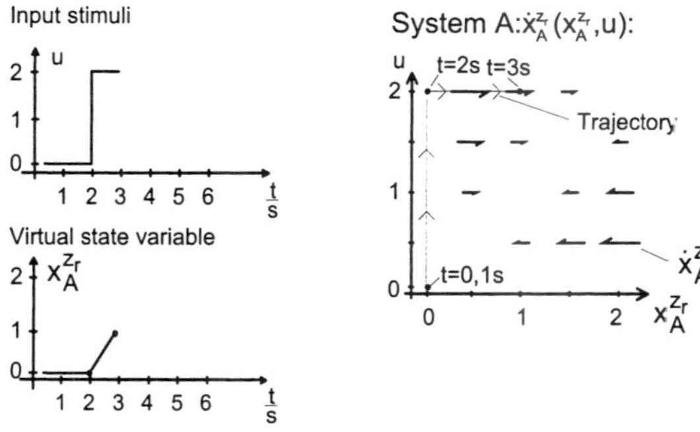

Figure 8.23 Input stimuli and trajectory for the intended point $[u = 2, x_A^{z_r} = 1]$

Assume a transient analysis starting in the DC operating point $[u = 0, x_A^{z_r} = 0]$ in the state space. The analysis will traverse through the state space along a trajectory and will end for the given example at the point $[u = 2, x_A^{z_r} = 1]$. The generation of an error stimulus inverts this process by trying to find a trajectory in reverse order starting from the intended point (the error state) and ending at a DC-operating point. The changes in u during the inversely running time t are used to compute the input stimulus function.

An algorithm for a simple full search is given in *Table 8.7*.

Table 8.7 Algorithm for input stimuli generation

```
Determine Transient Error Stimuli
  Start at the given point p in the extended state space.
  do while DC-equilibrium point not reached:
    Find all points {p_eq.state} with same state as actual point p.
    Find all points {p_time} with same u, which are predecessors
      in time of the actual point p.
    Sort {p_sort} = {p_eq.state} ∪ {p_time} according
      to minimal distance to a DC-equilibrium point.
    if distance to a DC-equilibrium point is lower than before then
      p = first of {p_sort}.
    else
      Backtrack.
    Store p in the trajectory.
    Store the sorted list for possible backtracks.
  Calculate the input stimuli using the trajectory.
```

8.4.4 Example: Log-Domain-Filter

This log-domain filter [35, 36] consists of a logarithmic compressor, a logarithmic integrator and an exponential expander. Thanks to the high compression ratio of internal signal voltages, log-domain filters are useful in low-power and especially low-voltage applications. Their operation principle is based on the exponential characteristics of bipolar transistors or weakly inverted MOSFETs. Even though, this principle can advantageously be implemented in Bipolar, BiCMOS, or CMOS technologies. Bipolar and BiCMOS filters are clearly superior in terms of high-frequency performance.

In log-domain circuits, the input current is logarithmically compressed by the nonlinear characteristic of a bipolar transistor into a voltage signal. After signal processing, the output current is regenerated by an exponential expansion of the internal voltage, done as well by a bipolar transistor. Between these two stages, the signal has to be processed in such a way that the total system performs in the specified way, i.e. as a linear filter in this example. Because of the logarithmic compression the signal voltage swing is very low. Therefore, the minimum supply voltage is almost independent of the processed signal. It is mainly determined by the sum of one junction voltage plus two saturation voltages of the transistors.

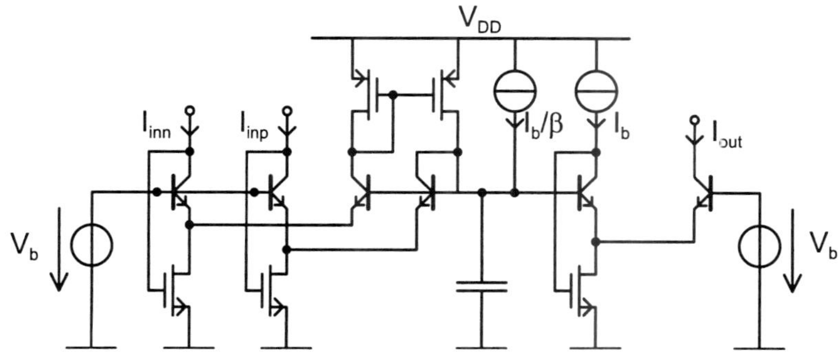

Figure 8.24 BiCMOS log-domain integrator

In *Figure 8.24*, the log-domain integrator schematic is shown. Such integrators can easily be connected to form regular filter structures of arbitrary order. Consequently, the design quality of the whole filter benefits from the improvement of the design verification of one integrator. Therefore, the work described hereafter focuses on the simplest example of a log-domain filter, that is, a 1st-order low-pass filter, which can be implemented easily by adding a negative feedback loop (I_{out} - I_{inn}) to the integrator of *Figure 8.24*.

8.4.4.1 Verification Results. Since a top-down design flow has not been used for this circuit, the formal verification tool can not be used in the manner of Section 1.1. However, the concept of formal verification can also be utilized for a bottom-up design flow. In this case, the tool is used to verify a nonlinear behavioral model, which has been implemented manually.

The logarithmic compressor model has two terminals: a single input current and an output voltage. The output voltage is given by the logarithm of the input current. Two parameters are used to fit the output characteristic to the behavior of the transistor netlist. All parasitic and second order effects, like input/output resistance, are left aside to get the ideal behavior. Along with an exponential expander model and a log-domain integrator model implemented in the same manner, the entire filter behavior is modeled.

The verification run compares the log-domain filter netlist versus this nonlinear behavior model. Since the model is a direct implementation of the log-domain filter block diagram, we expect small errors in the operating region.

The errors presented in *Figure 8.25* and *Figure 8.26* are plotted versus the input current I_{in} and the circuits main state variable, named uA_ld_int_basis. uA_ld_int_basis is an automatically generated name for the voltage across the capacitor in *Figure 8.24*. The DC error is smaller than 1% in a wide operating range. Moreover, the modified derivative error (*Figure 8.26*) remains lower than 0.3% during the whole verification

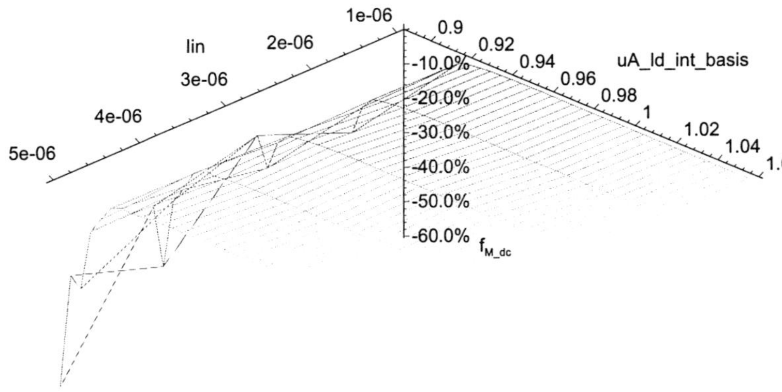

Figure 8.25 Netlist versus nonlinear behavioral model: modified DC error

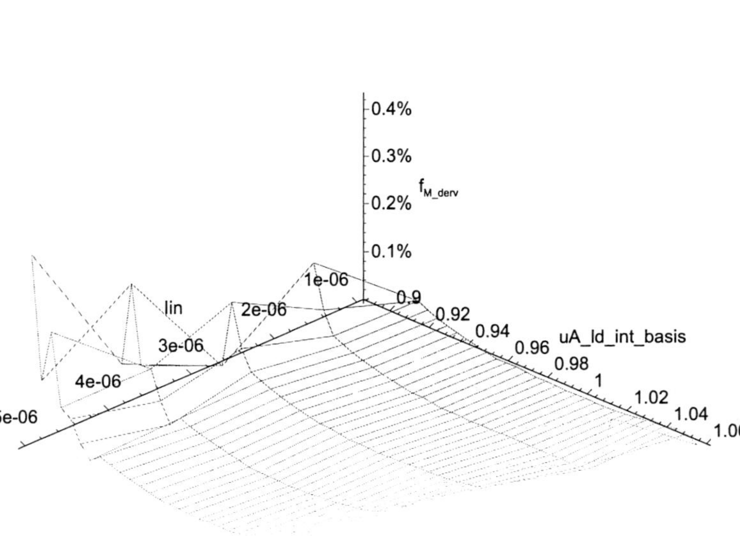

Figure 8.26 Netlist versus nonlinear behavioral model: modified derivative error

run. Assuming these small errors and the wide range of values for uA_ld_int_basis used in this verification run, the nonlinear behavioral model seems to be very good.

At the border of the operating region, the errors (especially the modified DC error) grow very fast. This effect is caused by an inaccurate modeling of the logarithmic compression and the exponential expansion. Due to the behavior of an exponential function, the rapid increase of the errors is natural.

8.5 CONCLUSIONS

In this chapter, algorithms for formal verification of analog systems circuits are presented. The algorithms compare two system descriptions on different levels of abstraction. They prove/disprove, that the systems have functionally similar input-output behavior. Additionally, in case of nonlinear dynamic circuits, an algorithm for generating transient stimuli is outlined to help the designer finding the design flaws with well known transient simulations. Some examples show the feasibility of the approaches.

All methods try to prove in a more or less graphical way the similarity of two system descriptions. Due to the large effort to do a mathematical proof, it is currently not possible to handle nonlinear dynamic circuits with tolerances. An other important step in the development of the formal verification methodology could be to restrict the proof to special properties of a system. For example: to prove that the gain of a system is always greater than a given value. This could be understood as model checking for analog circuits.

References

[1] Mentor Graphics: http://www.mentor.com/formalpro/

[2] Verplex Systems: http://www.verplex.com/product.html

[3] Avant! Corporation: http://www.avanticorp.com/product/1,1172,50,00.html

[4] P. Nussbaum, M. Hinners, L. Menevaut, "SimBoy: An Analog Simulator Interface for Automated Datasheet Generation", *Proceedings of the European Design Automation Conference* , pp. 37-41, Nov 1994.

[5] St. W. Director, P. Feldmann, K. Krishna, "Optimization of Parametric Yield: A Tutorial", *Proceedings of the Custom Integrated Circuits Conference* , pp. 3.1.1-3.1.8, Jan 1992.

[6] K. J. Antreich, H. E. Graeb, C. U. Wieser, "Circuit Analysis and Optimization Driven by Worst-Case Distances", *IEEE Transactions on Computer-Aided Design* , Vol. 13, No. 1, pp. 57-71, Jan 1994.

[7] Th. A. Henzinger, P.-H. Ho, "Algorithmic Analysis of Nonlinear Hybrid Systems", *International Conference on Computer Aided Verification* , pp. 225-238, June 1995.

[8] A. Puri, P. Varaiya, "Decidability of Hybrid Systems with Rectangular Differential Inclusions", *International Conference on Computer Aided Verification* , pp. 95-104, June 1994.

[9] St.W. Director, P. Feldmann, K. Krishna, "Optimization of Parametric Yield: A Tutorial", *Proc. Custom Integrated Circuits Conference*, pp. 3.1.1-3.1.8, 1992.

[10] A. Dharchoudhury, S.M. Kang, "Worst-Case Analysis and Optimization of VLSI Circuit Performances", *IEEE Trans. on Computer-Aided Design*, Vol. 14, No. 4, pp 481-192, 1993.

[11] K.J. Antreich, H.E. Graeb, C.U. Wieser, "Circuit Analysis and Optimization Driven by Worst-Case Distances", *IEEE Trans. on Computer-Aided Design*, Vol. 13, No.1, pp. 57-71, 1994.

[12] G.E. Mller-L., "Limit Parameters: The General Solution of the Worst-Case Problem of the Linearized Case", *IEEE Intern. Symp. on Circuits and Systems*, Vol. 3, pp. 2256-2259, 1990.

[13] St.W. Director, G.D. Hachtel, "The Simplical Approximation Approach to Design Centering", *IEEE Trans. on Circuits and Systems*, Vol. 24, No. 7, pp. 363-372, 1977.

[14] A. Levkovich, E. Zeheb, N. Cohen, "Frequency Response Envelopes of a Family of Uncertain Continuous-Time Systems", *IEEE Trans. on Circuits and Systems*, Vol. 42, No. 3, pp. 156-16, May 1995.

[15] S.P. Bhattacharyya, H. Chapellat, L.H. Keel, "Robust Control: The Parametric Approach", *Prentice Hall*, London, 1995 .

[16] P.-O. Gutman; C. Baril, L. Neumann, "An Algorithm for Computing Value Sets of Uncertain Transfer Functions in Factored Real Form ", *IEEE Trans. on Automatic Control*, Vol. 39, No. 6, pp. 1269-73, 1994.

[17] G. Gielen, W. Sansen, "Symbolic Analysis for Automated Design of Analog Integrated Circuits", *Kluwer Academic Publishers*, Boston, 1991.

[18] R.E. Moore, "Interval analysis", *Prentice Hall*, Englewood Cliffs, 1966.

[19] G. Alefeld, J. Herzberger, "Introduction to interval computations", *Academic Press*, New York, 1983.

[20] E.P. Oppenheimer, A.N. Michel, "Application of Interval Analysis Techniques to Linear Systems: Part I-III", *IEEE Trans. on Circuits and Systems*, Vol. 35, No. 9-10, 1988.

[21] E.G. Eszter, R.S. Sanchez Pena, "Value set boundary computation of uncertainty structures", *Proc. of the American Control Conference*, pp. 2210-13, 1992.

[22] J. O'Rourke, "Computational Geometry in C", *Cambridge University Press*, New York, 1993.

[23] D. Python, Ch. C. Enz, "An Antialiasing Filter Using Complementary MOS Transconductors Biased in the Tiode Region," *Deliverable 5.2.5, ESD-LPD Project "Alpins", Proc. IEEE Int. Symp. Circuits Syst.*, Vol. 2, pp. 184-187, June 1999.

[24] D. Python, Ch. C. Enz, "A 40μW, 75dB Dynamic Range, 70kHz Bandwidth Biquad based on complementary MOS Transconductors," *Deliverable 5.2.6, ESD-LPD Project "Alpins", Proc. European Solid-State Circuits Conference in Duisburg*, Sept. 1999.

[25] R. Krawczyk, "Interval Iteration for Including a Set of Solutions", *Computing 32*, pp. 13-31, 1984.

[26] R. Krawczyk, A. Neumaier, "Interval Newton operators for Function Strips", *Journal of Mathematical Analysis and Applications*, No. 124, pp. 52-72, 1987.

[27] R Krawczyk, "Properties of Interval Operators", *Computing 37*, pp.227-245, 1986.

[28] D. Oelschlgel, "Intervallmathematische Lösung nichtlinearer Gleichungssysteme", *Wissenschaftliche Zeitschrift TH Leuna-Merseburg 26*, PP.298-307, 1984, in German.

[29] L. Hedrich, E. Barke, "A Formal Approach to Nonlinear Analog Circuit Verification", *International Conference on Computer-Aided Design* , pp. 123-127, Nov 1995.

[30] F. H. Bursal, B. H. Tongue, "A New Method of Nonlinear System Identification Using Interpolated Cell Mapping", *Proc. of the American Control Conf* ., Vol. 4, pp. 3160-4, Jan 1992.

[31] S. Natarajan, "A systematic method for obtaining state equations using MNA", *IEE Proceedings-G* , Vol. 138, No. 3, pp. 341-345, Mar 1991.

[32] L. Fortuna, G. Nunnari and A. Gallo, "Model Order Reduction Techniques with Applications in Electrical Engineering", *Springer-Verlag London*, 1992.

[33] P. Feldmann, R.W. Freund, "Efficient Linear Circuits Analysis by Pade Approximation via the Lanczos Process", *IEEE Transactions on Computer Aided Design* , Vol. 14, No. 5, pp. 639-649, May 1995.

[34] P.E. Gill, W. Murray, M.H. Wright, "Practical Optimization", *Academic Press Inc*, London, 1981.

[35] M. Punzenberger, C. Enz, "A 1.2-V Low-Power BiCMOS Class AB Log-Domain Filter", *IEEE Journal of Solid-State Circuits*, Vol. 32, No. 12, 1997.

[36] M. Punzenberger, C. Enz, "A Compact Low-Power BiCMOS Log-Domain Filter", *IEEE Journal of Solid-State Circuits*, Vol. 33, No. 7, 1998.

9 A LOW POWER BICMOS 1 GHZ SUPER-REGENERATIVE TRANSCEIVER FOR ISM APPLICATIONS

Catherine Dehollain, Norbert Joehl, Michel Declercq,
Patrick Favre, Philippe Deval, Alexandre Vouilloz

Swiss Federal Institute of Technology EPFL, Lausanne, Switzerland
E-mail: catherine.dehollain@epfl.ch

This work has been performed during the European Esprit SUPREGE project (project ESD 25400) from the 1st December 1997 until the 31st March 2000. One of the main objectives of the SUPREGE project was to develop a new solution for micropower wireless data transmission over short distances. The frequency bands are the UHF ISM (Industrial Scientific and Medical) frequency bands (center frequency equal to 434 MHz and 868 MHz in Europe, and 916 MHz in the USA). As the ISM frequency bands are widely used, a half-duplex link between the receiver and the transmitter is often necessary. Typical applications are high-volume, low-cost and low-power data transmission systems such as:

- wireless computer peripherals
- distributed sensors
- biomedical telemetry circuits.

Most of those devices are battery-operated and require a supply voltage between 2 V and 3.2 V.

In practice, the target current consumption of the receiver has to be lower than 2 mA to extend the battery life. Simple systems use OOK (On-Off Keying) modulation. Data rates are between 1 kbit/s and 50 kbit/s. The typical transmission range is equal to 20 meters line-of-sight.

The micropower receiver described in this chapter, uses an original architecture based on the super-regeneration principle invented by Armstrong [1] in 1922. It is based upon the variation of the start-up time of an oscillator into which an RF signal

P. Wambacq et al. (eds.),
Low-Power Design Techniques and CAD Tools for Analog and RF Integrated Circuits, 193–213.

is injected. This technique was widely used in vacuum tubes circuits up to the fifties. It was progressively abandoned and replaced by the superheterodyne receiver with its superior frequency selectivity.

Nowadays, for short distance data transmission systems, the principle of super-regeneration allows a very simple architecture and appears to be particularly suited for micropower applications, compared to classical solutions such as the superheterodyne, the low IF (Intermediate Frequency) or the direct conversion receiver. Moreover, the super-regenerative receiver has a limited number of RF nodes, resulting in reduced supply current. Compared to a discrete-component version of a super-regenerative receiver, innovative analog integrated circuits techniques have improved the selectivity and sensitivity of the integrated super-regenerative receiver.

The external components required for the IC version of the super-regenerative receiver are limited to a resonator and some decoupling capacitors. In contrast, for classical solutions, it is often necessary to use external IF filters, like ceramic or SAW (Surface Acoustic Wave) filters, which are expensive components. The simple architecture of the super-regenerative receiver implies that its silicon area will be easily five times smaller than classical receivers such as the super-heterodyne receiver. Its very small area and its limited number of external components imply that its cost will be low compared to classical receivers.

The core of the super-regenerative receiver being an oscillator, the oscillator of the receiver can be re-used in the transmitter when adding a modulator. This is another important advantage compared to classical solutions.

The super-regenerative transceiver presented here has been developed by the Electronics Laboratory (LEG) of the Swiss Federal Institute of Technology, Lausanne (EPFL). It is worthwhile to note the numerous publications on super-regeneration ([2] to [12]) by the Electronics Laboratory of EPFL.

9.0.1 Explanation of the approach followed

First, the operating principles of the key blocks of the super-regenerative receiver have been thoroughly analyzed:

- RF oscillator,
- RF isolation amplifier
- LF quench signal generation

To end up with a circuit that features low power consumption, it was decided to use BiCMOS technology because it combines the best of bipolar and MOS technology.

Bipolar transistors have been used in the cross-coupled differential pair that realizes the negative conductance in the RF oscillator. At a given tail current, a bipolar differential pair has a higher transconductance than a MOS differential pair. As a result, the current consumption of the oscillator can be lowered by using bipolar transistors. Another advantage of bipolar transistors is their low 1/f noise corner, resulting in lower phase noise values close to the carrier.

MOS transistors have been used as current sources in the RF oscillator, the RF isolation amplifier, the envelope detector, and the transmitter circuits. CMOS technology has also been used in the bias control loop, the gm-C low-pass filter of the receiver and in the digital part of the circuit.

A first test circuit has been integrated in 0.8 μm AMS BiCMOS technology. It contains the isolation amplifier and a first version of the oscillator. Frequency selectivity measurements of this circuit showed that the excess loop gain of the oscillator signal should not exceed a certain threshold to obtain good receiver selectivity. This property has also been proven mathematically. On the other hand, the oscillator steady-state amplitude has to be high enough for proper operation of the subsequent envelope detector. For the measurements of this first circuit, an external amplifier and envelope detector were used to obtain good selectivity.

An improved oscillator circuit was implemented on a second test chip. This circuit provides a higher output voltage without increasing its excess loop gain and therefore selectivity is not affected. The oscillator output signal directly drives the on-chip envelope detector whose output signal is also used in a control loop to control the oscillator steady-state amplitude. Good selectivity and sensitivity have been obtained in this second circuit. An external quench signal was used to control the oscillator mode and by varying the shape of the quench signal, it has been verified that a sawtooth or a sinusoidal quench waveform results in better selectivity than a square wave quench signal.

Next, the quench generator was integrated together with the other receiver blocks. A first version of the transmitter was implemented on silicon as well.

Finally, the fourth integrated circuit (still in 0.8 μm AMS BiCMOS technology) contained the optimized receiver and transmitter. The receiver uses an optimized quench waveform and a power amplifier has been included in the transmitter. A Phase-Locked Loop controls the center frequency of both receiver and transmitter. Voltage and current reference circuits were included as well.

9.1 BASIC PRINCIPLES OF SUPER-REGENERATION

The principle of super-regeneration is based on the variation of the start-up time of the oscillator constituting the core of the super-regenerative receiver. The start-up time depends on the initial conditions of the oscillator circuit. By injecting an RF signal into the oscillator, the start-up time becomes dependent on the amplitude and frequency of the injected RF signal.

Without external signal injection, the process of building up of the oscillations is initiated by thermal noise. This process can be sped up by injecting an RF signal with frequency close to the natural frequency of the oscillator. The start-up time is found to be inversely proportional to the amplitude of the injected signal and to the frequency difference.

To show the influence of an injected signal upon the oscillator start-up time, the envelope of the output signal of the oscillator is shown in Figure 9.1 with and without an injected RF signal.

The basic block diagram of a super-regenerative radio receiver is shown in Figure 9.2. The amplitude modulated RF signal RFin is injected into the oscillator. The oscillator can be represented as a feedback loop constituted by an amplifier A and

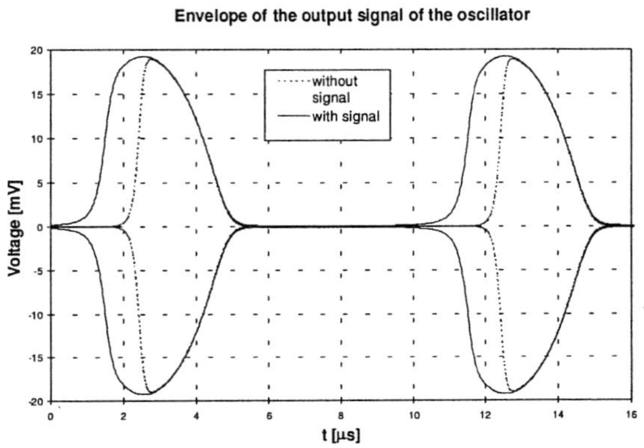

Figure 9.1 Envelope of the oscillator output signal for a 100 kHz sinusoidal quench signal.

a bandpass filter B. Compared to a classical oscillator circuit, two additional input signals are present: the quench signal and the RF input signal.

The quench signal controls the gain of the amplifier and thereby the overall transfer function.

The low-frequency quench signal periodically varies the loop gain AB of the oscillator.

Two operating modes can be distinguished:

- $|AB| < 1$ bandpass filter
- $|AB| > 1$ oscillator

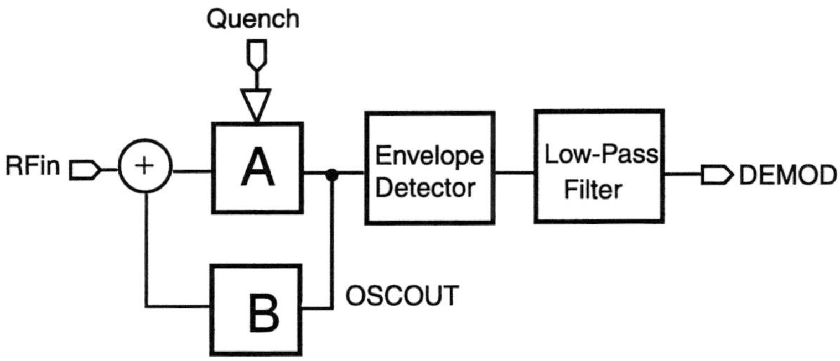

Figure 9.2 Simplified schematic of the super-regenerative receiver.

When the quench signal is originating from an independent generator, the receiver is called externally quenched. When the biasing of the oscillator is arranged in such a way that the oscillator output decreases by itself when reaching a given amplitude, the receiver is called self-quenched.

Addition of some passive components to the oscillator circuit can make it self-quenched. An externally quenched receiver requires an additional low-frequency oscillator circuit. By controlling the quench signal shape and amplitude, receiver selectivity can be optimized. This is not possible in the simpler self-quenched receiver. In order to benefit from this possibility, it was decided to go for an externally quenched receiver.

The easiest solution for demodulation of OOK modulated signals is to measure the lowpass filtered envelope of the oscillator output signal (Figure 9.2). The cut-off frequency of the lowpass filter must be lower than the quench frequency and higher than the bandwidth of the fundamental component of the digital data signal (typically half of the bit rate). As for every sampled system, the quench signal frequency must be at least twice as high as the digital data bandwidth. In practice, it will be chosen 5 to 10 times higher.

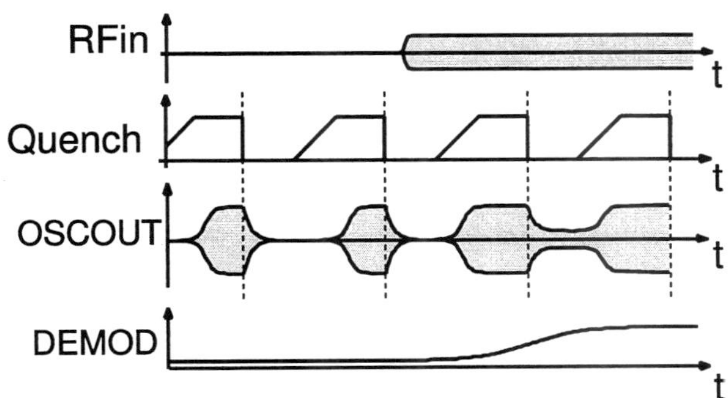

Figure 9.3 Main signals of the super-regenerative receiver.

The main signals of an externally quenched super-regenerative receiver are shown in Figure 9.3. The RF input signal RFin is 100 % amplitude modulated (OOK modulation) by the digital data signal. The quench signal restores the oscillation condition while rising. On the other hand, it cancels this condition when falling. The shape of the quench signal influences both receiver selectivity and sensitivity.

The quench signal rise time is long because it can be proven theoretically ([3], [5], [6], [12], [13], [14]) that the oscillator has to be turned on slowly to optimize receiver performance. On the other hand, the quench signal fall time is short in order to quickly dissipate the energy stored in the oscillator. The oscillator output signal is denoted by OSCOUT. When no input signal RFin is present $\{\text{data} = 0\}$, the building-up of the oscillations is slower than when RFin is present $\{\text{data} = 1\}$. As a result, the presence of the RFin signal influences the pulse width of the envelope of the oscillations. The envelope frequency is constant and equal to the frequency of the quench signal. The envelope is therefore a PWM (Pulse Width Modulated) signal.

Before the building-up of the oscillations when the loop gain AB < 1, the circuit behaves as a 2nd order bandpass filter with its bandwidth narrowing as the quench current increases.

The signal DEMOD represents the average value of the envelope of the oscillator output signal after the low-pass filter. The value of the demodulated signal is small when no RF input signal is applied; this case corresponds to $\{\text{data} = 0\}$. On the other hand, the presence of an RF signal, $\{\text{data} = 1\}$ results in a large value of the DEMOD signal. There is a time delay between the envelope of the RFin input signal and the DEMOD signal due to the demodulation and lowpass filtering.

In the case of the self-quenched receiver, oscillation stops when the output amplitude reaches a certain threshold. The quench frequency is proportional to the amplitude of the RF input signal. The effect of an RF input signal at the natural frequency of the oscillator is to increase the number of times the RF oscillator turns on and off per second without changing the pulse width of the envelope of the oscillations. This envelope becomes therefore a PDM (Pulse Density Modulated) signal.

9.2 SELECTION OF THE OSCILLATOR CIRCUIT

The RF oscillator is the key block of the super-regenerative receiver. In order to achieve maximum performance at low current, it was decided to use a harmonic oscillator instead of a regenerative one. In order to lower the sensitivity to supply voltage noise, a differential architecture was adopted.

9.2.1 LC oscillator

The harmonic oscillator is schematically shown in Figure 9.4, it comprises

- a parallel resonant RLC circuit
- an active negative impedance

The active circuit compensates for the resonator losses by implementing a negative resistance of value R_{neg}

$$R_{neg} = -\frac{1}{g_m} \tag{9.1}$$

The resonator losses are represented by the resistor R_0 in Figure 9.4. R_0 is related to the resonator quality factor Q by

$$Q = \frac{R_0}{\omega_0 L} \tag{9.2}$$

The use of a parallel resonant circuit yields low power consumption since the external AC current supplied to the resonator is a factor Q smaller than the current flowing through the inductor and capacitor. A higher quality factor Q results in a higher equivalent loss resistance R_0 requiring a lower value for the active transconductance g_m resulting in lower current consumption.

Figure 9.5 shows the simplified schematic diagram of the bipolar LC oscillator. The current sources Ibias and i_{qch} generate the DC bias and AC quench current. The total oscillator tail current Itail is equal to:

$$I_{tail} = I_{bias} + i_{qch} \tag{9.3}$$

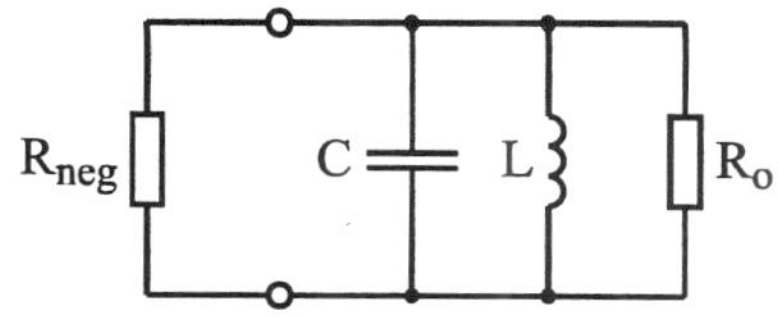

Figure 9.4 The LC oscillator.

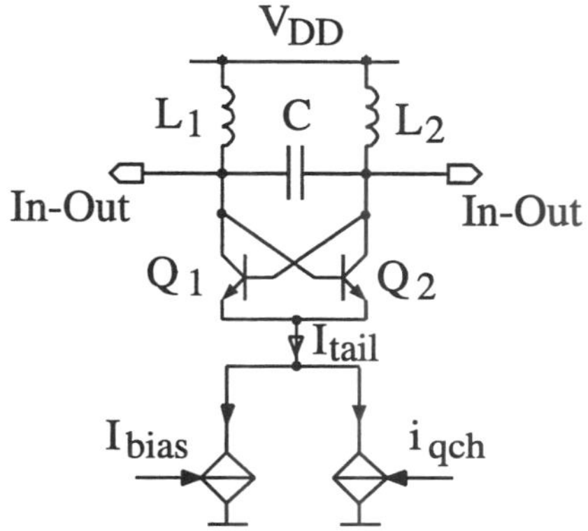

Figure 9.5 Simplified schematic of the oscillator.

The two inductors L_1 and L_2 each having value $\frac{1}{2}L$ and the capacitor C constitute the resonator. The resonator losses of this resonator, concentrated in a parallel equivalent loss resistance R_0 (not shown in Figure 9.5), are compensated by the negative impedance Rneg realized by the cross-coupled bipolar differential pair. Ignoring the base currents, G_{neg} the conductance undamping the parallel resonant circuit can be written as

$$G_{neg} = \frac{1}{R_{neg}} = -\frac{g_m}{2} = -\left(\frac{I_{bias} + i_{qch}}{4U_T}\right) \tag{9.4}$$

The use of bipolar transistors for the differential pair results in lower power consumption for a given conductance G_{neg}. The value of Gneg is proportional to the tail current I_{tail}.

It can be proven [3] that the oscillator circuit of Figure 9.5 behaves like a second order pass-band filter, which can also be considered as a selective amplifier, if :

$$\frac{1}{R_{neg}} + \frac{1}{R_0} > 0 \tag{9.5}$$

On the other hand, the circuit of Figure 9.5 behaves like an oscillator [3] if:

$$\frac{1}{R_{neg}} + \frac{1}{R_0} \leq 0 \tag{9.6}$$

The *critical current* I_{crit} is defined as the value of the total current for which the following holds:

$$\frac{1}{R_{neg}} + \frac{1}{R_O} = 0 \tag{9.7}$$

The critical current corresponds to the minimum tail current for which the oscillations start to build up.

The OOK modulated RF signal can be injected into the oscillator tank circuit. An input amplifier is required to transform the RFin input voltage signal delivered by the antenna into a current.

Ignoring the transistor and layout capacitances, the oscillator frequency f_0 is determined by the resonator inductance L and capacitance C and is equal to:

$$f_0 = \frac{1}{2\pi\sqrt{LC}} \tag{9.8}$$

9.2.2 Inductor choice

The following alternatives have been considered for the inductors in the oscillator circuit:

- Bond wire inductors
- Integrated spiral inductors
- External SMD inductors

The typical inductance of a bond wire is equal to 1 nH per mm length. Its quality factor is between 20 and 100. Accuracy is typically 5 % . A typical value required in the oscillator circuit would be 3 nH requiring a 3 mm bond wire. Bond wire size may become excessive compared to the size of the integrated circuit.

Integrated spiral inductors are realized on-chip by using two or more metal layers. An integrated inductor requires silicon area. The attainable quality factor in the frequency bands of interest (867 – 928 MHz) is lower than 8. The quality factor can be slightly increased by etching away the substrate underneath the spiral inductors. This post-processing increases the price of the integrated circuit.

The quality factor of good external SMD inductor is larger than 30. The combination of high quality factor, small size and low price makes the external SMD inductor the preferred choice for this application.

9.2.3 Oscillator operating modes

The oscillator circuit of Figure 9.5 behaves like a 2nd order bandpass filter if the losses of the LC resonator are partially compensated by the negative conductance G_{neg} of the cross-coupled differential pair.

This situation corresponds to the improvement of the quality factor, which has to be high enough in order to obtain a good selectivity. This mode is called the *regenerative mode*. The initial quality factor of the resonator Qini and the improved value Qimp are related as follows

$$Q_{imp} = Q_{ini}\left(\frac{R_{neg}}{R_{neg} + R_0}\right) \tag{9.9}$$

If the time, during which the oscillator tail current is larger than the critical current, is short enough to ensure that the output signal of the oscillator does not stabilize, then the amplitude of the demodulated output signal remains a linear function of the amplitude of the input signal RFin. This mode of operation, represented by the plain line in Figure 9.6 is called the *linear mode.*

If the time, during which the oscillator tail current is larger than the critical current, is long enough to ensure that the output signal of the oscillator stabilizes, then the amplitude of the demodulated output signal becomes a logarithmic function of the amplitude of the input carrier RFin. This mode of operation, represented by the dashed line in Figure 9.6, is called the *logarithmic mode.*

The logarithmic mode of operation has been chosen for this application in order to be able to measure and control the output steady-state amplitude of the oscillator.

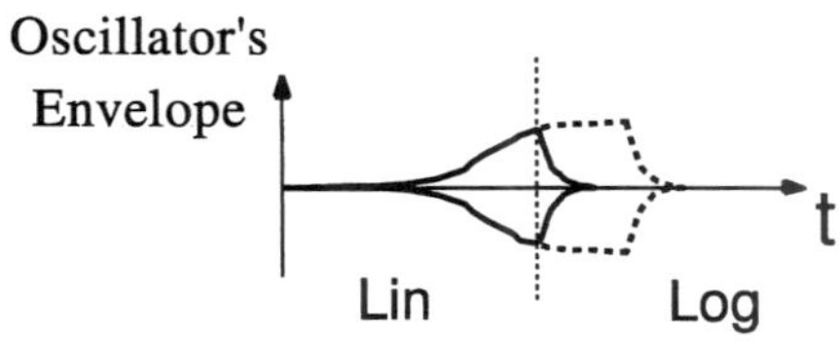

Figure 9.6 Linear mode (plain line) and logarithmic mode (dashed line).

9.3 RECEIVER PERFORMANCE OPTIMIZATION

The oscillator tail current, as defined by (9.3), is represented in Figure 9.7. The receiver behaves as a bandpass filter as long as the tail current is lower than the critical current. The tail current increases beyond the critical current until it reaches its final value. This final value is called the *start-up current* I_{start}. The difference between the critical current and the start-up current is equal to ΔI_{start}.

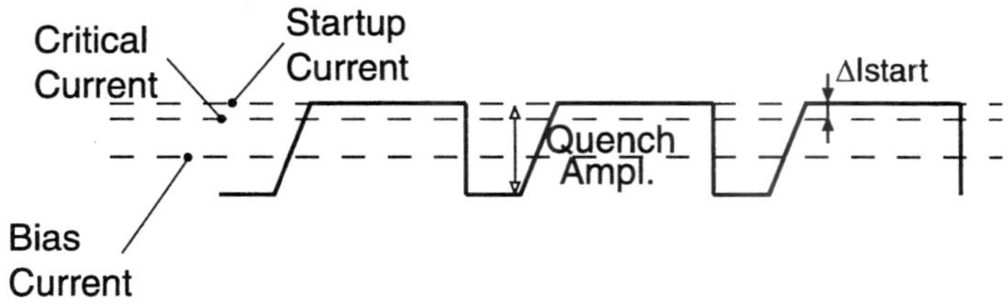

Figure 9.7 Oscillator tail current I_{tail} .

9.3.1 Adjustment of the oscillator tail current

What would be a good choice for the startup current? Let's define the parameter δ as

$$\delta = \frac{I_{start}}{I_{crit}} \tag{9.10}$$

It is clear that $\delta > 1$ is required for an increasing oscillator output signal amplitude. It can be shown [3] that the oscillator output steady-state amplitude is proportional to δ, whereas the selectivity is inversely proportional to δ. Figure 9.8 illustrates these relations.

In practice, it is not possible to choose a value of δ close to 1. This is due to the fact that the oscillator output signal drives the envelope detector, which requires a minimum voltage amplitude to work properly. Therefore, a trade-off must be made between frequency selectivity and output voltage amplitude. A solution to this problem will be proposed in the next section.

Good selectivity is obtained using a typical value $\delta = 1.01$ for the oscillator circuit shown in Figure 9.5. This value of δ results in a steady-state amplitude (differentially) around 10 mV. A bias control loop is used to automatically adjust the bias current, and in this way δ, by measuring the steady-state amplitude of the oscillator output signal.

Finally, the selectivity and the sensitivity performance of the receiver also depend on the characteristics of the quench signal like its frequency, waveform and amplitude [3], [5], [6].

Receiver performance benefits from a slowly rising quench signal. A sawtooth or a sinusoidal quench waveform is a much better choice than a square wave. A sawtooth quench signal is therefore generated by the on-chip quench generator.

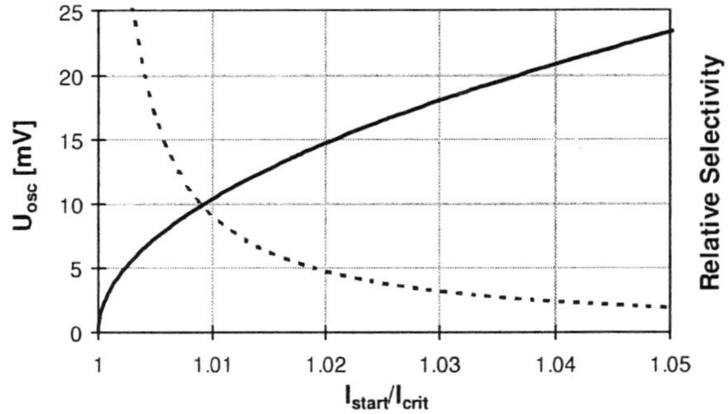

Figure 9.8 Output amplitude of the oscillator signal (plain line) and selectivity (dashed line).

9.3.2 Improved harmonic oscillator circuit

A capacitor voltage divider [2] has been introduced in the feedback loop of the oscillator as shown in Figure 9.9. Capacitor C2 has its value equal to nC1. The voltage amplification Av, which is defined as the ratio of the voltage V_{CC} between the 2 collectors and the voltage V_{BB} between the bases, is equal to:

$$A_v = \frac{V_{CC}}{V_{BB}} = 2\frac{C_2}{C_1} + 1 = 2n + 1 \tag{9.11}$$

The real part G_{neg} of the admittance Y_{in} seen between the two collector nodes is approximately equal to:

$$G_{in} = -\left(\frac{g_m}{4n+2}\right) \tag{9.12}$$

Comparison with expression (9.4) shows that the admittance is lowered by a factor (2n + 1). In order to have the same loop gain, the tail current of the differential pair must be increased by the same factor. This results in an output voltage V_{CC}that is (2n+1) larger compared to the oscillator circuit without capacitive divider. In this design n was chosen equal to 2.

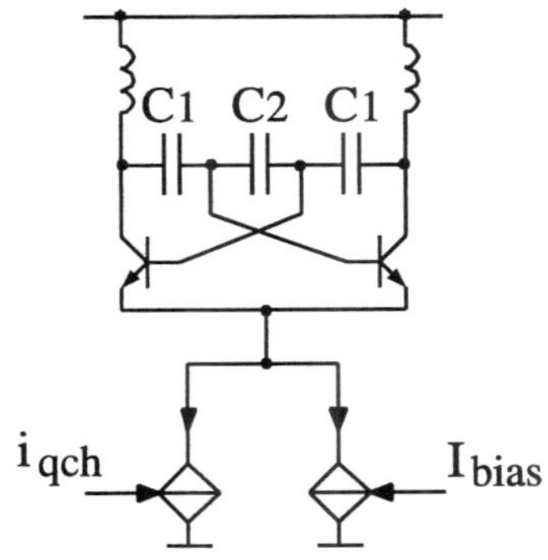

Figure 9.9 Improved oscillator circuit.

9.4 OVERALL RECEIVER ARCHITECTURE

The architecture of the receiver is represented in Figure 9.10. A LNA with differential input injects the RF signal into the oscillator. The LNA and the oscillator share the same parallel resonant LC load. The LNA isolates the LO signal from the antenna. The envelope detector also has differential inputs and outputs. The output voltage delivered by this circuit is proportional to the envelope of the oscillator output signal.

The output signal of the differential gm-C lowpass filter (shown in Figure 9.11) represents the average value of the envelope of the oscillator output signal. Since the quench frequency is dependent on the bit rate, the cut-off frequency of the lowpass filter is tuned accordingly to be equal to one fifth of the quench frequency. This gm-C filter implements a third order Butterworth lowpass transfer function. By varying the values of the transconductance (gm) and/or the values of the capacitors, it is possible to select 4 different values of the cut-off frequency. An output amplifier transforms the differential output signal from the gm-C filter in a single-ended output signal.

A sawtooth quench current is generated in a separate block. Figure 9.12 shows the quench current waveform. It becomes negative after completion of the sawtooth. This lowers the oscillator tail current resulting in faster turn-off of the oscillator signal. The energy in the parallel resonant tank circuit is dissipated rapidly by activating the MOS transistor across the tank. The quench frequency can be chosen equal to 50kHz, 100kHz, 200kHz or 500kHz.

The bias current of the oscillator is controlled through a loop that measures the output signal V_{env} of the envelope detector. Figure 9.12 shows the signals involved in this control loop. Signal V_{cmp} enables the comparison of the output amplitude Venv with its target value Vref. The bias current is corrected so as to make V_{env} equal to V_{ref}.

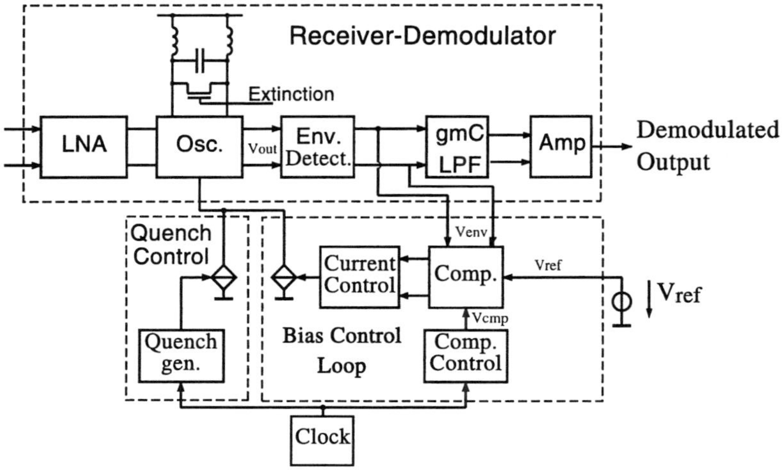

Figure 9.10 Complete receiver architecture.

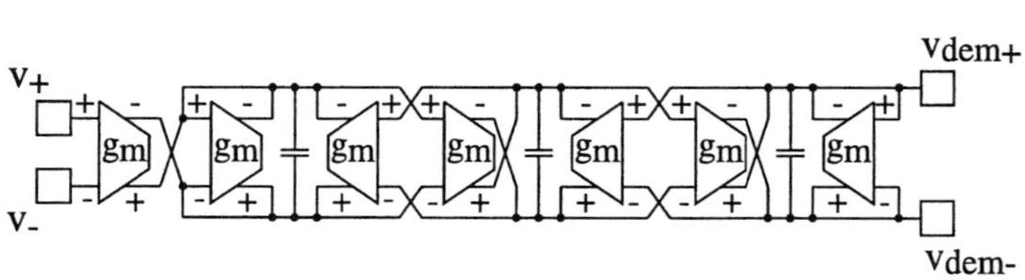

Figure 9.11 Third order gm-C lowpass filter.

9.4.1 Receiver isolation amplifier and oscillator schematics

The LNA isolation amplifier and the oscillator are shown in Figure 9.13. The LNA is constituted by transistors T_3 and $T_4$and cascode transistors T_5 and T_6. The latter im-

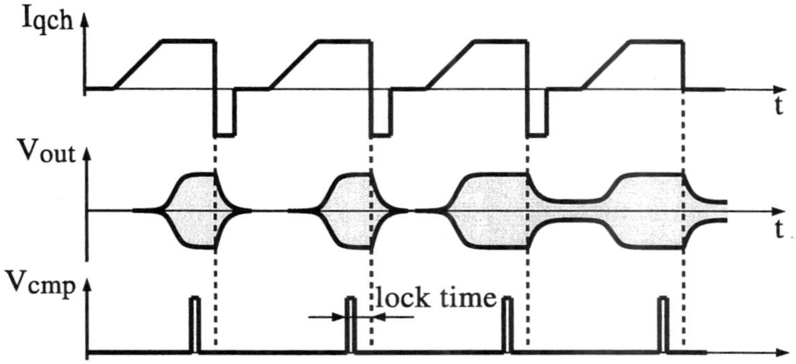

Figure 9.12 Signals involved in the bias control loop.

prove the isolation between antenna and oscillator. The antenna signal is capacitively coupled to the bases of T_3 and T_4.

The LNA is biased (current mirror T_7 – T_3 - T_4) by a current of 50 μA in each branch. This value has been chosen to achieve a good compromise between power consumption and sensitivity.

In transmit mode, the LNA is disabled by adjusting the base voltage of T_5 and T_6.

The cross-coupled differential pair T_1-T_2 constitutes the oscillator core that undamps the external LC resonator. Transistors T_1 and T_2 have a double-base structure to lower the base resistance, and thereby oscillator phase noise.

Transistor M_1 in parallel with the LC resonator is used to quickly dissipate the energy stored in the resonator at the end of each quench period.

The current consumption of the oscillator is equal to 310 μA for an output steady-state amplitude equal to 50 mV and an unloaded Q factor of the LC resonator equal to 50. The values of the two inductors of the LC resonator are equal to 2.7 nH each. For this output oscillator amplitude, the attenuated oscillator power injected into the antenna is equal to – 70 dBm.

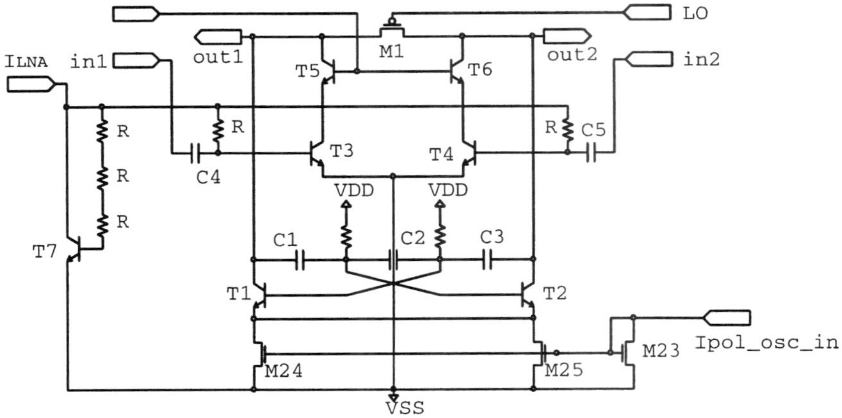

Figure 9.13 LNA Isolation Amplifier and Oscillator.

9.5 TRANSMITTER ARCHITECTURE

The transmitter architecture is shown in Figure 9.14. It employs a fully differential structure up to the antenna. The oscillator, an isolation amplifier and an open collector power amplifier constitute the building blocks of the transmitter. The isolation amplifier also drives the ECL prescaler of the PLL. The receiver oscillator is re-used for the transmitter. While transmitting the oscillator is not quenched. An isolation and power amplifier complete the transmitter. An external resistor fixes the output power.

The receiver LNA is turned off during transmission. The bias control loop, which controls the oscillator output amplitude, is active in both transmission and reception mode. This loop also guarantees that the prescaler receives sufficient input signal.

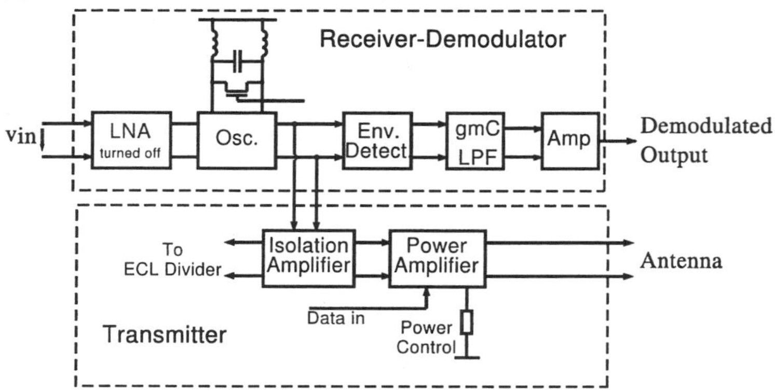

Figure 9.14 Transmitter architecture.

9.5.1 Transmitter Power Amplifier schematics

The schematic diagram of the power amplifier is shown in Figure 9.15. The output power depends on the value of an external resistor connected between ground and the PMOS current mirror input.

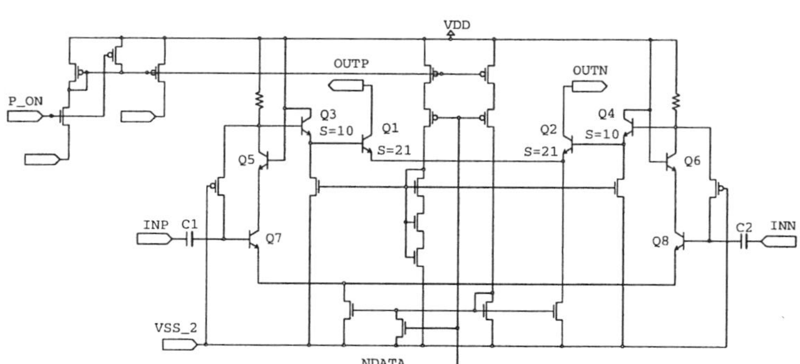

Figure 9.15 Power Amplifier.

The differential input signal V(INP,INN) is amplified by differential pair Q_7-Q_8 and their cascode transistors Q_5 and Q_6. A buffer amplifier with emitter followers Q_3 and Q_4 drives the final amplifier stage Q_1-Q_2.

OOK modulation is achieved by the NDATA signal that switches on and off the bias current of the amplifier. This power amplifier has been designed for a minimum output power of 1 mW in a 600 Ω differential load.

9.6 THE SAMPLED PHASE-LOCKED LOOP

A Phase-Locked Loop (PLL) controls the operating frequency of the receiver and transmitter.

Figure 9.16 shows the PLL block schematic diagram. The reference frequency is derived from a crystal oscillator using an external quartz at 6.78 MHz. The output signal of the RF oscillator of the receiver is applied to a fixed ECL (Emitter Coupled Logic) divider (division by 128). The output signal of this divider and the reference signal from the crystal oscillator are fed into a sequential phase comparator. The sequential phase comparator output is connected to an external loop filter that generates the VCO control voltage. The VCO voltage drives an external varactor diode that is part of the oscillator tank circuit. The sequential phase comparator output can be switched to high impedance state.

To insure proper operation of the ECL divider, a level detector measures the oscillator output level and a feedback loop guarantees that this level is larger than 40 mV during the ON periods of the PLL. During the time that the PLL is switched on, the oscillator is not quenched. On the other hand, during the OFF periods (also called memorization periods) the oscillator does not operate.

Figure 9.17 illustrates the operating principle of this PLL. Just after switching on the receiver, the oscillator is not quenched and the PLL runs continuously to quickly settle to its final frequency. The oscillator amplitude regulation operates at a 6.78MHz rate (equal to the frequency of the crystal oscillator) during this initialization period. Signal V_{cmp} is the sampling command signal of the bias control loop.

After PLL settling, the oscillator is periodically quenched and the PLL runs in a sampled mode of operation. After completion of the square wave the quench current Iqch becomes negative. The MOS transistor in parallel with the tank circuit is activated and ensures quick dissipation of the energy stored in the tank

During the OFF periods of the PLL, the phase states of both the sequential phase comparator and the ECL divider are retained. The control voltage V_{VCO} of the varactor diode is memorized by the low-passfilter during these OFF periods. However, while the PLL is "remembering", the VCO voltage will probably drop slightly due to leakage currents. This gives an initial frequency error the next time the PLL tries to lock. This error is compensated during the next ON period of the PLL.

The current consumption of the ECL divider is decreased from 2 mA to 0.25 mA during the memorization periods. The duty cycle of this sampled mode of operation is equal to 10 % .

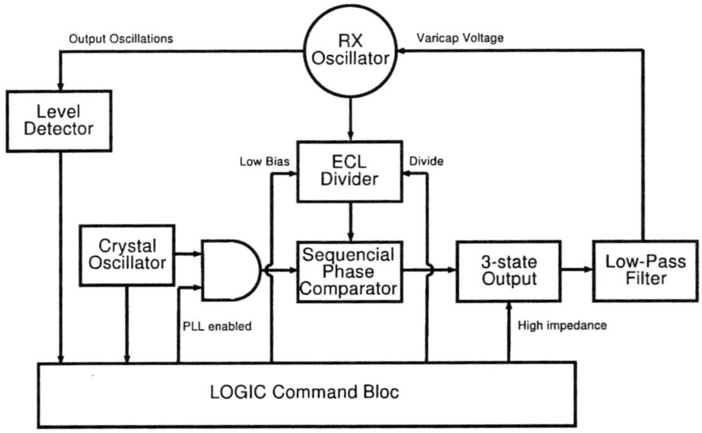

Figure 9.16 PLL block schematic diagram.

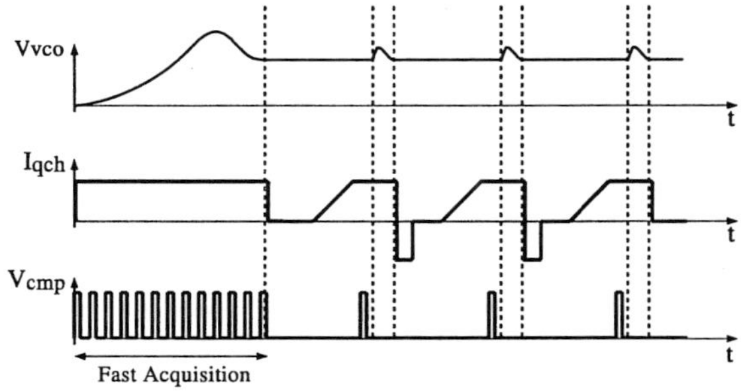

Figure 9.17 PLL timing diagram

9.7 THE COMPLETE TRANSCEIVER CIRCUIT AND ITS PERFORMANCE

The complete transceiver circuit has been realized in AMS BiCMOS 0.8 μm technology. A band gap circuit delivering a reference voltage for low battery detection was also included on the silicon.

Circuit size (including the pads) is 1.8 mm x 2.6 mm. The number of pads is equal to 42 and a CQFP44 package provided by Edgetek was used. A microphotography of the integrated circuit is shown in Figure 9.18.

The power amplifier of the transmitter was placed far away from the oscillator to lower the coupling between these blocks. This coupling could generate a positive feedback between the power amplifier and the oscillator that could corrupt the transmitter output spectrum.

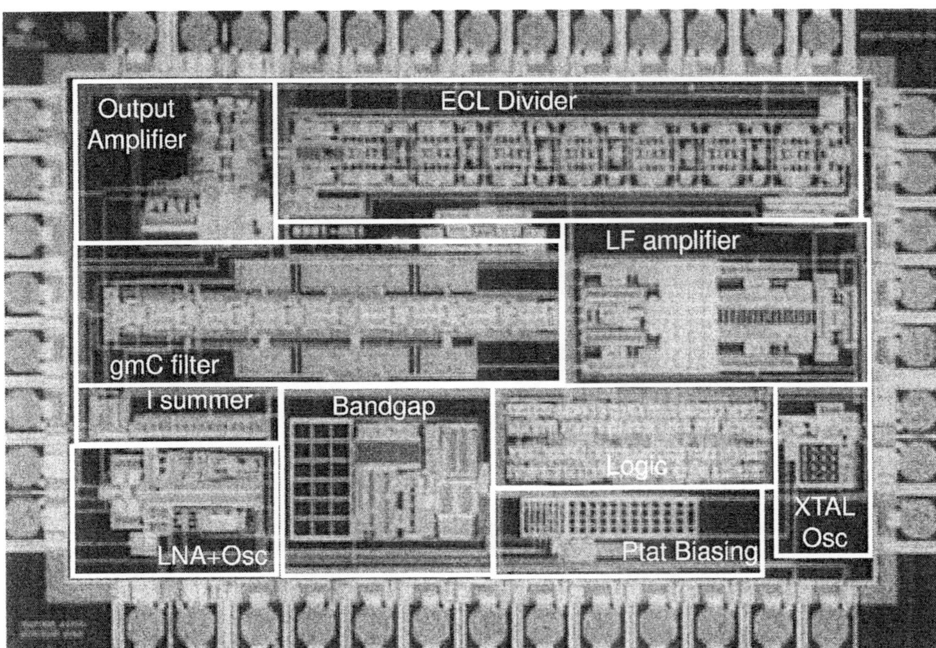

Figure 9.18 Microphotography of the circuit.

9.7.1 Measurement results

The circuit was measured using an LC resonator with an unloaded quality factor Q =50. An RF generator supplied the OOK modulated RF signal.

9.7.1.1 Receiver sensitivity. The sensitivity of the receiver has been measured at 868 MHz for the 4 different values of the quench frequencies (50, 100, 200 and 500 kHz). It is defined as the RF input level required for a Bit Error Rate, BER = 0.001. Figure 9.19 shows the results.

These measurements show that, first, the sensitivity increases with the quench frequency, with a peak at –105 dBm for a quench frequency equal to 200 kHz. Then, for a frequency equal to 500 kHz, the sensitivity decreases. These measurements are in accordance with the theory of the super-regenerative receiver ([3], [5], [6]) which proves that the sensitivity is best when the receiver works close to the linear mode, but still in the logarithmic mode in order to be able to control the output amplitude of the oscillator signal. When the quench frequency increases from 50 to 200 kHz, the receiver is more often turned on and off and therefore works closer to the linear mode of operation. A value of the quench frequency equal to 200 kHz corresponds to a quasi-optimum value for receiver sensitivity. When the quench frequency is increased to 500 kHz, the receiver is too close to the linear mode and as a consequence its sensitivity decreases.

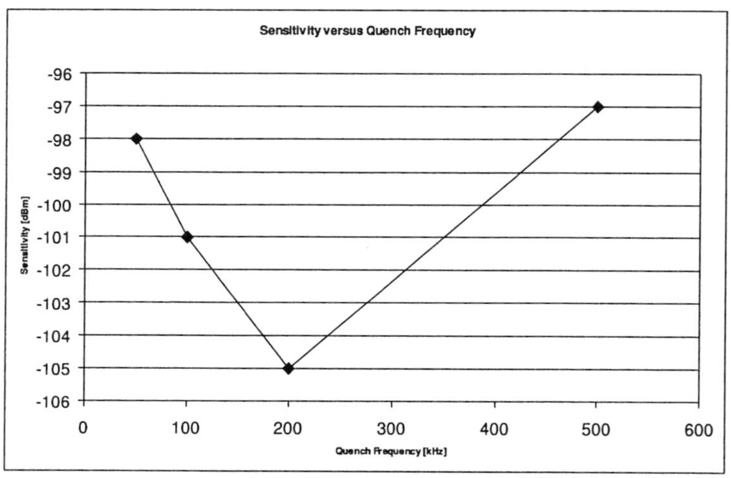

Figure 9.19 Receiver sensitivity at 868 MHz.

9.7.1.2 Receiver selectivity. The selectivity of the receiver has been measured at 868 MHz for the 4 different values of the quench frequencies (50, 100, 200 and 500 kHz). The measurement was performed by applying a modulated RF signal to the receiver at different RF frequencies and by adjusting its level for a constant demodulated data BER = 0.001. Quench frequencies between 50 and 200 kHz (curves Q50, Q100 and Q200) result in a constant –5 dB bandwidth of 150 kHz. This means that, for frequencies 75 kHz higher (or lower) than the center frequency (868 MHz), the level of the RF input signal has to be increased by 5 dB to obtain the same BER.

The maximum attenuation outside the frequency band is around 60 dB. At 500 kHz quench frequency, the –5 dB bandwidth increases to 400 kHz.

Compared to the results obtained with earlier versions of the receiver, Figure 9.20 shows that, for this last version, the selectivity is nearly constant at large distance from the center frequency. For the earlier versions, there was a sharp degradation of the receiver selectivity for signals more than 6 MHz away from the center frequency. It is recalled that in the literature, this phenomena is called *hangover*. In this new version, the shape of the quench current has been improved in order to avoid this problem. Indeed, a dip has been added when the quench goes down in order to turn off more the oscillator than in the previous circuit.

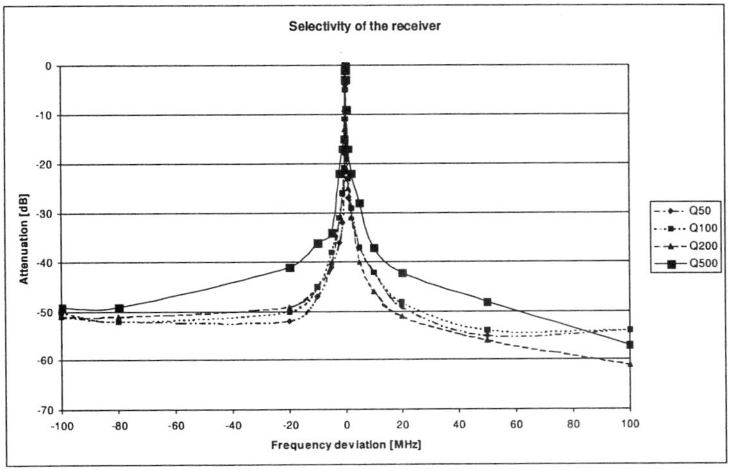

Figure 9.20 Receiver selectivity (200 MHz span).

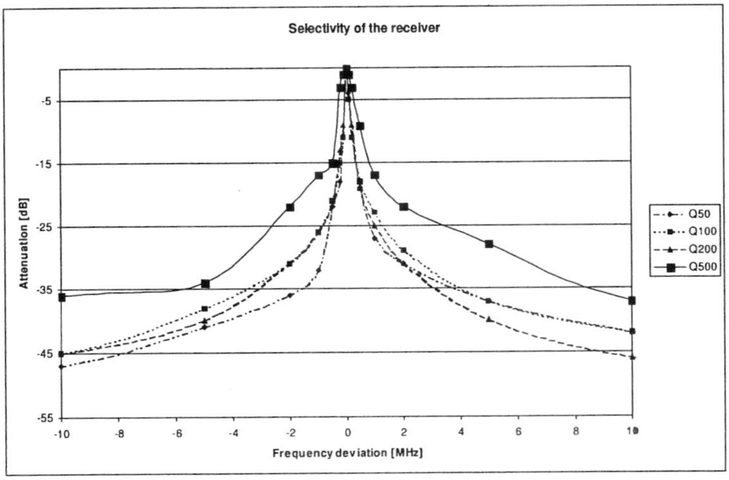

Figure 9.21 Receiver selectivity (20 MHz span).

9.7.1.3 Performance summary. Table 9.1 summarizes the transceiver performance. Transmitter supply current was measured for 0 dBm output power at 868

MHz. The receiver supply current was measured at 868 MHz. The external resonator had a quality factor Q = 50. The maximum operating frequency is package-limited and given for a 44-pin CQFP package. By using chip-on-board technology, the maximum operating frequency can be increased to 1.5 GHz.

Table 9.1 Transceiver performance

Minimum operating voltage	2.4 V
Operating frequency	300 - 1000 MHz
Receiver supply current at 2.5 V	1.5 mA
Transmitter supply current at 2.5 V	6 mA
Sensitivity, 10 kbit/s, BER = 0.001	- 105 dBm
Selectivity, –5 dB, 200 kHz quench frequency	150 kHz
Maximum data rate at 500 kHz quench frequency	150 kbit/s

9.8 CONCLUSION

A low-voltage low-power transceiver for ISM applications has been presented. The sensitivity of the super-regenerative receiver is comparable to that of classical super-heterodyne receivers. Frequency selectivity is sufficient for short-range wireless data transmission in the ISM frequency bands, but not good enough for GSM applications.

The results obtained show that the super-regenerative principle is a very good solution for short-range wireless data communication.

The same architecture could be designed for higher ISM frequency bands (e.g. the 2.45 GHz frequency band) by using a more advanced technology, such that the AMS SiGe BiCMOS 0.8 μm technology or the STMicroelectronics SiGe BiCMOS 0.35 μm.

References

[1] E. H. Armstrong, "Some recent developments of regenerative circuits", Proc. Inst. Radio Eng., Vol. 10, August 1922, pp. 244-260.

[2] P. Favre, N. Joehl, M. Declercq, C. Dehollain and P. Deval, "A 2V, 600 μA, 1 GHz, BiCMOS super-regenerative receiver", in ISSCC Digest of Technical Papers, San Francisco 1998, Vol. 41, pp. 128-129.

[3] A. Vouilloz, M. Declercq and C. Dehollain, "Selectivity and sensitivity performances of superregenerative receivers", in CD-ROM and Proceeding of ISCAS 98' Conference, Monterey (USA), 1998

[4] N. Joehl, P. Favre, P. Deval, A. Vouilloz, C. Dehollain and M. Declercq, "A BiCMOS Micropower 1 GHz Super-Regenerative Receiver", Proceeding of URSI ISSSE 98' Conference, Pisa (Italy) 1998, pp. 74-78.

[5] P. Favre, N. Joehl, A. Vouilloz, P. Deval, C. Dehollain and M. Declercq, "A 2V 600 uA 1 GHz BiCMOS Super-Regenerative Receiver for ISM Applications", IEEE Journal of Solid-State Circuits, Vol. 33, n0.12, December 1998, pp. 2186- 2196.

[6] A. Vouilloz, C. Dehollain and M. Declercq, "Modelisation and Simulation of Integrated Super-Regenerative Receivers", CD-ROM and Proceeding of ICECS 99' Conference, Pafos (Cyprus), 5-8 September 1999, Vol.1, pp. 521-524.

[7] A. Koukab, M. Declercq, and C. Dehollain, "Analysis and Simulation of Substrate Coupling in RF IC", Proceedings of the International Workshop on Low Power RF integrated Circuits, Lausanne (Switzerland), 19-20 October 1999, pp. not numbered.

[8] P. Favre, N. Joehl, M. Declercq, C. Dehollain and P. Deval, "A PLL stabilized Super-Regenerative Transceiver", Proceedings of the International Workshop on Low Power RF integrated Circuits, Lausanne (Switzerland), 19-20 October 1999, pp. not numbered.

[9] A. Vouilloz, C. Dehollain and M. Declercq, "A Low Power CMOS Super-Regenerative Receiver at 1 GHz", CD-ROM and Proceedings of CICC 2000' Conference, Orlando (USA), 21-24 May 2000.

[10] C. Dehollain, M. Declercq, N. Joehl, P. Favre, P. Deval, P. Chênes, Sultan, G. Bellini, P. Besseux, V. Valence, R. Magnin, "Low Power Wireless Data Transmission Systems at Short Distance for ISM Applications", Proceeding of ESD LPD 2000' Workshop, Rapallo (Italy), 25 and 28 July 2000, ISBN 90-5326-036-6, pp. 213-243.

[11] P. Favre, N. Joehl, P. Deval, M. Declercq and C. Dehollain, "A Low Power 1 GHz Super-Regenerative Transceiver with time-shared PLL control", Proceedings of the European Solid-State Circuits Conference (ESSCIRC), Stockholm (Sweden), 19-21 September 2000, pp. 84-87.

[12] A. Vouilloz, M. Declercq and C. Dehollain, "A Low Power CMOS Super-Regenerative Receiver at 1 GHz", to be published in March 2001 in the IEEE Journal of Solid-State Circuits, special issue on the CICC 2000 Conference.

[13] W. E. Bradley, "Superregenerative", Electronics, September 1948, pp. 96-98.

[14] A. Hazeltine, D. Richman and B.D. Loughlin, "Superregenerator design", Electronics, Sept. 1948, pp. 99-102.

[15] N. Joehl, C. Dehollain, P. Favre, P. Deval and M. Declercq, "A Low Power 1 GHz Super-Regenerative Transceiver with time-shared PLL control", to be published in July 2001 in the IEEE Journal of Solid-State Circuits, special issue on the ESSCIRC 2000 Conference.

10 LOW-VOLTAGE SWITCHED-CAPACITOR FILTERS

Andrea Baschirotto

Universita' degli Studi di Lecce, Italia

E-mail: andrea.baschirotto@unile.it

Recently the interest for low-power low-voltage integrated systems has consistently increased. This is due to the growing popularity of portable equipment and to the reduction of the possible supply voltage allowed by the modern scaled-down IC technologies. Figure 10.1 shows the foreseen reduction of the supply voltage for the next years.

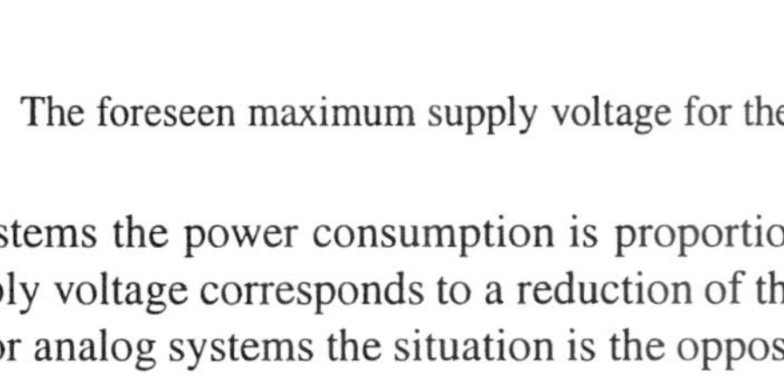

Figure 10.1 The foreseen maximum supply voltage for the coming years

For the digital systems the power consumption is proportional to V_{DD}^2. Thus the reduction of the supply voltage corresponds to a reduction of the power consumption. On the other hand, for analog systems the situation is the opposite. In fact, for a given supply voltage V_{DD}, the maximum possible voltage swing (SW) in an analog system is about equal to

$$SW = V_{DD} - 2V_{OV} \tag{10.1}$$

with V_{OV} the upper and lower saturation voltage of the output stage. The power consumption (P) can then be obtained by multiplying V_{DD} by the total current I:

P. Wambacq et al. (eds.),
Low-Power Design Techniques and CAD Tools for Analog and RF Integrated Circuits, 215–250.

$$P = V_{DD} I \tag{10.2}$$

The noise (N) is kT/C-limited and is inversely proportional to a part of the current αI

$$N^2 \propto \frac{1}{\alpha I} \propto \frac{V_{DD}}{\alpha P} \tag{10.3}$$

The analog system dynamic range (DR) can then be written as follows

$$DR \propto \frac{SW^2}{N^2} \propto \frac{[V_{DD} - 2V_{OV}]^2}{1/\alpha I} = \alpha P \frac{[V_{DD} - 2V_{OV}]^2}{V_{DD}} \tag{10.4}$$

A given dynamic range requires power consumption

$$P \propto \frac{1}{\alpha} DR \frac{V_{DD}}{[V_{DD} - 2V_{OV}]^2} \tag{10.5}$$

The analog power consumption increases for lower supply voltages, as qualitatively shown in Figure 10.2.

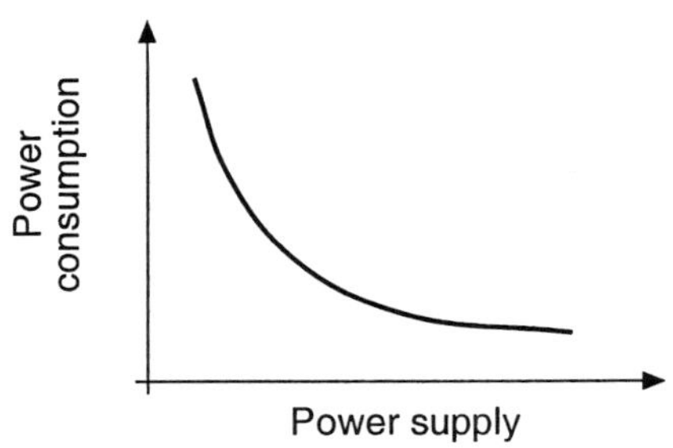

Figure 10.2 The relation between supply voltage and power consumption

This relation is confirmed in Table 1, where the performance of various analog $\Sigma\Delta$ modulators is shown. The modulators are thermal noise limited and can be compared using the following figure of merit (F):

$$F = \frac{4kTDR^2BW}{P} \tag{10.6}$$

It appears that systems operating at 5V (even when realized in technologies that are not state-of-the-art) outperform more recent low-voltage implementations. Despite the higher power consumption, low-voltage analog circuits are required in future mixed-signal systems. Therefore specific low-voltage solutions have to be developed. In

Table 10.1 ΣΔ modulator comparison

Reference	*Year*	V_{DD} *[V]*	*DR [dB]*	*B[kHz]*	*P [mW]*	*F [x 10^6]*
Peluso	1998	0.9	77	16	0.040	330
Williams	1994	5	104	50	47	445
Nys	1997	5	77	16	40	800

analog systems, SC techniques provide an efficient solution (in terms of performance accuracy and dynamic range) for the implementation of several analog functions.

However, supply voltage reduction should not degrade the analog circuit performance. It is fundamental to achieve a maximum output swing (rail-to-rail) to maintain a sufficiently large dynamic range. As the supply voltage decreases, the noise level remains to first order constant, while the signal swing decreases more than linearly. From the technological point of view CMOS technology seems to be more appropriate than BiCMOS, for the following two reasons:

- scaled-down CMOS technologies feature threshold voltages as low as 0.4 – 0.5 V. This allows the MOS devices to operate at a bias voltage (V_{GS}) comparable to that of bipolar devices (V_{BE} = 0.7V). Therefore bipolar devices are no more clearly advantageous in term of voltage compliance;

- SC circuits are usually included in mixed-signal systems where the digital part is much larger than the analog one. The use of a standard CMOS process (without low-threshold voltage option) is often mandatory for cost reasons [3, 4].

As a consequence, in the following only standard CMOS technology will be considered.

The classical structure of an SC integrator embedded in a closed loop topology is shown in Figure 10.3, where it is preceded and followed by a similar stage. An SC integrator is constituted by an input branch, consisting of four switches and a capacitor, and by an integrator, realized by an opamp with a feedback capacitor.

Regarding the opamp operation, this can be achieved with an appropriate design using a supply voltage as low as V_{TH}+ $2V_{OV}$, as described in the following sections. However such an opamp can only be used with some modifications at system level. More generally speaking, the problem of the design of low-voltage SC circuits in standard CMOS technology can be described by two conceptually different questions:

- for a given V_{DD}, using standard SC structures (no charge pump, no switched-opamp), which performance (maximum sampling frequency, output signal swing, etc.) can be achieved?

- for a rail-to-rail signal swing, which is the minimum supply voltage, adopting any design technique?

The first question relates to applications requiring which SC circuits operating at a given supply voltage, not necessarily the theoretical limit of the analog technique.

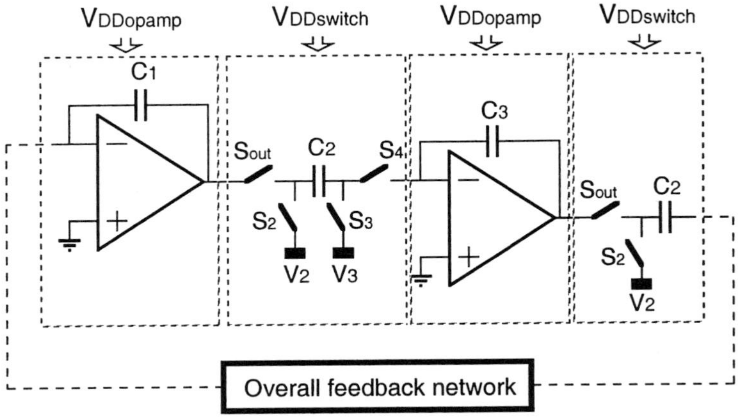

Figure 10.3 Typical SC integrator

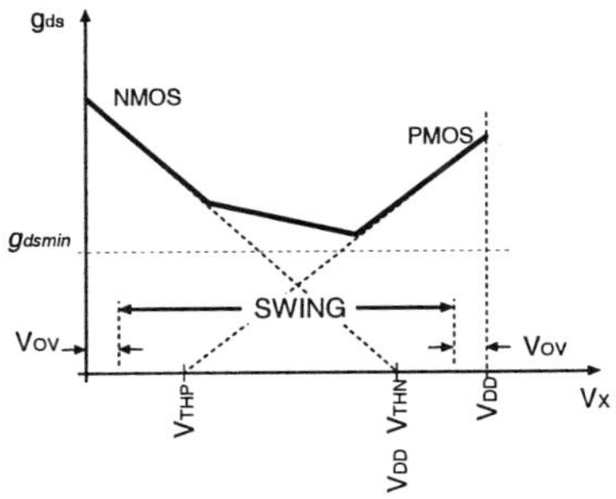

Figure 10.4 Switch conductance at $V_{DD} = 5$ V

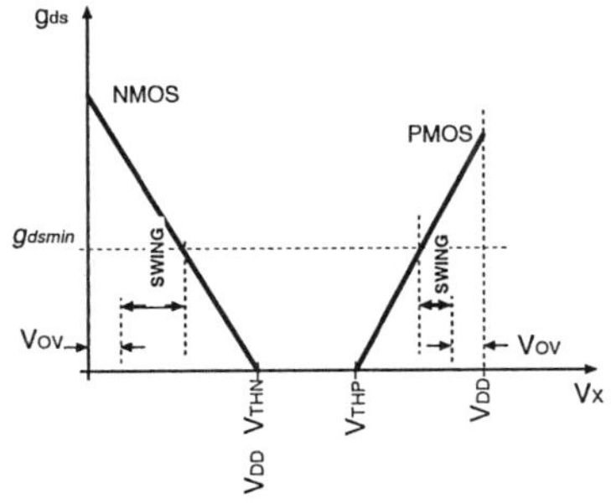

Figure 10.5 Switch conductance at $V_{DD} = 1$ V

The second question corresponds to the research towards new solutions, developed in order to achieve the maximum performance at the lowest supply voltage.

This chapter is organized as follows: section 10.1 replies to the first question and addresses the achievable performance at a given supply voltage using standard SC structures. For the second question section 10.2 provides three different approaches at system level. Section 10.3 deals with circuit solutions for the basic building blocks. Experimental results from a design example are proposed in section 10.4, while section 10.5 addresses open issues and future developments.

10.1 STANDARD STRUCTURES FOR LOW-VOLTAGE SC CIRCUITS

Fundamental limitations of low-voltage SC circuits originate from the switches. From Figure 10.5, it follows that low-voltage circuits using standard complementary switches suffer from reduced signal swing. In addition, this signal swing is only possible in two regions: one is closed to ground and the other is closed to V_{DD}. When operating in these regions, the use of complementary switches is no longer advantageous, and single MOS switches (NMOS-only or PMOS only) can be used. An NMOS-only switch is appropriate when the signal's DC level is close to ground, while a PMOS-only switch is used when the signal's DC level is close to V_{DD}. This however makes the possible signal swing to depend on the power supply. In fact, let's consider the case of an NMOS-only switch connected to a signal biased at V_{DC} and with a signal swing VSW. The resulting switch conductance is shown in Figure 10.6.

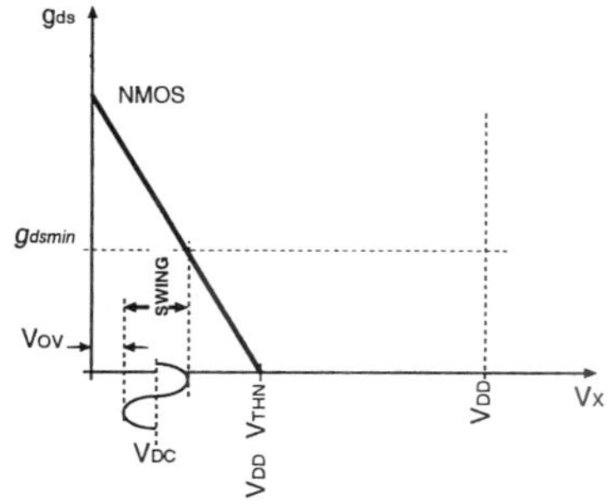

Figure 10.6 Possible swing using an NMOS-only switch

In this case the minimum supply voltage V_{DDmin} is given by:

$$V_{DD\,\min} = V_{DC} + V_{SW} + V_{THN}\left(V_{DC} + V_{SW}\right) + V_{OV} \tag{10.7}$$

Note hat the NMOS threshold voltage V_{THN} depends on the voltage the switch is connected to, i.e. (V_{DC}+ V_{OV}) due to the body effect. As a consequence, V_{DDmin} depends directly and indirectly on V_{SW}. This low-voltage SC design approach has been adopted in the design of a Sample & Hold circuit, the schematic of which is shown in Figure 10.7. It presents a pseudo-differential (PD, i.e. two single-ended structures driven with opposite signals) double-sample (DS, i.e. the input is sampled during both clock phases) structure.

The operating principle is the following. During phase 1 capacitors C_{1P} and C_{1M} sample the input signal, referred to V_{DD}. During phase 2 capacitors C_{1P} and C_{1M} are connected in the opamp feedback loop, producing the output sample.

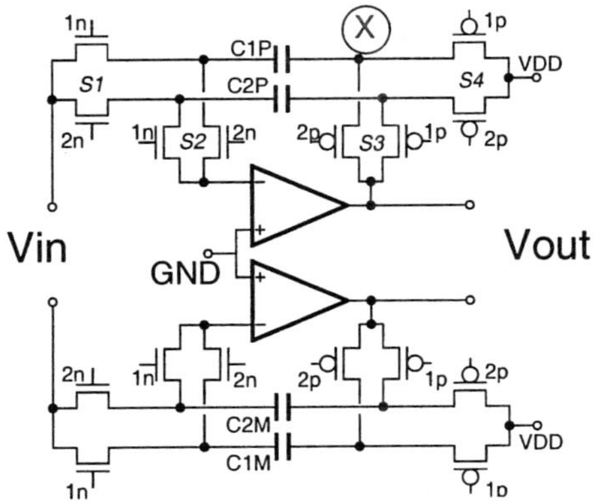

Figure 10.7 PD-DS Sample & Hold circuit

This structure offers the following advantages. The pseudo-differential structure does not require a Common-Mode Feedback circuit, a critical block for low-voltage circuits. The DS structure avoids any opamp-reset phase which causes in single-sampled structures large output steps. Slew-rate requirements are then relaxed, also because the opamp has to charge only the output load, and not the feedback capacitor. Furthermore the opamp always operates with feedback factor equal to one, achieving maximum speed of response. Finally a negligible droop rate is expected assuming an opamp MOS input device with zero input current. Switch operations are guaranteed by proper control of the voltage at the node where the switches are connected to, as shown in Figure 10.8.

The opamp input DC voltage is set to ground by the feedback action. Switch S2 is then realized with a single NMOS device. Switch S4 is connected to V_{DD}, and then it is realized using a PMOS device. The DC component of the input voltage V_{in_DC} is close to ground: this allows the realization of S1 with a single NMOS device. The opamp output DC voltage (V_{out_DC}) is then fixed at the value:

$$V_{out\_DC} = V_{DD} - V_{in\_DC} \tag{10.8}$$

which is close to V_{DD}. S3 is then realized with a PMOS device. The minimum supply voltage required by the structure is then determined by the correct operation of switches S1 and S3, and is given by:

$$V_{DDn} > V_{in\_DC} + V_{SW} + V_{THN}\left(V_{in\_DC} + V_{SW}\right) + V_{OV} \tag{10.9}$$

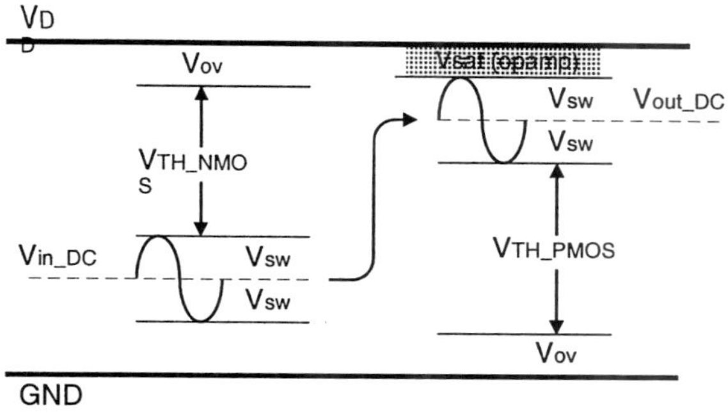

Figure 10.8 Voltage drop across the switches

$$V_{DDp} > V_{sat} + 2V_{SW} + V_{THP}\left(V_{out\_DC} - V_{SW}\right) + V_{OV} \tag{10.10}$$

where V_{SW} is the peak value of the single-ended signal amplitude, V_{THN} and V_{THP} are the maximum values of the NMOS and PMOS threshold voltages obtained for the body effect evaluated at the maximum level of the signal swing. V_{sat} is the minimum distance from V_{DD} for the opamp output node before it enters in the saturation region. As previously anticipated, V_{DD} depends on the signal swing directly and indirectly, through the dependence of the V_{TH} on the signal value. These aspects can be studied by plotting the available output swing vs. the power supply V_{DD}. In the proposed design the following values have been used: V_{OV} = 50 mV, V_{sat} = 80 mV. The minimum supply voltage required for a given differential output swing is shown in Figure 10.9. The line with the stars indicates the technological typical case for both NMOS and PMOS, while all the other lines indicate all the possible combinations of NMOS and PMOS worst case situations for the 0.5μm CMOS technology used. For the typical case 600 mV_{pp} output swing is possible at 1.2 V supply voltage.

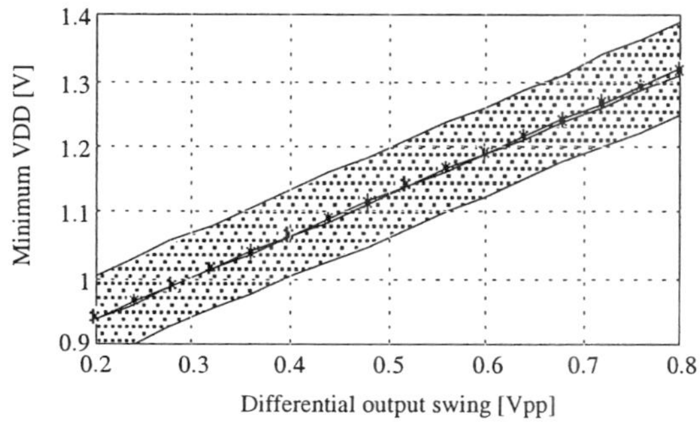

Figure 10.9 Minimum supply voltage versus output swing

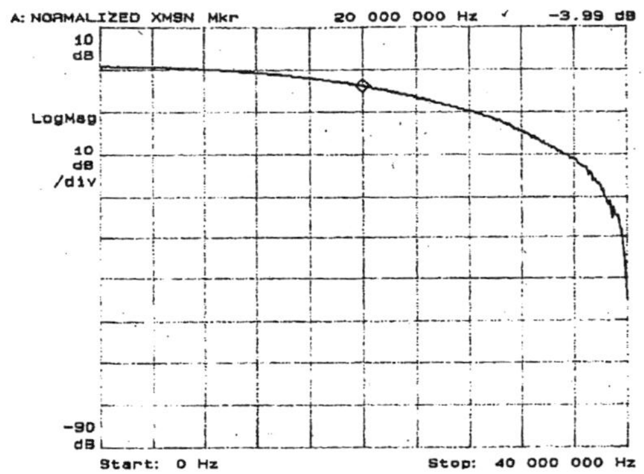

Figure 10.10 Sample & Hold frequency response

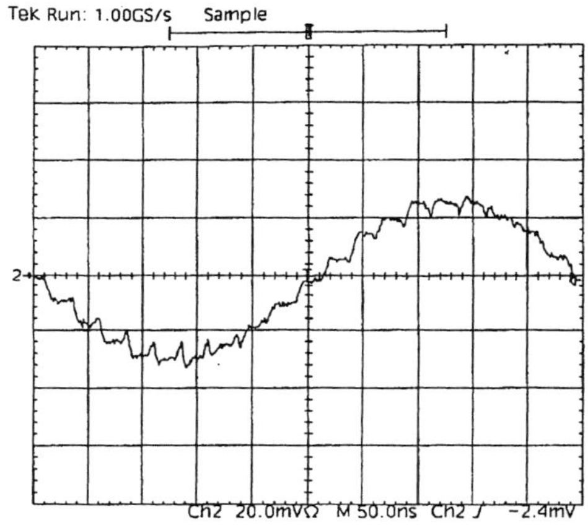

Figure 10.11 Output differential waveform

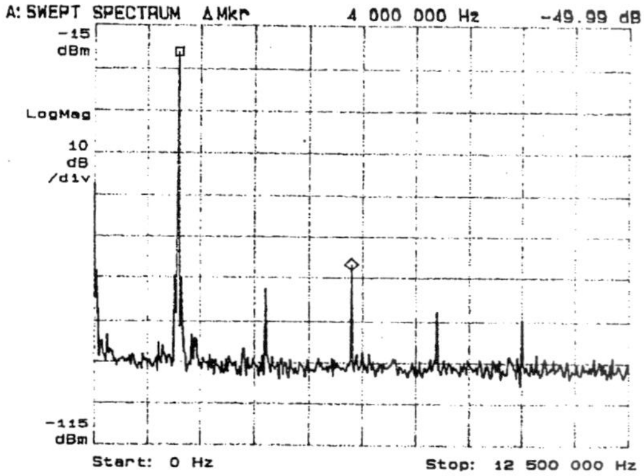

Figure 10.12 Output spectrum

The S& H prototype has been realized in a standard 0.5μm CMOS technology (V_{THN} = 0.60 V, V_{THP} = 0.66 V) with a chip area of 0.15 mm^2. The sampling capacitor value is 1 pF; the switch sizes are 30 μm / 0.5 μm for the NMOS and 50 μm / 0.5 μm for the PMOS section. The S& H circuit has been characterized at 40 MHz sampling frequency. Figure 10.10 shows the S& H frequency response, in agreement with the expected sin(x)/x behavior. Figure 10.11 shows the differential output waveform for an input signal of 600 mV_{pp} amplitude at 2 MHz. The output signal spectrum is shown in Figure 10.12, where a -50dB THD is observed. Figure 10.12 shows the THD vs. the input signal amplitude for 1MHz and 2 MHz input signal frequency. Up to 600 mV_{pp} signal amplitude the S& H exhibits a THD lower than –50 dB, i.e. better than 8-bit accuracy. For comparison the THD obtained with a single-ended (SE) signal is shown as well, showing clearly the improvement obtained by using the PD solution. Figure 10.14 shows the THD as a function of frequency for a 600 mV_{pp} input signal. It can be seen that 8-bit accuracy is achieved up to 3 MHz.

The circuit operates on a supply voltage as low as 1 V consuming 500μW. The maximum sampling frequency at 1 V is 20 MHz and the measured THD for a 1MHz input signal is shown in Figure 10.13, giving 8-bit accuracy for a signal of 350 mV_{pp}. Although pseudo-differential, the circuit exhibits a CMR higher than 41 dB up to 5 MHz input frequency as shown in Figure 10.15. Table 10.2 summarizes the performance of this solution. The main advantage is the high sampling frequency at the cost of a reduced and technology-dependent signal swing.

10.2 IMPROVED SOLUTIONS FOR LOW-VOLTAGE SC CIRCUITS

As shown in Figure 10.3, the switches and opamp sections of a SC circuit may operate at different supply voltages. Three different approaches have then been presented in literature for low supply voltage operation:

- on-chip voltage multiplier for all circuits

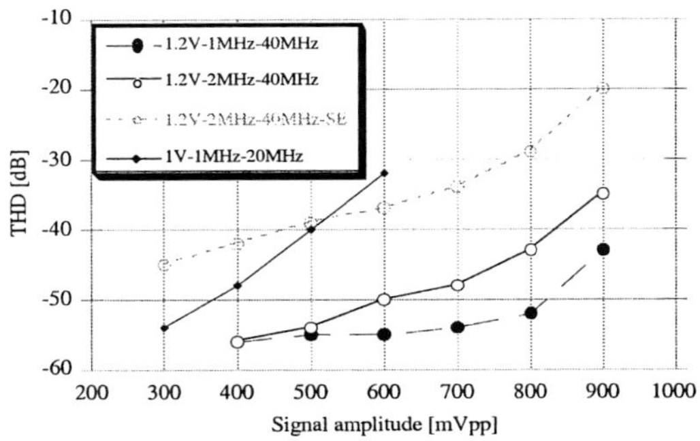

Figure 10.13 THD vs. input signal amplitude

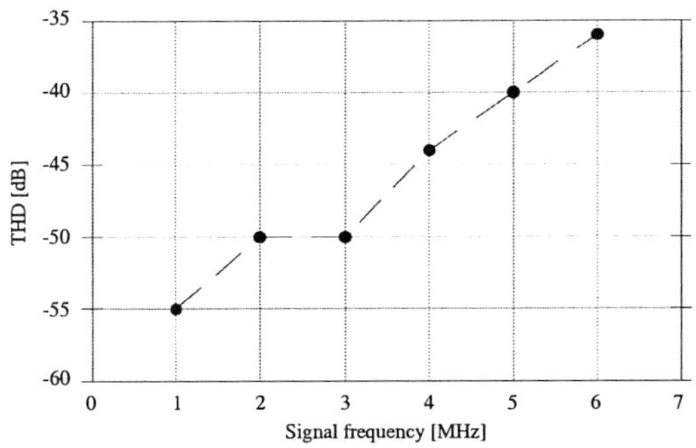

Figure 10.14 THD vs. input signal frequency

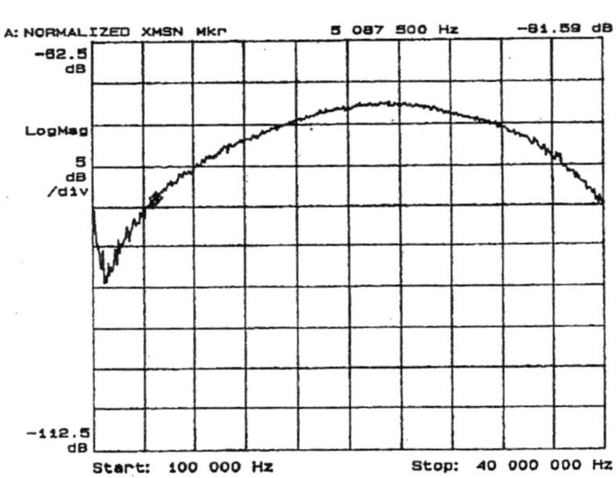

Figure 10.15 Common-mode Rejection (-20 dBm source, 20 dB probe attenuation)

Table 10.2 Sample & Hold performance

Technology	0.5 μm CMOS
Chip area	0.15 mm^2
Supply voltage	1.2 1.0 V
Power consumption	1.2 1.0 mW
Sampling frequency	40 20 MHz
THD 1% (@ 2MHz)	825 500 mV$_{pp}$
CMRR (f < 5 MHz)	> 41 dB

- on-chip clock voltage multiplier
- switched-opamp techniques

Only the third approach uses one a single low-voltage supply for all the blocks.

The different approaches will now be presented in more detail. They use different supply voltages for the switch and the opamp sections. Each solution has its own specific problems and limitations. For all of them, however, it is mandatory for the SC filter to deliver rail-to-rail output swing. This condition can be met by setting the opamp output DC voltage to $^1\!/_2 V_{DD}$.

10.2.1 On-chip supply voltage multiplier

If the SC designer wants to re-use his "high-voltage" circuits, the only possible design approach is to generate on-chip an auxiliary supply voltage V_{DDmult} powering the complete SC filter. In this way the SC filter is designed using the available analog cells for opamps and switches, without applying any low-voltage modifications and operate from the multiplied supplied voltage (i.e. with $V_{DDswitch} = V_{DDopamp} = V_{DDmult}$). A possible circuit solution for an on-chip voltage multiplier is shown in Figure 10.16.

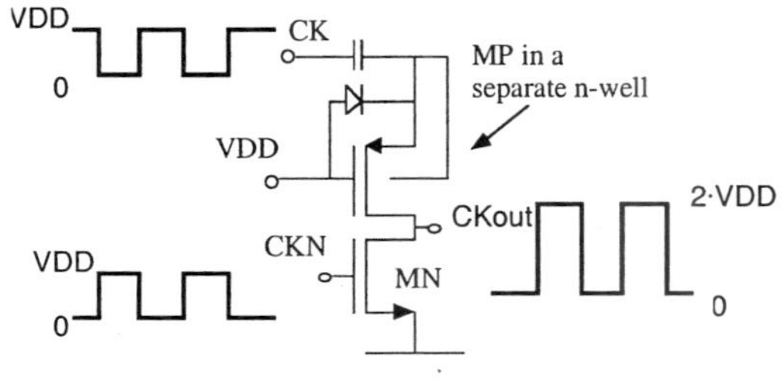

Figure 10.16 A clock voltage multiplier circuit

Correct switch operation is guaranteed by using the higher supply voltage by adopting complementary transmission gates ensuring a minimum conductance for all possible voltages they are connected to (between 0 and V_{DD}). When the opamps are operating at a higher supply voltage, they can deliver a larger signal swing, resulting in a larger dynamic range. E.g., [5] shows a solution where 4 V_{pp} signal amplitude has been achieved at -80dB THD, with a dynamic range larger than 80 dB using a 1.5V supply voltage.

The on-chip supply voltage multiplier suffers from some several limitations:

- technology limited: scaled-down technologies show lower values for the maximum electrical field strength between gate and channel (for gate oxide breakdown) and between drain and source (for hot electron damage); this results in an absolute limit for the value of the multiplied supply voltage
- external capacitor(s) are usually required, implying additional cost and board space
- conversion efficiency of the charge-pump cannot be 100 % thereby limiting the application of this approach in battery operated portable systems

These limitations make this solution the least feasible for future applications and it will not be discussed any further.

10.2.2 On-chip clock voltage multiplier

A second and more feasible alternative to operate low-voltage SC filters is the use of on-chip voltage multiplier circuit powering the clock generation circuits [6-12] that drive the switches, while the opamps operate from the low-supply voltage. In this case, the voltage multiplier has only to drive the capacitive load of the switch gates, while it is not required to supply any DC current to the opamp. No external capacitor is then required and the SC filter is fully integrated. Using this design approach the switches can operate as in standard SC circuit operating at a higher supply voltage. The opamp has to be properly designed in order to operate from the reduced supply voltage. In particular the opamp input DC voltage should be $V_{in\_DC} = 0$ (this will explained later), while the opamp output DC voltage is required to be $V_{out\_DC} = \frac{1}{2}V_{DD}$ to achieve rail-to-rail output swing. These DC levels are not equal and so a voltage level shifting circuit is required. Such a level shift can be efficiently implemented using SC techniques. In this way the operation is possible due to the full functionality of the switches at any input voltage using the multiplied clock supply. In a typical SC integrator Figure 10.3), applying $V_2 = V_{out\_DC}$ and $V_3 = V_{in\_DC}$ gives the proper DC voltage at the opamp input and output nodes.

This design approach, like the previous one, suffers from the technology limitation associated with the gate oxide breakdown. In addition, a potential problem arising from choosing V_{in_DC} equal to ground is the possible charge loss during the transient, as shown in Figure 10.17.

The inverting input node of the opamp is always connected to a PN junction (shown in Figure 10.18) associated with the drain/source of the NMOS switch S_4 in Figure 10.3. This junction is normally reverse biased. Due to negative charge transfer, clock feedthrough and the finite response time of the opamp, negative voltage spikes can occur as shown in Figure 10.17.

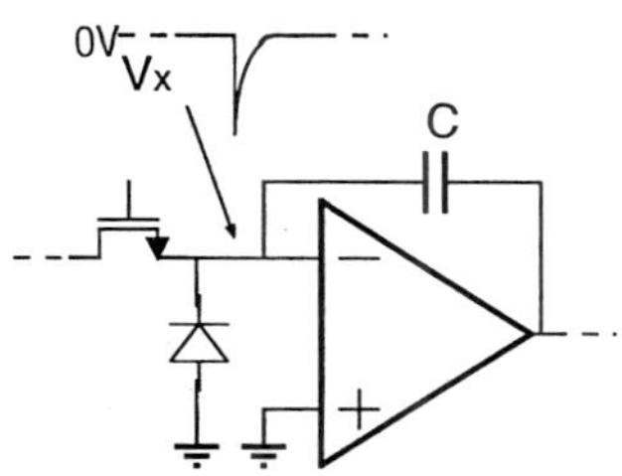

Figure 10.17 Excursion of node X during operation

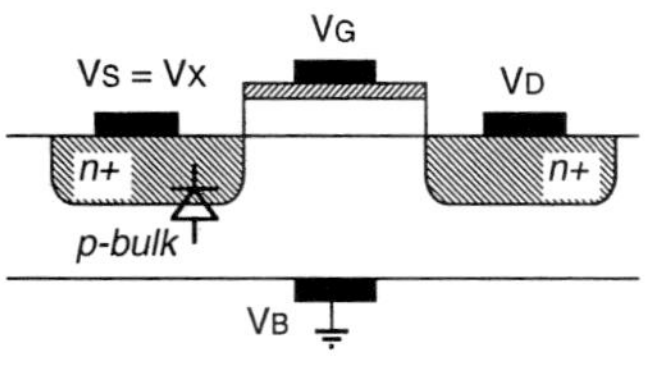

Figure 10.18 NMOS switch structure

These negative spikes can forward bias the bulk diode resulting in charge leakage. E.g., at a temperature of 100 °C, a 500 mV spike of 5 ns duration results in a 0.5 mV voltage error on a 1 pF integration capacitor. The spike amplitude is proportional to the signal level (limited by the opamp supply voltage) and to the clock amplitude (i.e. the clock supply voltage). This problem is therefore difficult to solve for a 5 V SC filter, while it becomes progressively less important as the opamp and clock supply voltage are reduced. At low supply voltage, the contribution of clock feedthrough dominates. Precautions are still required to obtain good results.

Despite the problems mentioned above, this approach is very popular since it allows the filter to operate at high sampling frequency. Using this approach, a 20 Ms/s sampling rate has been reported in a pipelined A/D converter [11]. This design solution can be improved by driving all the switches with a fixed V_{OV}. In this case a constant switch conductance is ensured and this also reduces signal-dependent distortion. It however requires a specific charge-pump circuit for each switch, increasing area, power consumption and noise injection.

10.2.3 Switched-opamp techniques

In order to avoid any kind of voltage multiplier, a third approach called 'Switched-OpAmp' (SOA) technique has been presented in literature [12]. The SC circuits developed with this solution require a dedicated design for both opamp and switches. The basic considerations leading to the SOA technique are the following:

- the optimum condition for the switches driven with a low supply voltage is to be connected either to ground or to V_{DD}. Switch S4 in Figure 10.3 is connected to virtual ground. As a consequence the opamp input DC voltage has to be either ground or V_{DD}

- biasing the opamp input node at ground or V_{DD} also allows reduction of the required opamp supply voltage. On the other hand, biasing the opamp output at $1/2V_{DD}$ allows to achieve rail-to-rail output swing
- proper operation of switch S_{out} connected to the opamp output node is not guaranteed if the supply voltage is reduced below some minimum value. Its functionality has then to be guaranteed in some different way.

The use of the 'Switched-OpAmp' solution satisfies all these requirements. The idea is to use an opamp featuring a tri-state condition. AS a result the critical output switch S_{out} is no longer required and can be eliminated. The critical problem is then moved to the opamp design. Figure 10.19 shows the resulting SOA SC integrator [14-16]: the function of the critical switch S_{out} is replaced by turning on and off the opamp biasing through switch S_a.

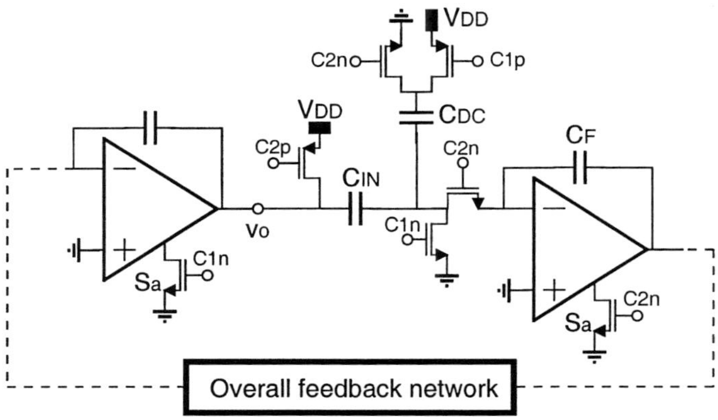

Figure 10.19 Switched-Opamp SC integrator

The opamp input DC voltage is set to ground: this relaxes the opamp supply requirements and guarantees correct operation for S_4 of Figure 10.3. An important characteristic of this scheme is that all the switches are connected either to ground (realizable with a single NMOS device) or to V_{DD} (realizable with a single PMOS device). In this way all the switches are driven with the maximum overdrive, i.e. $V_{DD}-V_{TH}$. The minimum supply voltage required for proper operation of the switches is then given by:

$$V_{DD\,\min} = V_{TH} + V_{OV} \tag{10.11}$$

where V_{TH} is the larger of the two threshold voltages (N–type and P–type). The value of V_{DDmin} is of the same order as the minimum supply voltage required for the operation of digital CMOS circuits.

As previously described, the SOA technique also requires level shifting techniques due to the difference between the opamp input and output DC voltages. This is efficiently implemented in the scheme of Figure 10.19 with the switched-capacitor CDC that gives a fixed charge injection into virtual ground. The charge balance at the opamp input node implies that

$$-C_{IN}V_{out\_DC} - C_{DC}V_{DD} = C_{IN}\left(V_{in\_DC} - V_{DD}\right) + C_{DC}V_{in\_DC} = 0 \quad (10.12)$$

Since V_{in_DC} is set to ground, the opamp output DC voltage V_{out_DC} should be fixed at:

$$V_{out\_DC} = V_{DD}\frac{C_{IN}}{C_{DC}} \quad (10.13)$$

To set V_{out_DC} to ½V_{DD} it is necessary to choose C_{DC}= $2C_{IN}$. This guarantees optimum operation of the circuits shown in Figure 10.19.

This concept can be also illustrated by the equivalent continuous-time scheme of Figure 10.20 A DC current balance at the virtual ground allows having the virtual ground at 0 V and the opamp output node at ½V_{DD}. This scheme however requires a negative supply voltage –V_{DD}. The use of a negative resistor –R_{DC} should allow to use just one supply, as shown in Figure 10.21. This is however impossible for real continuous-time circuits, while for SC circuit a negative impedance can be obtained using proper phasing of the clock driving the switches.

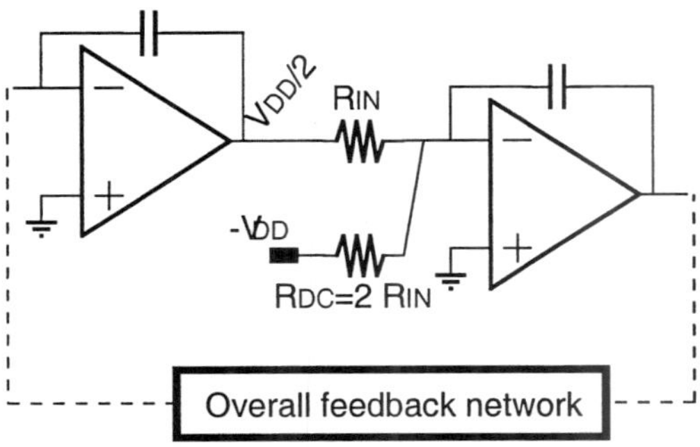

Figure 10.20 Continuous time equivalent circuit with a negative supply

The effect of capacitor C_{DC} is limited to the DC voltage level, while C_{DC} does not no affect the signal transfer function H(z) given by:

$$H\left(z\right) = \frac{C_{IN}}{C_F}\frac{z^{-1/2}}{1 - z^{-1}} \quad (10.14)$$

The main problems of the SOA approach are the following:

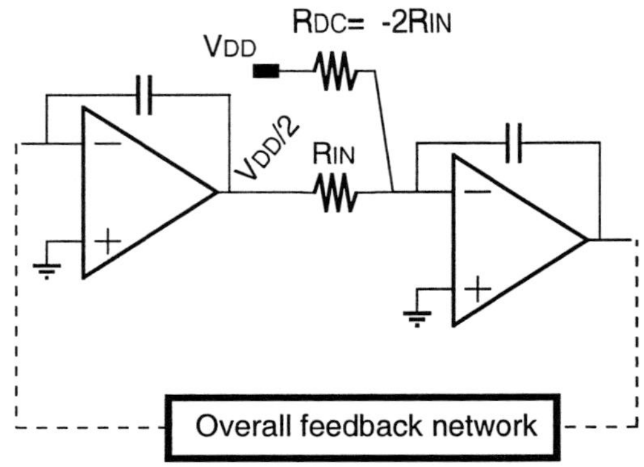

Figure 10.21 Continuous time equivalent circuit with a negative impedance

- only the non-inverting and delayed SC integrator has been up to now proposed in literature. Thus, a sign change must be properly implemented to close the basic two-integrators loop and to build high-order filters. This problem is still open for the single ended structure and the only proven solution is the use of an additional inverting stage [13]
- inaccuracy in the value of CDC gives additional offset at the output node which limits the output swing
- noise and other signals present on the power supply are injected into the signal path

A fully differential structure can solve all of the above problems since

- a fully-differential architecture provides both signal polarities at each node, which allows to build high order structures without any extra elements (e.g., inverting stage);
- any disturbance (offset or noise) injected by C_{DC} results in a common mode signal, which is largely rejected by the fully-differential operation.

Despite these advantages, fully–differential structures have a drawback: they require a common-mode feedback (CMFB) circuit, which becomes critical at low supply voltage as discussed in the following sections. In addition to this, the SOA design approach still suffers for the following open problems:

- SOA structures require an opamp that is turned on and off. The opamp turn-on time limits the maximum sampling frequency
- charge loss at the virtual ground node due to spikes is still possible. A SOA structure is less vulnerable to charge loss than the clock multiplier architecture, since the clock amplitude is lower

- the output signal of a SOA structure is available only during one clock phase, because during the other clock phases the output is set to zero. If the output signal is read as a continuous-time waveform the zero-output phase has two effects: a gain loss of a factor 2, and increased distortion. The distortion is due to the large output steps resulting in slew-rate-limited signal transients and glitches. When the SOA integrator precedes a sampled-data system (like an ADC) the SOA output signal can be sampled during one clock phase and the problem is no longer present

- the input coupling switch has to cope with the entire voltage swing and is therefore still critical: only AC coupling through a capacitor appears a good solution, up to now.

10.3 CIRCUIT LEVEL CONSIDERATIONS

In this section the circuit design of the basic blocks necessary to build low-voltage SC filters is discussed. All the blocks to be designed refer to the clock multiplier and to the SOA approach, since for the first approach the cells are designed in a standard way.

10.3.1 Opamp design

The MOS transistor output impedance of scaled-down technologies decreases and, as a consequence, also the achievable gain-per-stage decreases. In addition at low supply voltage, stacked configurations (like a cascode) are not possible.

In this situation a sufficiently large opamp gain can be achieved by adopting multistage structures that, for stability reasons, tend to have relatively small bandwidth compared to single stage structures. Moreover, these structures suffer from increased turn-on time when used in a SOA architecture.

On the other hand they enable, to the first order, the separate optimization of the different stages. In this way it is possible to operate with V_{in_DC} set to ground and V_{out_DC} set to $^1/_2V_{DD}$.

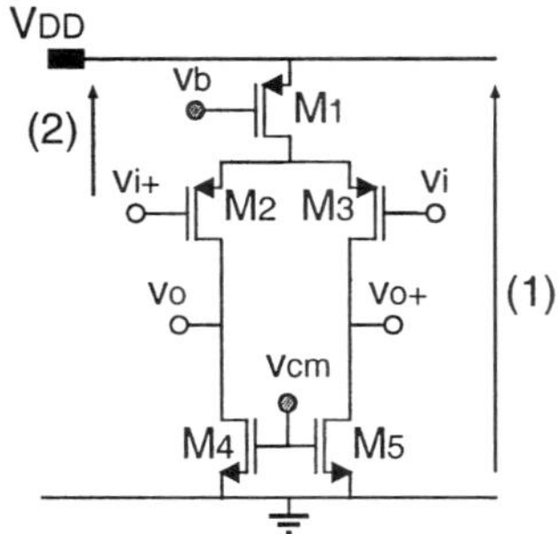

Figure 10.22 Differential input stage

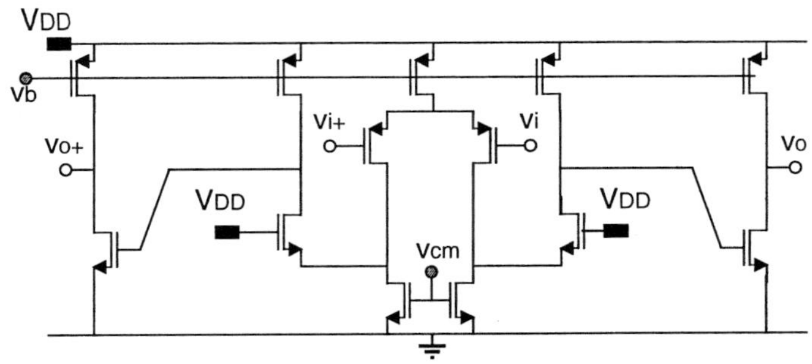

Figure 10.23 Complete two-stage amplifier

Regarding the input stage, the best solution to reduce noise coupling and offset appears to be the differential structure. The simplest differential input stage is shown in Figure 10.22 employing PMOS input transistors with NMOS loads. NMOS input switches have to be used. The minimum supply voltage for correct operation (V_{DDmin}) is given by:

$$V_{DD\,\min} = \max\{3V_{ov} + V_{outpp}, V_{in\_DC} + V_{THp} + 2V_{ov}\} \tag{10.15}$$

The first condition is forced by condition that the three stacked devices M_1, M_3 and M_5 have to operate in the saturation region ($V_{DS} > V_{OV}$), assuming an output swing V_{outpp}. The second condition is forced by the sum of V_{GS} drop across the input device plus the V_{DS} of the PMOS current source. In the case where $V_{TH} > V_{OV} + V_{outpp}$ and $V_{in\_DC} = 0$, the theoretical minimum supply voltage V_{DDmin} obtained is

$$V_{DD\,\min} = V_{TH} + 2V_{ov} \tag{10.16}$$

Both the above conditions can be satisfied by the complete differential amplifier shown in Figure 10.23. The folding structure has been used to ensure that the drains of M_4 and M_5 are biased slightly higher than one Vov above ground. This keeps all the devices (with the possible exception of the cascode device itself) in the saturation region all the time. This is due the small voltage swing present at the source of the cascode devices and the gate of the NMOS device in the output stage. The minimum supply voltage V_{DDmin} for this circuit is still given by $V_{TH} + 2V_{OV}$, assuming that V_{TH} is the largest of the NMOS and the PMOS threshold voltages.

The structure of Figure 10.23 corresponds to a fully differential amplifier. When designing a single–ended structure, a current mirror must be implemented to realize the differential-to-single ended conversion. The use of a MOS connected as a diode in the signal path - typical of classical current mirror topologies - is not possible since it would increase the minimum supply voltage. Figure 10.24 shows a possible circuit capable to operate at the minimum supply voltage $V_{DDmin} = V_{TH} + 2V_{OV}$: a low-voltage current mirror is used (indicated with dashed line).

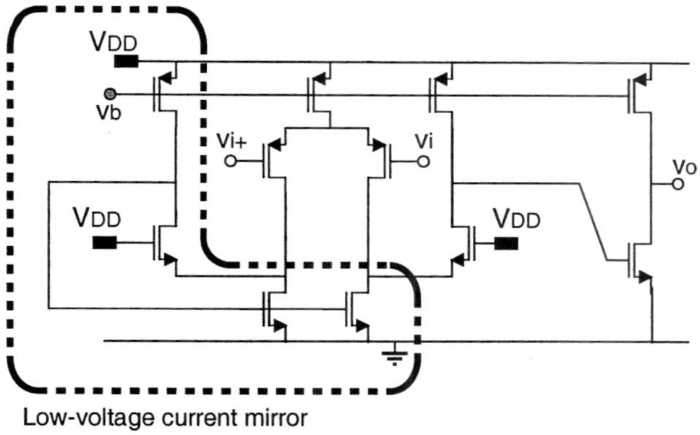

Figure 10.24 Figure 10.24 - Single-ended opamp with low-voltage current mirror

As an alternative to the above class-A input stage, a class-AB input stage has been proposed in literature [18]. This stage is based upon the input differential pair shown in Figure 10.25. Applying a differential signal to this circuit gives two equal output currents (I_{out}). The complete class-AB opamp schematic is shown in Figure 10.26. It comprises the low-voltage current mirror shown in Figure 10.24, and is able to operate with the $V_{DDmin} = V_{TH} + 2V_{OV}$. This stage can be used when a large capacitive load is present and power consumption must be reduced. However, as is usually the case in a class-AB stage, its structure is relatively complicated which results in increased noise and offset.

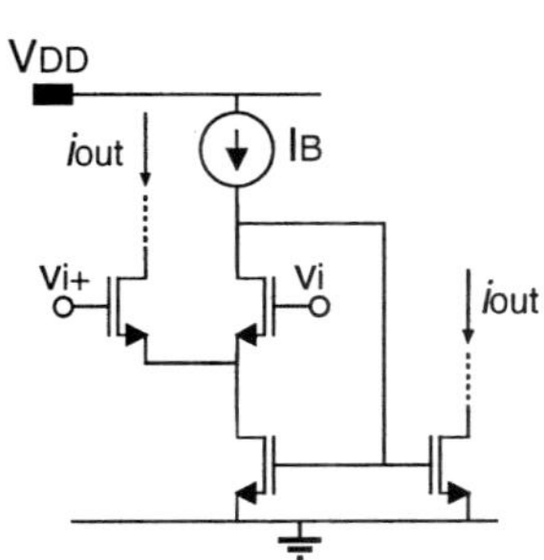

Figure 10.25 Class-AB input pair

A key feature parameter in SOA architectures is the turn-on time of the opamp. This parameter limits the maximum sampling frequency and should be minimized. The main limitations are found to be:

- turn-on time of current sources
- re-charging of the compensation capacitance

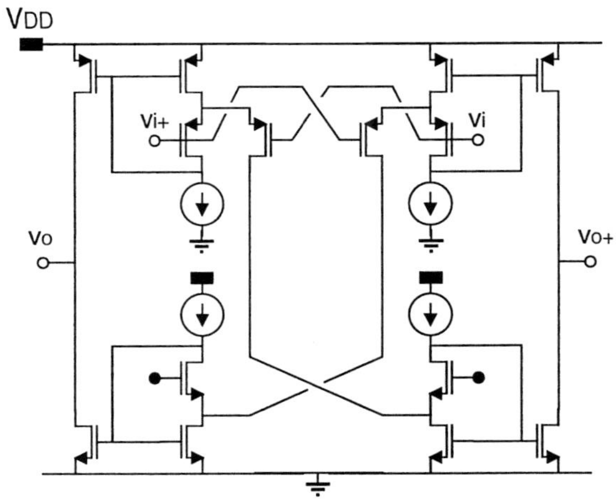

Figure 10.26 Class-AB opamp

In early SOA realizations [12], the opamp was fully turned off as shown in Figure 10.27, i.e. by shorting to V_{DD} the gate of the PMOS current sources.

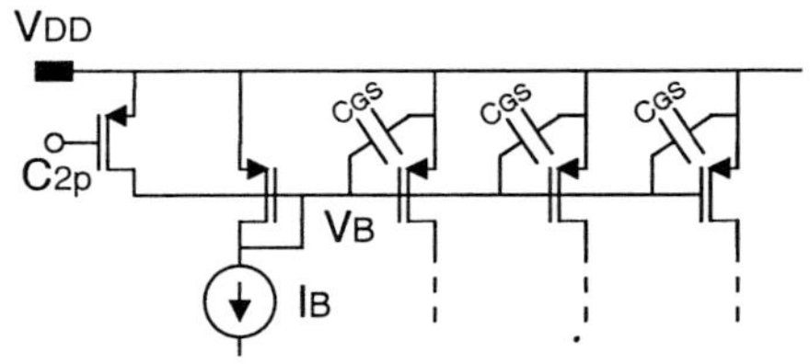

Figure 10.27 Current source turn-off scheme

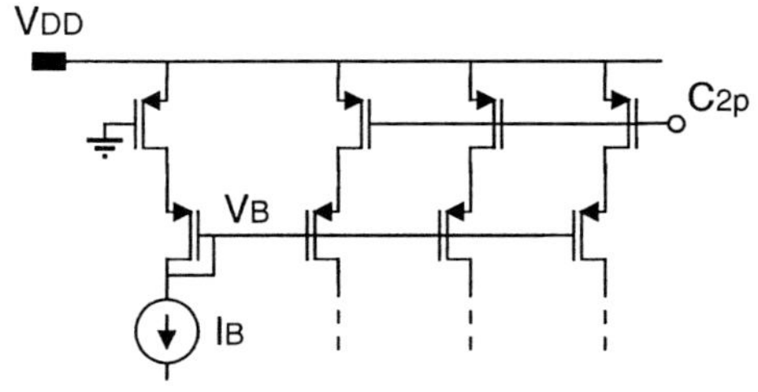

Figure 10.28 Improved current source turn-off scheme

This approach results in a long turn-on time since all the capacitances of the gates connected to V_B must be charged and discharged with the constant bias current I_B. A possible improvement is given in Figure 10.28. In this scheme the current sources are turned-off through MOS series switches. Only the current paths are opened, while the gate capacitances are kept charged and they do not require extra time to be recharged.

The opamp turn-on time can also be reduced by turning off only specific sections of the full opamp, maintaining the rest of the structure active and ready to operate at the turn-on instant. For instance, in the opamp shown in Figure 10.29 [15-16] only the second stage is turned off, while the first stage is kept active even during the off phase. Table 10.3 shows the opamp performance.

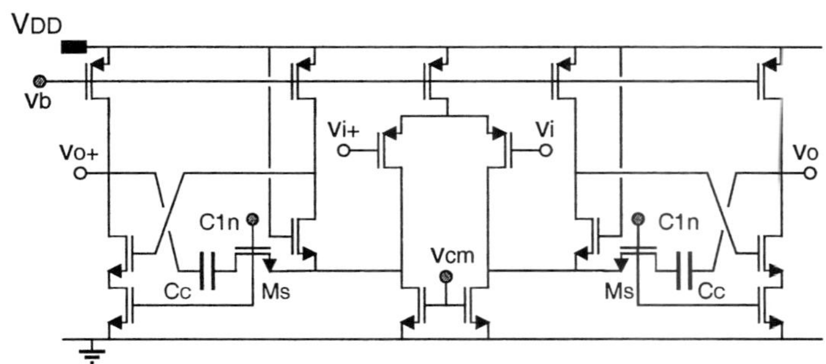

Figure 10.29 Switched-OpAmp structure

Table 10.3 Opamp performance

Technology	0.5 μm CMOS
Supply voltage	1 V
Power consumption	80 μW
Open loop gain	> 75 dB
Unity gain bandwidth	30 MHz
Switching time	< 100 ns

The second reason for increased turn-time is the re-charging of the compensation capacitance C_C. Actually the compensation capacitance C_C voltage changes considerably during the off-phase when the output voltage is tied to V_{DD}. Capacitance C_C must then be re-charged during the next on-phase. Isolating Cc during the off-phase can reduce this effect. Figure 10.29 shows a possible realization [15-16], where C_C is connected to the source of M6, while switch MS is inserted in series with it. During the off-phase, these switches disconnect C_C that preserves its charge. Switch M_S can properly operate at the same V_{DDmin} required by the opamp by having the source voltage of M6 one V_{OV} voltage higher than ground.

10.3.2 Common-mode feedback (CMFB) design

For a fully differential topology an additional CMFB circuit is required to control the output DC voltage. Both continuous–time and sampled–domain (dynamic) approaches can be used.

In a continuous–time solution, the inputs of the CMFB circuit are DC coupled to the opamp output nodes whose DC voltage is equal to $\frac{1}{2}V_{DD}$. At low supply voltage $\frac{1}{2}V_{DD}$ is lower than V_{TH}. No MOS gate can therefore be directly connected to the opamp output node. Figure 10.30 shows a circuit using a passive level shift circuit to circumvent this limitation. In this case a trade–off exists between the amount of signal present at the CMFB circuit input and the amount of level shift. In addition the resistive level shift decreases output stage gain. Finally, this scheme requires a minimum supply voltage

$V_{DDmin} = V_{TH} + 3V_{OV}$, i.e. larger than the one needed by the rest of the filter. This means that the CMFB circuit has now become the limiting factor for the V_{DD} reduction.

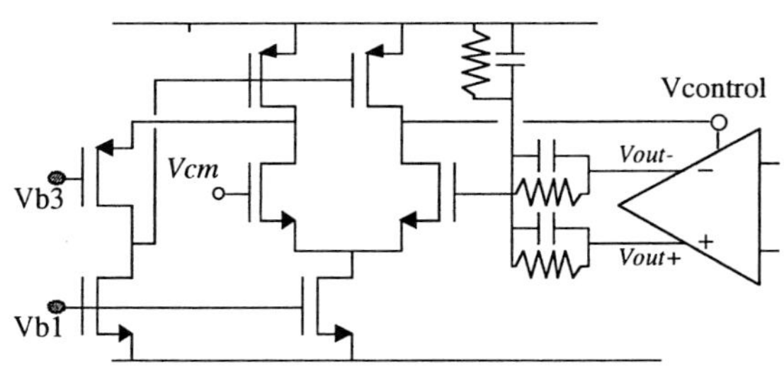

Figure 10.30 Low-voltage continuous-time CMFB circuit

In a dynamic CMFB circuit the key problem is to properly turn on and off the switches. This could be easily done using a voltage multiplier and for this case no further discussion is necessary. On the other hand problems arise for the SOA approach. A possible solution for the dynamic CMFB using a single-ended switched-opamp integrator is shown in Figure 10.31.

The opamp used in the integrator has a PMOS input stage with an input common-mode voltage equal to ground. Furthermore all the switches are connected to either ground or V_{DD}. The CMFB steady state occurs when the following condition is met:

$$C_P\left(V_{DD} - V_{out+}\right) + C_M\left(V_{DD} - V_{out-}\right) + C_{CM}\left(0 - V_{DD}\right) = 0 \qquad (10.17)$$

By choosing $C_{CM} = C_M = C_P$, the opamp common mode output voltage is set to $\frac{1}{2}V_{DD}$. The CMFB integrator adds a pole in the CM loop. To improve CM loop stability and bandwidth, continuous-time feed-forward capacitors C_{FF} are included in the CM loop to create a zero. They are disconnected during the differential opamp reset phase and stay charged. Notice that this solution requires a V_{DDmin} equal to $V_{TH} + 2V_{OV}$, exactly the same as the V_{DDmin} required by the rest of the filter. Another

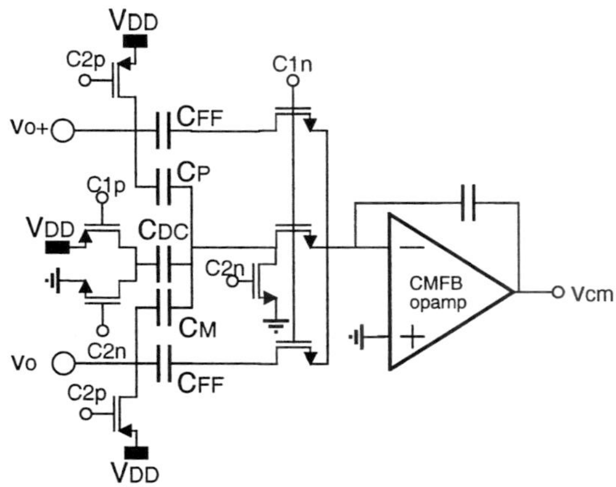

Figure 10.31 Low-voltage SOA sampled-data CMFB circuit

solution for a low-voltage CMFB circuit is shown in Figure 10.32. It has no opamp in the loop and can operate at higher frequency.

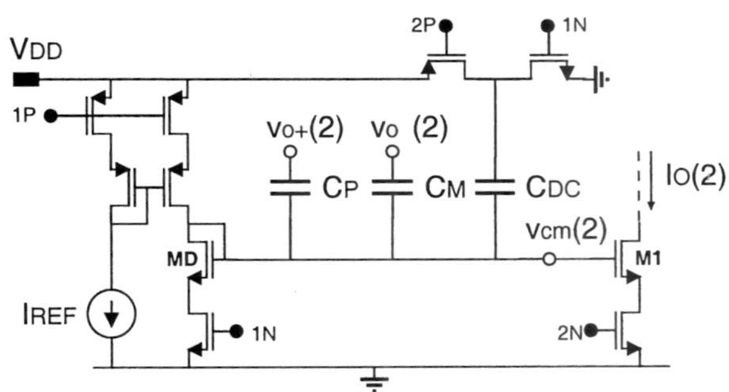

Figure 10.32 Proposed low-voltage CMFB circuit

The circuit operates as follows. The main differential opamp is assumed to operate during phase 2, while during phase 1 it is turned off and its output nodes are connected to V_{DD}. During phase 1 capacitor C_{DC} is connected between ground and node $V_{CM}(1)$ and thus charged to $V_{CM}(1)$; V_{CM} assumes the ideal value obtained with a replica bias branch including the diode-connected MOS device M_C in the current mirror active only during phase 1. Capacitors C_P and C_M are both charged to the $V_{CM}(1) - V_{DD}$.

During phase 2 (which corresponds to the active phase of the switched opamp) the current mirror M_D is turned off and the charge conservation law can be applied to node V_{CM}.

By choosing $C_1 = C_2 = C_{DC}$ the output opamp common-mode voltage is set to $^1/_2$ V_{DD}. In addition capacitors C_P and C_M are properly charged in order to operate like a battery between the output nodes and the control voltage V_{CM} that can be used to bias the opamp. For instance the output current Io of transistor M_1 can be used to bias the first stage of the switched-opamp.

Notice that all the switches used in the proposed scheme are connected to either ground (implemented with only NMOS devices) or to V_{DD} (implemented with only PMOS devices). This means that the proposed CMFB scheme can operate from a single supply voltage as low as $V_{TH}+ 2V_{OV}$, which is the same limit found for the main differential opamp.

10.3.3 Reduction of charge loss due to voltage spikes

Setting $V_{in\_DC} = 0$ could make the SOA structure suffer from the possible charge loss during the turn-on transient. In fact, at the moment of turn-on (the feedback loop is not yet active), two voltage steps are applied to the injecting capacitor C_{DC} and C_{IN}, as shown in Figure 10.33.

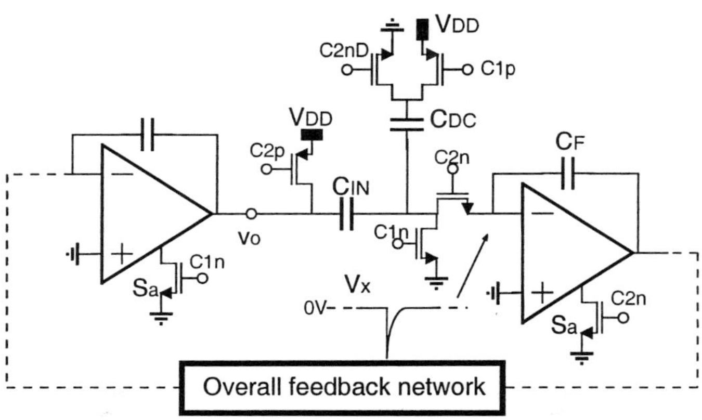

Figure 10.33 Switched-opamp integrator

Capacitor C_{DC} is switched from V_{DD} to ground and it corresponds to be driven with a negative step. Thus it tends to decrease the voltage V_x. On the other hand, capacitor C_{IN} is switched from V_{out} to V_{DD}, i.e. it is driven with a positive step that tends to increase voltage V_x. The result of the two steps gives a spike on V_x. Considering that a negative spike on V_x corresponds to a possible charge loss across the substrate diode, the negative spike can be reduced ensuring that the positive contribution from C_{IN} is injected before the negative contribution from C_{DC}. Appropriate sizing of the switches and the use of a delayed clock phase help to achieve this goal.

Considering the linear model of the SC integrator during the on-phase shown in Figure 10.34 where the two capacitors are connected simultaneously, the excursion of voltage V_x can be described in the time domain by the equation:

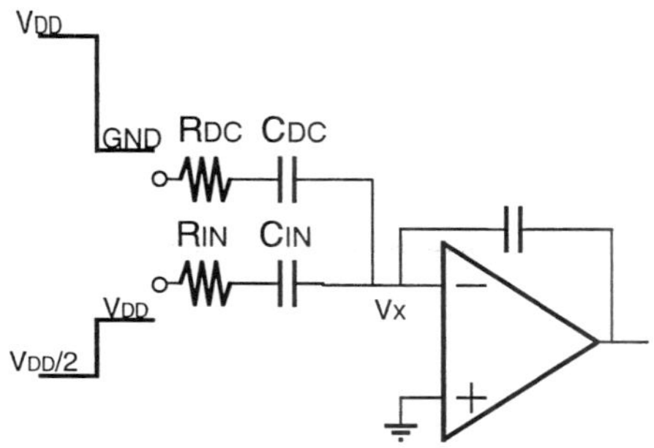

Figure 10.34 Continuous time equivalent of the integrator after turn on edge

$$V_x(t) = V_{DD}\frac{2R_{IN} - R_{DC}}{2(R_{IN} + R_{DC})}\exp\left(-\frac{R_{IN} + R_{DC}}{3R_{IN}R_{DC}C_{DC}}t\right) \tag{10.18}$$

where R_{IN} and R_{DC} are the on-resistance of the switches connected to C_{IN} and C_{DC} respectively. Therefore the switch size can be chosen in order to have $2R_{IN} > R_{DC}$, which corresponds to make V_x move in the positive direction. In this case the large size of these switches is not critical for the charge injection because they are not connected to the virtual ground. Figure 10.35 shows an improved version using delayed clock phases, a scheme usually adopted to reduce signal dependent distortion. In fact, in addition to the longer time constant with C_{DC}, this capacitor is also driven by delayed phase 2D.

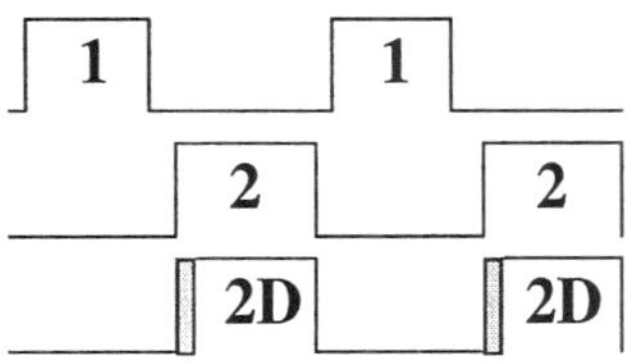

Figure 10.35 Proposed clock phase timing

10.4 A SOA SC FILTER DESIGN EXAMPLE

Using the concept mentioned above, a band-pass switched-opamp filter is reported. The band-pass response solves the problem associated with the input DC coupling that is still open for the SOA approach. The band-pass biquad cell prototype has been realized using a standard 0.5μm CMOS technology (V_{THP} = 0.7 V, V_{THN} = 0.65 V). The band-pass filter center frequency is designed to be one fourth of the sampling frequency and its quality factor equals Q= 7. Figure 10.32 shows the filter structure. It consists of two SOA differential integrators that operate during opposite clock phases. The chip area is about 0.15 mm^2. A unit capacitance of 0.25pF has been used.

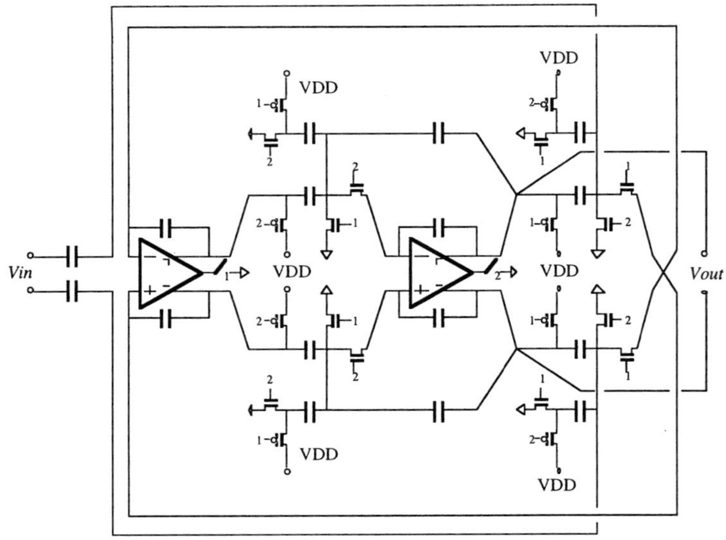

Figure 10.36 Prototype band-pass filter architecture

The prototype has been fully characterized at V_{DD}= 1 V, which is slightly larger than V_{TH}+ $2V_{OV}$
(Vov $\approx$ 100mV). Under this condition, the filter power consumption equals 160μW. The filter frequency response for clock frequencies from 1.8 to 4.2 MHz is shown in Figure 10.37.Some peaking (2 dB) occurs for sampling frequencies up to 3.4 MHz. For higher sampling frequencies, this effect, which is due to incomplete opamp settling (limited in particular by the opamp turn-on time) , becomes more pronounced. At a sampling frequency f_s= 1.8 MHz, the total output noise equals 800 μV_{RMS}.

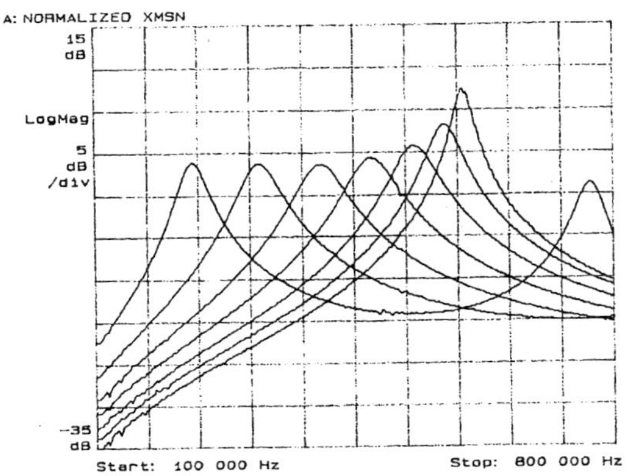

Figure 10.37 Frequency response (V_{DD} = 1V)

The immunity to power supply noise has been measured in terms of power supply rejection ratio (PSRR) in the pass band, and in terms of power supply rejection (PSR) up twice the sampling frequency. Figure 10.38 shows the PSRR in the pass band (300kHz - 600kHz): it is higher than 46dB. For frequencies up to $2f_s$ the PSR is higher than 45dB.

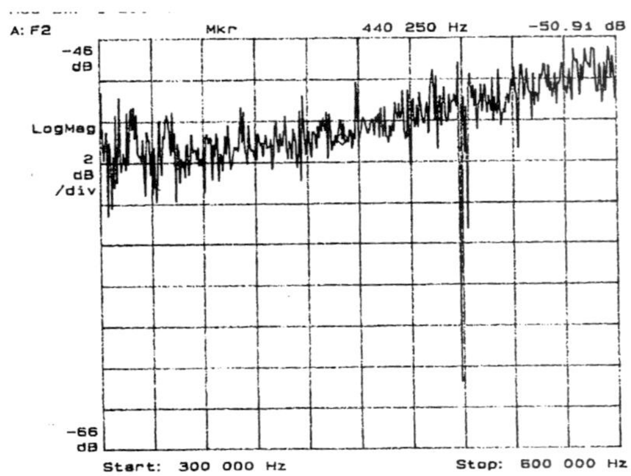

Figure 10.38 PSSR measurement results

These measurement results clearly show that the differential solution guarantees good immunity to power supply noise, even if the noise is directly injected in the signal path. The single-ended and the differential output signals are shown in Figure 10.39-b, respectively. The single-ended waveform illustrates the switched-opamp mechanism that is based on the fact that during one clock phase the output signal is not available and the output node is tied to the power supply. Figure 10.39 also shows the signal settling for signal amplitude as large as 1.6 V_{pp}. In this situation (since the output common-mode is set slightly higher than $\frac{1}{2}V_{DD}$) the peak signal value is within 100 mV from the supply voltage.

The 3rd order intermodulation (IM) equals 3% for two input signals of 500mV_{pp} each, while 1% IM occurs for two input signals of 400mV_{pp}, as shown in Figure 10.41. The dynamic range for 3% IM is about 50dB. The THD is characterized at a frequency f_{in} within the pass band. This results in the folding of the third harmonic into the pass band at (f_s-3f_{in}). The 1% THD corresponds to a 725 mV_{pp} input signal and the 3% THD to a 1.1 V_{pp}input signal, as shown in Figure 10.37.

The above mentioned linearity measurement results are obtained from the complete output waveform (i.e. reading the output waveform in a continuous-time manner) and produce worse results compared to the case where the output signal is sampled only during its "valid" output phase. During the reset phase very large voltage steps (that can be almost equal to V_{DD}) occur at the opamp output and the result is a slew rate limited signal with some glitches.

These effects degrade the linearity of the continuous-time waveform. They, however, have no effect on the sampled version of the output signal, provided that it settles within the clock phase. For instance for the case of two 700mV_{pp} signals, the output

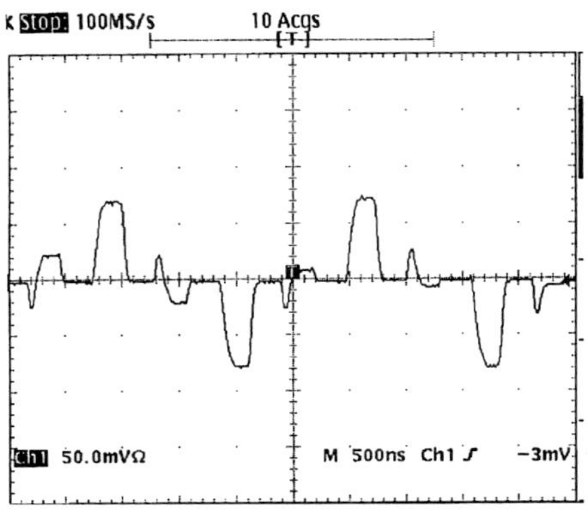

Figure 10.39 Output single-ended waveform with 1.6Vpp differential input signal

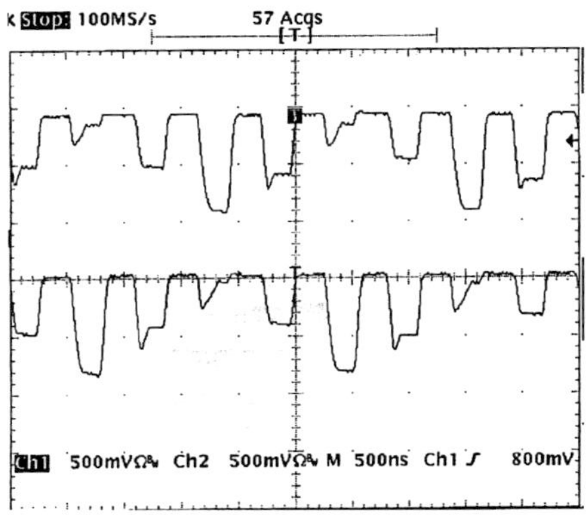

Figure 10.40 Output differential waveform with 1.6Vpp differential input signal

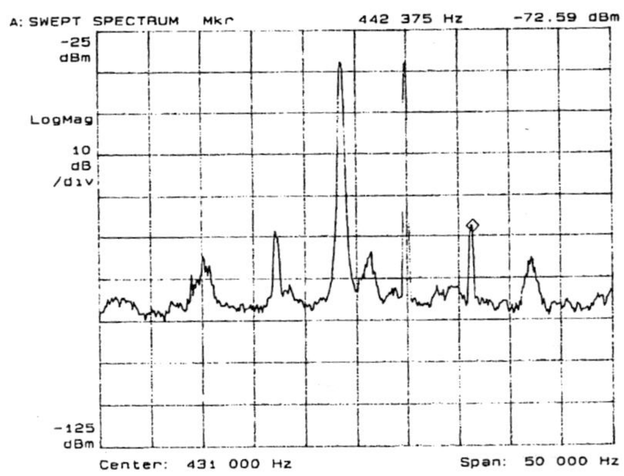

Figure 10.41 1% IM measurement

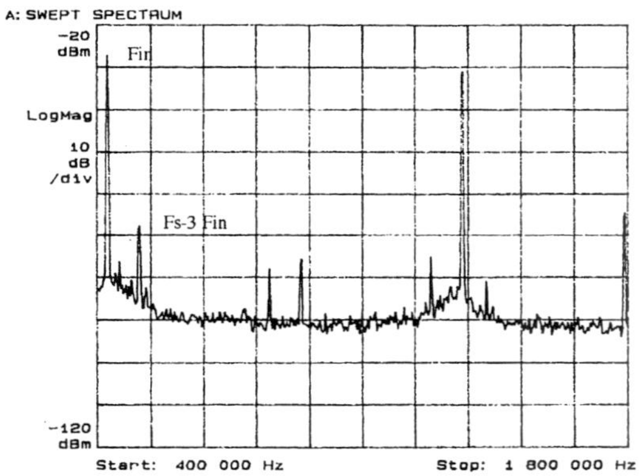

Figure 10.42 1% THD measurement

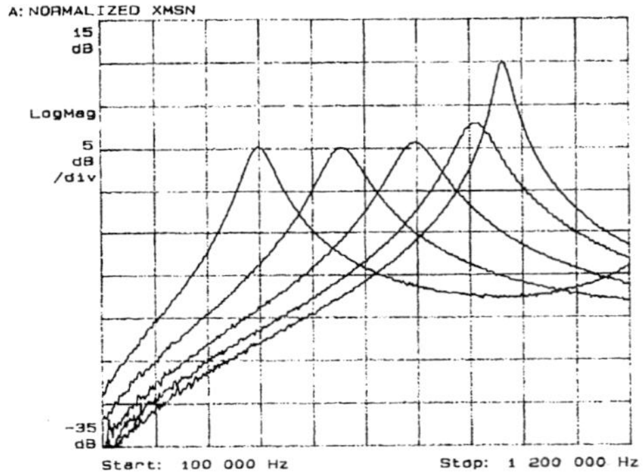

Figure 10.43 Frequency response at $V_{DD} = 0.9V$

signal was sampled and an FFT analysis has been performed. This results in a 1.8% IM, while for the continuous-time waveform the IM is 4% .

Finally the filter is still fully functional at a supply voltage as low as 0.9 V (i.e. leaving only 200mV for $2V_{OV}$) for $f_s < 3.2$ MHz. The frequency response for sampling frequencies between 500 kHz and 3 MHz is shown in Figure 10.43. Table 10.4 summarizes the filter performance.

Table 10.4 Filter performance

Technology	0.5 μm CMOS
Chip area	0.15 mm^2
Supply voltage	1 V
Power Consumption	160 μW
Sample frequency f_s	1.8 MHz
Centre frequency f_0	435 kHz ($\frac{1}{4}f_s$)
Quality factor Q	6.6
Maximum output swing	1.6 V_{pp}
THD 1%	725 mV_{pp}
THD 3%	1.1 V_{pp}
IM3 1%	800 mV_{pp}
IM3 3%	1.0 V_{pp}
Dynamic range (IM3 3%)	50 dB
PSRR	45 dB

10.5 OPEN ISSUES AND FUTURE DEVELOPMENTS

Several circuit implementations have shown the feasibility SC systems operating at low supply voltages. However, there remains room for further improvement. In this section two issues are addressed:

- switched-opamp with a series-switch
- further supply voltage reduction

10.5.1 Active switched-opamp series switch

One of the deficiencies of the SOA technique is the implementation of the series switch connected to the input signal that can have rail-to-rail swing. The input signal cannot be directly connected to a gate since its DC voltage is at $\frac{1}{2}V_{DD}$ being higher than the threshold voltage V_{TH}.

A possible solution consists in connecting the input signal to passive impedance connected to some kind of virtual ground. Figure 10.44 shows the concept of this solution: it consists of a switched buffer, implemented with a switched opamp in inverting configuration.

It can be easily verified that the relation between the in and output DC voltage is:

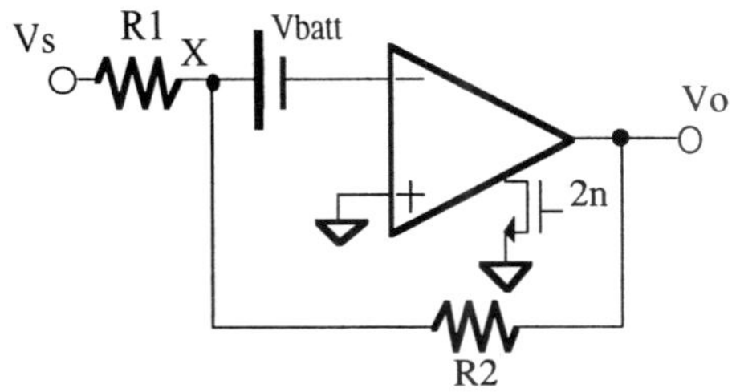

Figure 10.44 Concept of the series switch

$$V_{out\_DC} = \left(1 + \frac{R_2}{R_1}\right) V_{BAT} - \frac{R_2}{R_1} V_{s\_DC} \tag{10.19}$$

The values presented earlier $\{V_{batt} = 0, V_{in\_DC} = 0, V_{out\_DC} = \frac{1}{2}V_{DD}\}$ require an input signal DC level $V_{s\_DC} = -\frac{1}{2}V_{DD}$, which is a value that cannot be delivered by a similar previous stage.

If, on the other hand we choose $V_{batt} = \frac{1}{2}V_{DD}$, node X acts like a virtual ground at $\frac{1}{2}V_{DD}$.

If we choose $R_1 = R_2$, then

$$V_{out\_DC} = V_{DD} - V_{s\_DC} \tag{10.20}$$

The DC level of in and output signal are equal for $V_{s\_DC} = \frac{1}{2}V_{DD}$. This is exactly the value required for rail-to-rail swing. Figure 10.45 shows the complete circuit. In this circuit the level shifting voltage V_{batt} has been implemented with SC techniques: during phase 1 capacitor C_1 is charged to V_{DD}, while capacitor C_2 is uncharged since both terminals are connected to V_{DD}. During phase 2 capacitors C_1 and C_2 are connected in parallel. Choosing $C_1 = C_2$, yields a voltage equal to $\frac{1}{2}V_{DD}$ across both capacitors. During phase 2 no charge is added to C_1 and C_2, acting therefore like a $\frac{1}{2}V_{DD}$ level shifter between the opamp input node (at ground potential) and node X. During phase 2 the opamp is switched on forcing V_X to $\frac{1}{2}V_{DD}$. The feedback network (R_1-R_2) is active and the output voltage $V_o(2)$ follows $V_s(2)$ with gain equal to -1. The value of $V_o(2)$ is sampled on C_s, which is the input capacitor of the following stage in which it injects its charge during the next phase 1.

The circuit was simulated at a supply voltage of 1 V and its performance evaluated in terms of THD for an input signal of 87.9 kHz and a 1 MHz clock frequency. Figure 10.46 shows the output waveform for a signal of $1.8V_{pp}$. Figure 10.47 shows the harmonic distortion as a function of differential input signal amplitude. At 1.8 V_{pp} THD is lower than –60 dB (0.1 %).

10.5.2 Opamp supply voltage reduction

The possibilities for supply voltage reduction are considered in this section. In section 3 the minimum opamp supply voltage has been shown to be $V_{TH} + 2V_{OV}$.

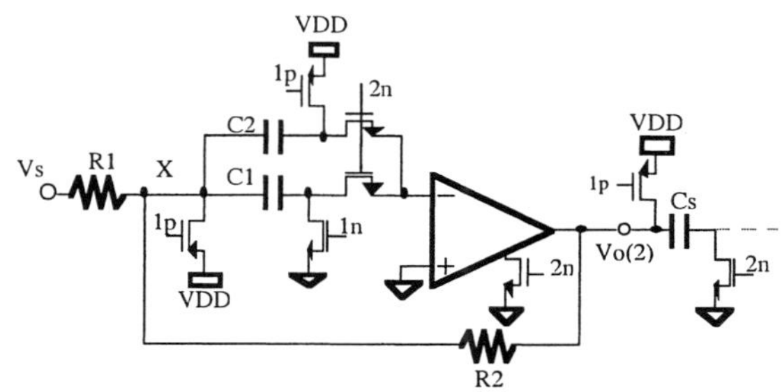

Figure 10.45 Complete switched-opamp buffer

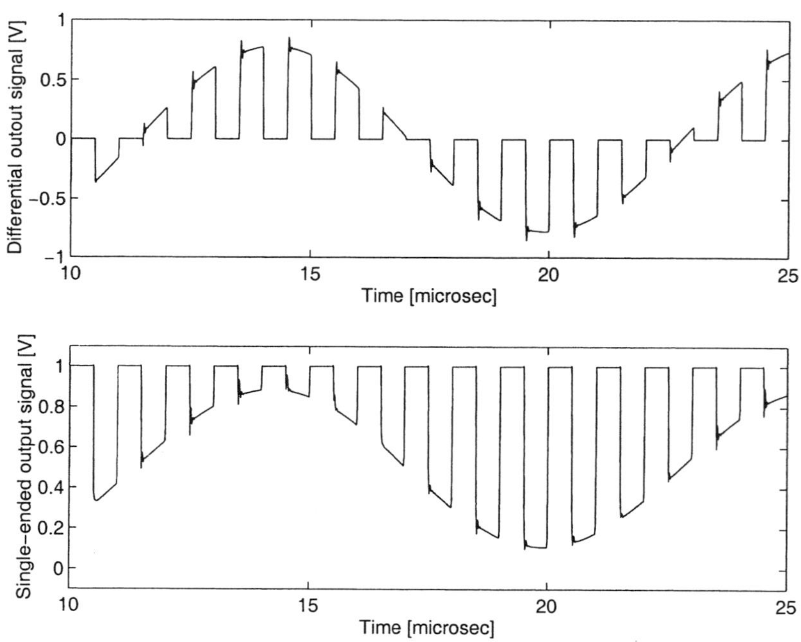

Figure 10.46 single-ended signal)

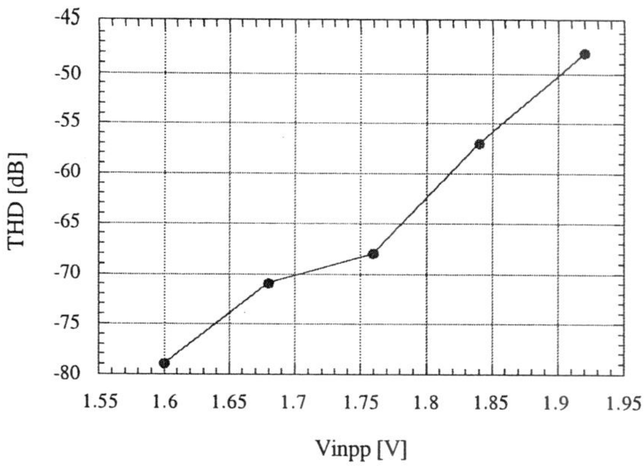

Figure 10.47 THD vs. differential signal amplitude

This value can be lowered when more specific structures are used, with consequences at both system and circuit level.

For the input stage, a pseudo-differential structure as shown in Figure 10.48 can be used [19]. The current source M_1 (Figure 10.22) has been eliminated. This reduces the minimum supply voltage to $V_{TH}+ V_{OV}$. This approach, however, reduces both CMRR and PSRR.

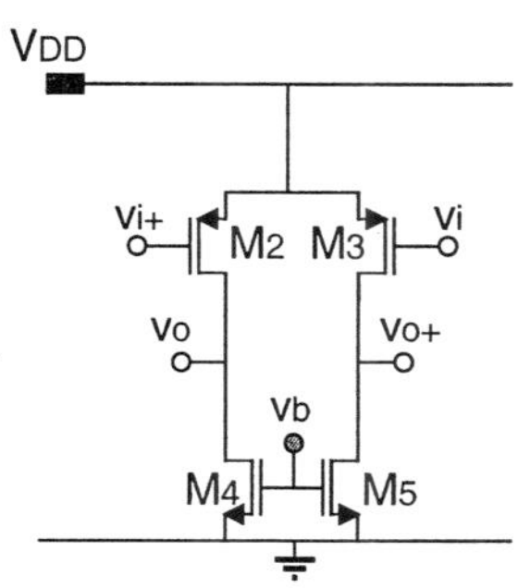

Figure 10.48 Pseudo-differential input pair

For the output stage, in literature a SC-coupling of the stages has been proposed, as shown in Figure 10.49. It consists of precharging a capacitor to be used as a level shifter between the input and the output stages. With this approach the input voltage of the output stage can be set to V_{DD}. This reduces the minimum supply voltage for the output stage to $V_{TH}+ V_{OV}$.

From these considerations, a SC filter operating from a $V_{DDmin}= V_{TH}+ V_{OV}$ is possible. The opamp supply voltage limit has now become equal to the supply voltage of the digital circuits.

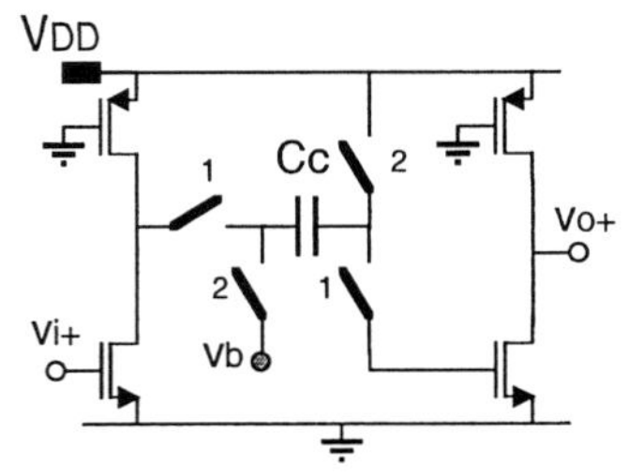

Figure 10.49 Complete two-stage single-ended opamp with ac-coupling

The use of a pseudo-differential structure is also advantageous since it does not require common-mode feedback The pseudo-differential opamp has the same gain for both differential and common–mode input signals. Therefore, the overall feedback present at the filter level is active for both differential and common–mode signals. This is equivalent to two single–ended filters operating out of phase.

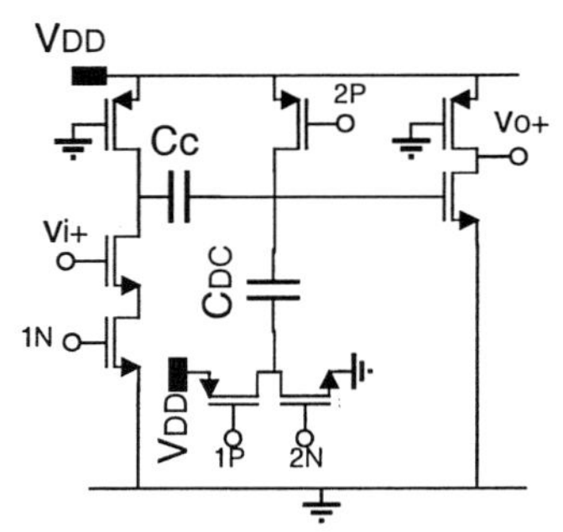

Figure 10.50 An ultra-low-voltage opamp

10.6 CONCLUSIONS

The realization of low-voltage SC circuits operating down to 1 V supply voltage in a standard CMOS technology has been discussed. The possible approaches to achieve low-voltage designs have been described and compared. Table 10.5 summarizes the results.

Between the three possibilities, the switched-opamp approach seems to be the most promising for low-voltage and low-power systems. However, the SOA approach requires the re-design of the entire SC system and still has a number of open issues:

- DC coupling
- Opamp turn-on time
- CMFB implementation

Table 10.5 Comparison of different approaches for the realization of low-voltage SC filters

	Standard design	*Supply voltage multiplier*	*Clock voltage multiplier*	*Switched OpAmp*
Switch supply voltage	V_{DD}	V_{DDmult}	V_{DDmult}	V_{DD}
Opamp supply voltage	V_{DD}	V_{DDmult}	V_{DD}	V_{DD}
New opamp design	Yes	No	No/Yes	Yes
New switch design	No	No	No	Yes
Output swing	—	+	-	-
Voltage spike charge loss	+	+	- -	-
Gate Break-down (VGS limit)	+	-	-	+
Hot electron (VDS limit)	+	-	+	+
Sampling freq. limitation	+	+	+	-
Power consumption	+	–	-	+
External components	No	Yes	No	No
Continuous waveform Gain	1	1	1	1/2
Return-to-zero distortion	+	+	+	- -
Input coupling	+	+	+	- - (?)

In any case it seems possible to realize 1 V SC filters with performance similar to those achieved with circuits operating at higher supply voltages. The price to be paid for similar performance appears to be higher power consumption.

References

[1] A. Matsuzawa, "Low-voltage and low-power circuit design for mixed analog/digital systems in portable equipment", IEEE J. of Solid-State Circuits, Vol. SC-29, no.4, pp. 470-480, April 1994.

[2] R. Castello, F. Montecchi, F. Rezzi, and A. Baschirotto, "Low-voltage analog filter", IEEE Transactions on Circuits and Systems - II - Nov. 1995 - pp. 827-840.

[3] Y. Matsuya, and J. Tamada, "1V power supply low-power consumption A/D conversion technique with swing suppression noise shaping", IEEE Journal of Solid-State Circuits, pp. 1524-1530, Dec. 1994.

[4] T. Adachi, A. Ishikawa, A. Barlow, K. Takasuda, "A 1.4V switched-capacitor filter", IEEE 1990 Custom Integrated Circuits Conference, pp. 8.2.1 - 8.2.4.

[5] G. Nicollini, A. Nagari, P. Confalonieri, and C. Crippa, "A -80dB THD, $4V_{pp}$ switched-capacitor filter for a 1.5V battery-operated systems", IEEE Journal of Solid-State Circuits, Vol. SC-31, no.8, pp. 1214-1219, August 1996.

[6] J. F. Dickson, "On-chip high-voltage generation in MNOS integrated circuits using an improved voltage multiplier", technique, IEEE J. of Solid-State Circuits, Vol. SC-11, no.3, pp.374-378, June 1976.

[7] Y. Nakagone, H. Tanaka, K. Takeuchi, E.Kume, Y. Watanabe, T. Kaga, Y. Kawamoto, F. Murai, R. Izawa, D. Hisamoto, T. Kisu, T. Nishida, E. Takeda, K. Itoh, "An experimental 1.5V 64Mb DRAM", IEEE JSSC, Vol. 26, no. 4, pp. 465-472, April 1992.

[8] F. Krummenacher, H. Pinier, A. Guillaume, "Higher sampling frequency in SC circuits by on-chip clock voltage multiplier", 1983 European Solid-State Circuits Conference (ESSCIRC 83), pp. 123-126.

[9] R. Castello, and L. Tomasini "A 1.5V high-performance switched-capacitor filters in BiCMOS technology", IEEE Journal of Solid-State Circuits, Vol. SC-26, no.7, pp. 930-936, July 1991.

[10] Jieh-Tsorng, Yueh-Huang Chang, Kuen-Long Chang, "1.2V CMOS Switched-Capacitor circuits", IEEE International Solid State Circuits Conference , February 1996 , pp. 388-389.

[11] T.B. Cho, and P.R. Gray, "A 10b, 20Ms/s 35mW pipeline A/D converter", IEEE J. of Solid-State Circuits, Vol. SC-30, no. 3, pp. 166-172, March 1995.

[12] J. Crols, and M. Steyaert, "Switched-Opamp: an approach to realize full CMOS switched-capacitor circuits at very low power supply voltages", IEEE Journal of Solid-State Circuits, Vol. SC-29, no.8, pp. 936-942, August 1994.

[13] V. Peluso, M. Steyaert, and W. Sansen, "A switched-opamp 1.5V-100μW $\Delta\Sigma$ modulator with 12 bits dynamic range", European Solid-State Circuits Conference (ESSCIRC 96) , Sept. 1996, pp. 256-259.

[14] A. Baschirotto, R. Castello, and F. Montecchi, "Design strategy for low-voltage SC circuits", IEE Electronics Letters, 3rd March 1994, Vol. 30, pp. 378-379.

[15] A. Baschirotto, R. Castello, "A 1V 1.8MHz CMOS Switched-opamp SC filter with rail-to-rail output swing", IEEE Journal of Solid-State Circuits, December 1997, pp. 1979-1986.

[16] A. Baschirotto, R. Castello, "A 1V 1.8MHz CMOS Switched-opamp SC filter with rail-to-rail output swing", IEEE International Solid-State Circuits Conference (ISSCC '97) - San Francisco, February 1997, Digest of technical papers, pp. 58-59.

[17] R. Castello, A. Baschirotto, and A. Nagari, "Low voltage, switched capacitance circuit employing switched operational amplifiers with maximized voltage swing", European Patent EP689286A1, Japanese Patent JP8130422A2, US Patent US5745002.

[18] V. Peluso, P. Vancorenland, A. Marques, M. Steyaert, W. Sansen, "A 900mV 40μW Switched-opamp DS modulator with 77dB dynamic range", IEEE International Solid-State Circuits Conference, February 1998, pp. 68-69.

[19] G. Nicollini, F. Moretti, M. Conti, "High-frequency fully-differential filter using operational amplifiers without common-mode feedback ", IEEE Journal of Solid-State Circuits, Vol. SC-24, no.3, June 1989, pp.803-813.

11 CMOS LOW-NOISE AMPLIFIER DESIGN

Francesco Svelto, Enrico Sacchi,
Francesco Gatta, Danilo Manstretta, Rinaldo Castello

Universita' di Pavia, Italy
E-mail: fsvelto@ele.unipv.it

The Low Noise Amplifier (LNA) is usually the first block in a radio receiver as shown in Figure 11.1; it amplifies the antenna signal to a suitable level for the first mixer. The overall noise factor F of a cascade of multiple stages with individual gain G_i and noise factor F_i is given by [1]

$$F = F_1 + \frac{F_2 - 1}{G_1} + \frac{F_3 - 1}{G_1 G_2} + \frac{F_4 - 1}{G_1 G_2 G_2} + \dots \tag{11.1}$$

Clearly the overall receiver noise figure is dominated by the first stage, which apart from a passive band-pass filter that may precede the receiver is constituted by the LNA.

A low LNA noise figure (typically NF $<$ 3 dB) means a minimum degradation of the signal-to-noise ratio of the wanted signal resulting in high sensitivity. The gain of the LNA relaxes the noise figure requirements for the subsequent stages (mixers, IF stage).

The LNA design is very critical because of its placing in the receiver path; what goes wrong in the LNA cannot be compensated for afterwards. Together with the VCO, the LNA is the most challenging to realize in a CMOS technology. It requires,

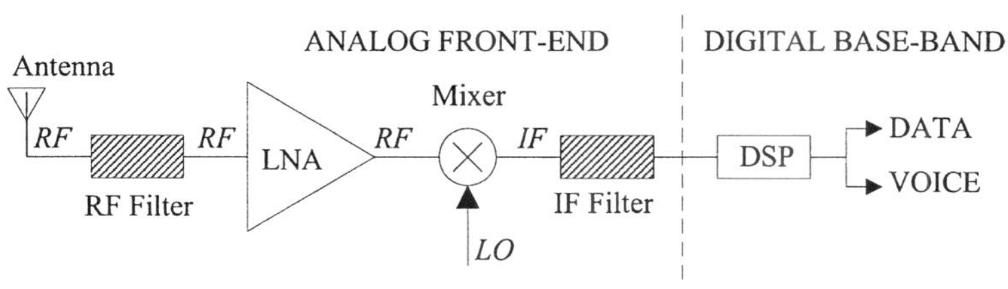

Figure 11.1 A simplified radio receiver block diagram

P. Wambacq et al. (eds.),
Low-Power Design Techniques and CAD Tools for Analog and RF Integrated Circuits, 251–265.

in fact, both high quality passive (integrated inductors) and active (high f_T MOS transistors) devices. Layout and package parasitic inductance and capacitance as well as substrate coupling have to be taken into account during the design phase, to avoid degradation of the receiver performance or even instability.

The design of a LNA operating at 900 MHz not only requires the optimization of the classical parameters like gain, noise figure, and DC current consumption, but also excellent matching of the LNA input circuitry to its source impedance.

In this chapter the importance of good input matching will be highlighted and different input matching networks will be discussed. Noise optimization and the realization of a variable gain amplifier are addressed as well. Finally, a particular LNA based on such a network will be analyzed in more detail.

11.1 LNA INPUT MATCHING

The LNA has to provide an accurate (usually 50 Ω) input impedance to correctly terminate an unknown length of transmission line delivering the antenna signal to its input. Insertion of a passive (helical or SAW) pre-selection filter between antenna and LNA makes this constraint even more important, because the transfer function of this filter is strongly dependent on the load impedance.

In the design of high frequency circuits, it is convenient to think and calculate in terms of incident, transmitted and reflected (scattered) voltage waves rather than port voltages or currents [2]. Scattering (S) parameters are used to characterize both active and passive components. They are usually measured with a network analyzer [3].

Recent CAD tools used for RFIC design like *SpectreRF* and *ADS* not only offer conventional Spice-like analysis modes like *AC, NOISE*, and *TRANSIENT* analysis, but also provide *s parameter*-analysis for small signal and *Harmonic Balancing* modes for large-signal high-frequency analysis.

Since it is signal *power* and not voltage that counts, LNA gain has to be expressed as power gain with matched in-and output. A good input matching is fundamental, because it allows the maximum power transfer from the antenna (RF source) to the LNA. Failing to do so, results in a partial reflection of the power received from the antenna. This results in lower receiver sensitivity.

Given the rather severe minimum detectable signal level specifications of various communication systems, (GSM, for example, requires a *sensitivity* [1] of -102 dBm), this may imply that a receiver design is no longer compatible with an existing standard.

11.1.1 Reflection Coefficient

Figure 11.2 shows a transmission line of characteristic impedance Z_0 terminated by a load impedance Z_L. When $Z_L \neq Z_0$, reflections will occur at the end of the line where the load is located.

The resulting voltage and current on each point of the line can be found by superposition of the incident and reflected voltage and current [2].

[1] Sensitivity: the minimum signal power required at the receiver input yielding sufficiently high SNR or sufficiently low Bit Error Rate (BER) in the particular radio system [2]

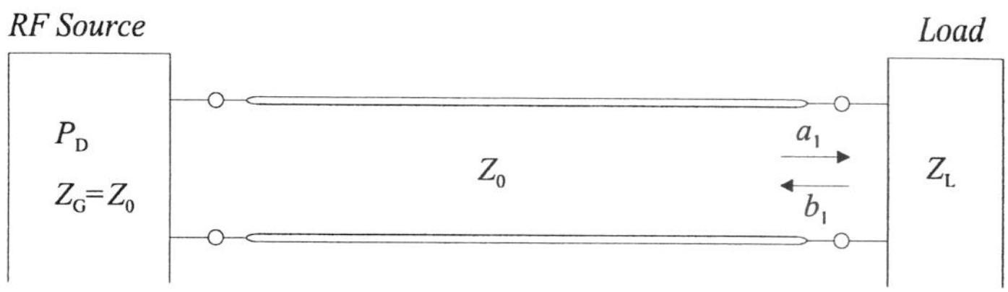

Figure 11.2 Transmission line terminated with impedance Z_L

The ratio of reflected voltage V_R to incident voltage V_I at the load is called the reflection coefficient Γ; it is a complex number that has both magnitude ρ and phase ϕ.

$$\Gamma = \rho \exp(j\phi) = \frac{V_R}{V_I} = \frac{I_R}{I_I} \tag{11.2}$$

It is possible to express the load impedance Z_L in terms of the characteristic line impedance $Z_0$and the (measured) reflection coefficient Γ.

$$Z_L = Z_0 \frac{1+\Gamma}{1-\Gamma} \tag{11.3}$$

11.1.2 Scattering parameters

Two-port networks can be characterized in various ways. Four parameters are sufficient to characterize the relation between input and output voltage and current. At low frequencies, two-ports are usually characterized by their impedance (Z) or admittance (Y) parameters. The use of these parameters requires, however, the availability of true current or voltage sources and open and short circuits very close to the device under test.

At high frequencies (where device size is no longer negligible to the wavelength λ), it becomes virtually impossible to have true voltage and current sources, and open and short circuits close to the device are hard to implement. Measuring voltage and current using probes will disturb the device under test, and possibly make it oscillate or even destroy it [3].

For these reasons at high frequencies, different parameters are required that can be accurately measured and do not require low or high-impedance sources. The so-called scattering (S) parameters comply with these requirements. They are based upon a measurement system of characteristic impedance Z_0, usually 50 or 75 Ω, that measures incident (a_1) and reflected (b_1) voltage waves at the in and output ports using directional couplers that do not have to be placed close to the device under test. Ports 1 and 2 of the device under test are always terminated by the characteristic impedance Z_0.

The (S) parameters have a direct relation with the reflection and transmission coefficients of the two-port and the (S) parameters of a number of cascaded two-ports is rather straightforward to calculate.

Table 11.1 Relation between Z_L, S_{11}, return loss and normalized transmitted power

Z_L [Ω]	$\Gamma = S_{11}$	Return loss [dB]	P_T/P_D	*10 log*(P_T/P_D) [dB]
0	-1	0	0	-∞
16.7	-0.5	6	0.75	-1.25
41	-0.1	20	0.99	-0.46
50.0	0	∞	1	0
61	0.1	20	0.99	-0.46
150	0.5	6	0.75	-1.25
∞	1	0	0	-∞
-j50	-j	0	0	-∞
j50	j	0	0	-∞

For the transmission line shown in Figure 11.2, the reflection coefficient at the load (port 1), which could be the LNA input, is equal to S_{11}. We can now rewrite (11.2) as

$$\Gamma = \rho \exp(j\phi) = S_{11} = \frac{b_1}{a_1} \tag{11.4}$$

The load impedance can be rewritten as

$$Z_L = Z_0 \frac{1+\Gamma}{1-\Gamma} = Z_0 \frac{1+S_{11}}{1-S_{11}} \tag{11.5}$$

It is straightforward to show that $|S_{11}|^2$represents the ratio between the reflected and incident power at port 1:

$$|S_{11}|^2 = \frac{|b_1|^2}{|a_1|^2} = \frac{|b_1|^2}{2} \cdot \frac{2}{|a_1|^2} = \frac{P_{refl}}{P_{inc}} = \frac{P_{refl}}{P_D} \tag{11.6}$$

where $P_D$is the *available source power* which is actually the power of the incident wave P_{inc} traveling from the RF source towards the load.

The so-called return loss (RL) is a positive scalar quantity defined as

$$RL = -20 \log |S_{11}| \tag{11.7}$$

Table 11.1 illustrates some typical values for a system where $Z_0 = R_S = 50$ Ω. Real load impedance values give real values for Γ. Purely reactive loads, result in total reflection, since no energy can be dissipated (only stored) in such a component.

11.1.3 Typical LNA Input Matching Networks

The input impedance of a MOS transistor is capacitive. This impedance has to be matched to the 50 Ω source impedance for maximum power transfer. It is not easy to realize such a loss-less network using on-chip passive components. The LNA noise figure could be dominated by the matching networks noise. This can be interpreted as insertion of a resistive divider in front of the LNA.

Many solutions have been presented in literature. The four most interesting ones are shown in Figure 11.3, in a single-ended version. The performance trade-off between the precision of the input impedance and the noise contribution of each stage is now briefly discussed.

Figure 11.3a shows a resistive implementation; a cheap solution in several ways. This broadband input matching provides a good input matching up to the frequency where the transistor input capacitance starts to become noticeable. This simple solution seriously degrades the LNA noise figure. The matching resistor is a noisy device. Its presence increases the noise power present at the LNA input. Moreover, it dissipates part of the incoming signal, resulting in less input signal power.

Another solution is the use of a common-gate MOS stage. This approach is shown in Figure 11.3.b and seems to be a good solution. The input impedance seen at the source is $1/g_{\rm m}$ and it is sufficient to correctly design the MOS and its DC biasing current to provide a good, broadband input matching. The problem of this stage, however, is that its minimum noise figure is equal to 2.2 dB, assuming Γ = 2/3, and can be higher with the technology scaling (where γ increases) [7]. This is a fundamental limit.

The *shunt and series feedback* topology is illustrated in Figure11.3c; this is another broadband input matching network used in many early discrete-component wide-band amplifiers. It is rarely used in IC implementations, since resistor tolerances are reflected in the input impedance. Moreover this kind of input architecture has a higher power consumption (see [7] and [8]).

The most commonly used input matching network, potentially the best, employs inductive source degeneration [4], [9] as illustrated in Figure11.3d; it is possible to show that for a given power consumption, this input stage has the lowest achievable noise figure.

The input impedance synthesized by this input network can be calculated using the small-signal equivalent circuit of Figure 11.4. The voltage at the input node is given by:

$$v_{in} = v_{GS}[1 + (g_m + sC_{GS})\, sL_S] + sL_G i_{in} \tag{11.8}$$

Substituting the gate to source voltage by

$$v_{GS} = \frac{i_{in}}{sC_{GS}} \tag{11.9}$$

yields for the input impedance

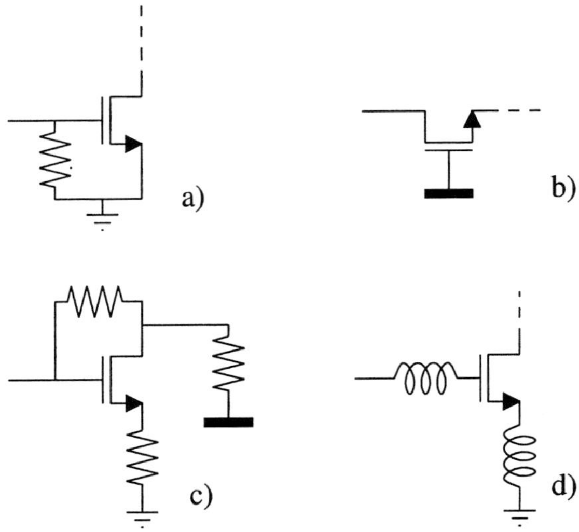

Figure 11.3 Various matching networks [4].

$$Z_{in} \equiv \frac{v_{in}}{i_{in}} = \frac{g_m}{C_{GS}} L_S + \frac{1}{sC_{GS}} + s\left(L_S + L_G\right) = \omega_T L_S + \frac{1}{sC_{GS}} + s\left(L_S + L_G\right) \tag{11.10}$$

A real impedance of 50 Ω can be obtained when the following conditions are met:

$$Im[Z_{in}] = 0 \rightarrow \omega_0 = \sqrt{\frac{1}{C_{GS}\left(L_S + L_G\right)}} \tag{11.11}$$

$$Re[Z_{in}] = 50 \rightarrow \omega_T L_S = 50 \tag{11.12}$$

Then, at the resonant frequency ω_0, given by (11.11) the input impedance equals

$$Z_{in} = \omega_T L_S \tag{11.13}$$

Including also the effect of gate-drain capacitance Eq. (11.6) changes, but a good input matching can still be obtained. At the resonant frequency, the input impedance is

$$Z_{in} \approx \omega_{Teff} L_S \tag{11.14}$$

with the effective transition frequency ω_{Teff} given by:

$$\omega_{Teff} \equiv \frac{\omega_T}{1 + 2\frac{C_{GD}}{C_{GS}}} \tag{11.15}$$

This represents the effective ω_T of the MOS device, degraded by the gate-drain capacitance.

The input stage using inductive source degeneration features several important properties: it is the only one that provides, at a given frequency, the correct real input impedance without the introduction of resistors, which would degrade the noise figure. Moreover, the transconductance of the stage has been decoupled from the transistor transconductance g_m.

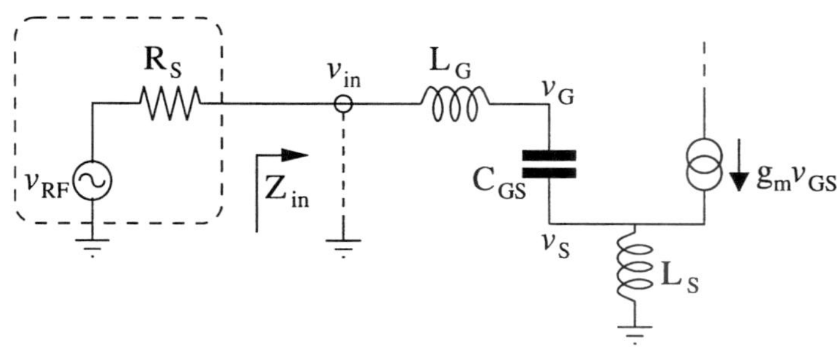

Figure 11.4 Equivalent circuit to evaluate the transconductance gain.

Using (11.13), the gate-source voltage can be written as:

$$v_{GS} = j\omega_0 C_{GS} \frac{v_{RF}}{R_S + Z_{in}} = j\omega_0 C_{GS} \frac{v_{RF}}{R_S + \omega_T L_S} \tag{11.16}$$

We now define the quality factor Q_{in} of the input matching network

$$Q_{in} \equiv \left| \frac{v_{GS}}{v_{RF}} \right| = \frac{1}{\omega_0 C_{GS} \left(R_S + \omega_T L_S \right)} \tag{11.17}$$

with R_s the source impedance equal to 50 Ω. The transconductance of the input stage can now be written as:

$$G_m = \frac{I_{out}}{v_{RF}} = g_m Q_{in} = \frac{\omega_T}{\omega_0} \frac{1}{R_S \left(1 + \frac{\omega_T L_S}{R_0}\right)} = \frac{\omega_T}{2\omega_0 R_S} \tag{11.18}$$

More precisely, taking also into account the effect of the gate-drain capacitance of the MOS transistor, (11.18) changes into

$$G_m \cong \frac{\omega_{Teff}}{2\omega_0 R_S} \tag{11.19}$$

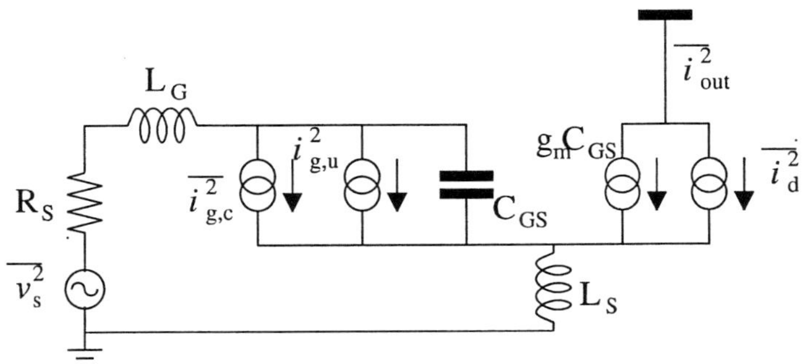

Figure 11.5 Small-signal model of LNA input stage.

11.2 LNA NOISE OPTIMIZATION

We will now address the minimization of the LNA noise figure. Figure 11.5 shows the small signal equivalent circuit with the noise sources of the input stage. Besides the thermal noise source in the drain current, the induced gate noise source has been included. This latter source represents the noise current induced into the gate by the variation of the potential along the channel caused by thermal noise. The induced gate noise is therefore partially correlated with the drain thermal noise (with a correlation coefficient c, equal to 0.4 j in long channel devices). The equivalent circuit shows two induced gate noise current sources, one correlated to the channel thermal noise ($i_{g,c}^2$) and one uncorrelated ($i_{g,u}^2$).

The noise figure of this configuration can be analyzed considering the output noise current. We have split this output noise current into three different terms: i_{Rs}^2 (output noise due to the driving resistance), i_{n1}^2 (output noise due to thermal noise and to the correlated induced gate noise), i_{n2}^2 (output noise due to uncorrelated induced gate noise)

$$\overline{i_{Rs}^2} = \frac{4KTg_m^2}{\omega_0 R_S C_{GS}^2} = 4KT\left(\frac{\omega_T}{\omega_0}\right)^2 \frac{1}{R_S} \tag{11.20}$$

$$\overline{i_{n1}^2} = KTg_{d0}\left\{\frac{\delta\alpha^2}{5\gamma}|c|^2 + [1+2|c|]Q_{in}\sqrt{\frac{\delta\alpha^2}{5\gamma}}\right\} = KTg_{d0}A \tag{11.21}$$

$$\overline{i_{n2}^2} = KTg_{d0}\left\{\frac{\delta\alpha^2}{5\gamma}\left(1-|c|^2\right)\left(1+4Q_{in}^2\right)\right\} = KTg_{d0}B \tag{11.22}$$

Using equations (11.18), (11.20), (11.21) and (11.22), the following expression for the noise figure is derived:

$$NF = 1 + \gamma R_S g_{d0}\,(A+B)\left(\frac{\omega_0}{\omega_T}\right)^2 = 1 + \frac{1}{2}\frac{\gamma}{\alpha}\frac{1}{Q_{in}}(A+B)\left(\frac{\omega_0}{\omega_T}\right)^2 \tag{11.23}$$

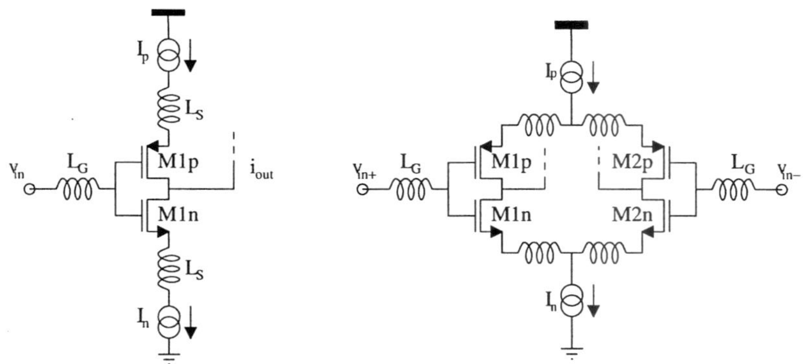

Figure 11.6 single ended and differential

The contribution of the drain noise can be lowered by increasing Q_{in} but at the expense of an increase in the contribution of the induced gate noise, since the current is injected into a higher impedance. Therefore, for a given source impedance value, there exists an optimum device width, i.e. an optimum Q_{in} for which the two contributions are equal and the NF has a minimum.

It can be shown that the noise figure can be lowered by increasing the biasing current or by increasing the source resistance [4]. The former increases power consumption, whereas the latter makes the LNA input more sensitive to noise coupling from the other blocks in the receiver. A different approach to lower the LNA noise figure is discussed in the next paragraph. A new topology, in which an inductively degenerated PMOS input stage is shunted together with a NMOS one using the same current, is proposed. It will be shown that in deep sub-micron processes this topology provides the same performance as the NMOS-only input stage while consuming only half of the current.

11.2.1 The P-N MOS Input Stage with Inductive Degeneration

One of the results of technology downscaling is that PMOS and NMOS devices will show more and more similar features. PMOS devices suffer from lower carrier mobility with respect to NMOS devices. On the other hand, due to the technology downscaling, NMOS devices are working closer and closer to the carrier velocity saturation limit, when operating at the typical biasing currents used in CMOS LNAs. This means that the transition frequency of PMOS devices will come closer to that of NMOS transistors [10]. For this reason, a possible way to improve the performance of the input stage with inductive source degeneration would be the use of PMOS and NMOS devices in a parallel configuration [11], to better exploit the current re-use feature (Figure11.6).

To understand the concept behind the new topology, let us start with the classical NMOS LNA, shown in Figure 11.7a. By doubling its current, we get figure 11.7b. The input impedance at resonance is not affected provided that

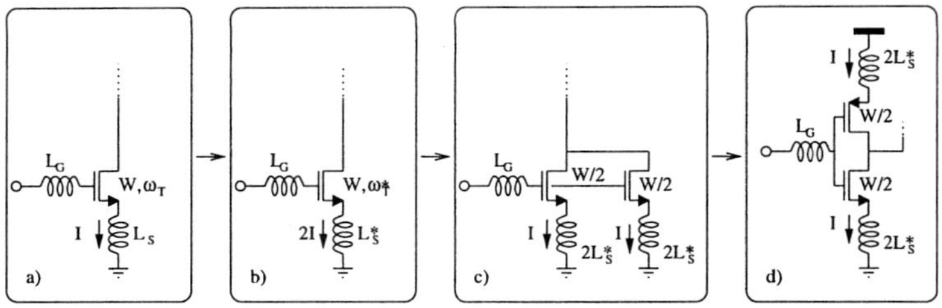

Figure 11.7 Transformation of the NMOS-only input stage (a) into the P-N input stage (d).

Table 11.2 Comparison between simulated optimum NMOS and P-N MOS differential LNA (f = 900 MHz).

Type	Techno	ω_{TN}/ω_{TP}	Rnoise [W]	NF [dB]
NMOS-only	0.35 μm		22	1.58
P-N	0.35 μm	2.5	24	1.7
NMOS-only	0.18 μm		8.8	.7
P-N	0.18 μm	1.7	9	.72

$$\omega_T L_S = \omega_T^* L_S^* \tag{11.24}$$

The circuit of Figure 11.7c, made of two NMOS transistors with half the gate width connected in parallel and with the same total biasing current, is equivalent to that of Figure 11.7b.

Let us assume, to gain an intuitive insight, that NMOS and PMOS devices have the same transition frequency $\omega_{TN} = \omega_{TP}$. One NMOS can then be substituted by a PMOS, as shown in Figure 11.7d, yielding the PMOS-NMOS input stage that achieves the same noise figure as the single NMOS stage of Figure 11.7b, but at only half the current.

The same beneficial effect is obtained on the linearity, since the devices of Figure 11.7b and the one in Figure11.7d have the same overdrive voltage.

The assumption used in this section ($\omega_{TN} = \omega_{TP}$) is not true, but useful for an intuitive explanation of the idea. More quantitative conclusions can be derived on the achievable noise and small signal parameters once the actual transition frequencies and noise coefficients of the NMOS and PMOS transistors are used. Simulations were carried out for the two possible LNA configurations (NMOS-only, with 16mA differential, and $P - N$shunted, with 8mA differential), in a 0.35 μm and 0.18 μm CMOS technology, using the Philips MM9 device models. The results are summarized in Table 11.2.

From Table 11.2 the following conclusions can be derived: in a 0.35 μm CMOS technology, the new proposed P-N shunted input stage, working at 900 MHz, has an

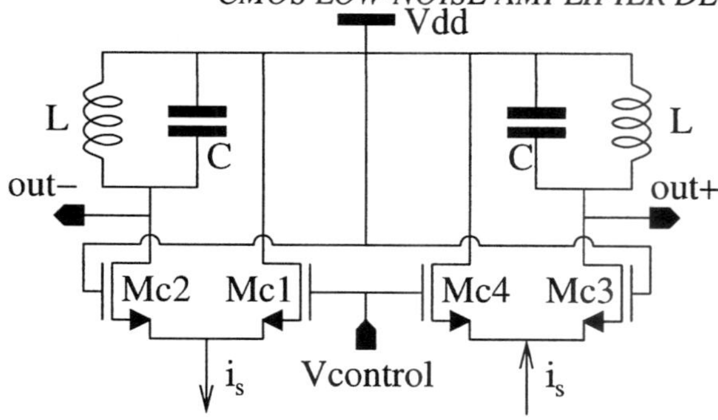

Figure 11.8 A detail of the differential variable gain cascode.

input noise resistance 10% higher than the NMOS-only counterpart, while consuming half the current. This difference reduces to about 3% in a 0.18 μm CMOS technology.

11.3 THE VARIABLE GAIN AMPLIFIER

In many wireless systems, the LNA is required to implement a variable gain function; this is because the signal received by the antenna has a very wide dynamic range (e.g. -102 dBm to -24 dBm in GSM). When the signal power is high, maximum LNA gain will lead to distortion and intermodulation. To cope with this problem automatic gain control (AGC) has to be provided. One of the places where this can be implemented is the LNA.

Generally, when the gain is lowered, the LNA noise figure is degraded. This is not a problem, since the gain will only be lowered for high input signal levels. In this case, a sufficient signal to noise ratio is still guaranteed.

11.3.1 Variable Gain Implementation

A possible solution to achieve a voltage-controlled gain consists in the design of a "differential cascode stage". This means that, by a control voltage, it is possible to steer the LNA output signal between a parallel resonant load and the power supply. Figure 11.8 illustrates this principle.

When the signal current i_s flows through **Mc2** and **Mc3**, gain is maximum. When, on the contrary, all the signal currents flows in **Mc1** and **Mc4**, the gain is zero. The voltage V_{control} determines the fraction of the signal current flowing into the power supply.

Actually only the control voltage V_{control} is external to the chip, while the gate bias voltage of **Mc2** and **Mc4** is internally generated mostly to avoid common mode oscillations and to improve immunity to external disturbances when the LNA works in maximum gain configuration.

11.4 THE COMPLETE LNA

Figure 11.9 shows the schematic of the LNA that was integrated in a 0.35 μm CMOS technology. It was designed to have the performance reported in table 11.3:

Table 11.3 Performance figures

Operating frequency	900 MHz
S_{21}	> 16 dB
Noise Figure	< 2.5 dB
Input IP3	> -5 dBm
Current consumption @ 2.7 V	8 mA

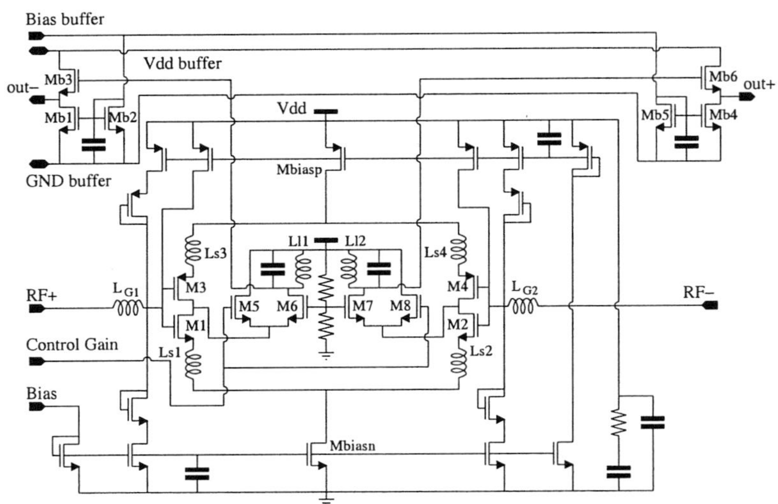

Figure 11.9 Complete schematic of the realized LNA.

Transistors M_1-M_2 with inductors L_{S1} and L_{S2} together with transistors M_3-M_4 and inductors L_{S3} and L_{S4} constitute the inductively degenerated *P-N* stage. Transistors M_{5-8} constitute the cascode stage that implements the variable gain function. The gate inductances L_{G1} and L_{G2} are the series connection of on-chip spiral inductors, bond wires and external SMD inductors, used to fine-tune the input matching circuit. This solution was chosen to limit the noise generated by the on-chip inductors.

The device size and component values of the LNA are given below:

M_{1-4}: $W = 400\ \mu$m, $L = 0.35\ \mu$m;

M_{5-8}: $W = 400\ \mu$m, $L = 0.35\ \mu$m;

$L_{G1\text{-}2}$: 3.1 nH (external inductors added to obtain the correct resonant frequency);

L_{S1-4}: 1.6 nH (the value of these inductors sets the input impedance and LNA gain);

Ll1, Ll2: 7.4 nH (its Q-factor and the value of the load inductance set the maximum gain);

M_{BIASN}: $W = 1200\ \mu$m, $L = 0.7\ \mu$m;

M_{BIASP}: $W = 700\ \mu$m, $L = 0.7\ \mu$m.

These transistor sizes yield 4 mA flowing in each input NMOS transistor, 3 mA flowing in each of the two input PMOS transistors. The difference of 1 mA biases the cascode stage.

11.4.1 Output Buffer

To measure and test the LNA in a 50 Ω system, an output buffer was required. We designed a classical source-follower and, to realize an output impedance of 50Ω considerable current was required. (19 mA for each output buffer). The biasing circuit of the output buffer, constituted by transistors M_{b1}, M_{b2}, M_{b3}, M_{b4}, M_{b5} and M_{b6} is separated from the LNA biasing circuit to allow a correct evaluation of the LNA current consumption.

11.4.2 Measurement Results

The LNA shown in Figure 11.9 has been realized in a 0.35 μm CMOS RF technology with on-chip spiral inductors, realized using a thick metal 4 layer. All the inductors shown in Figure 11.9 are on-chip, apart from the gate inductor. The latter is constituted by the series connection of:

- 4.1 nH on-chip,
- 3 nH bond-wire,
- 7 nH external SMD, used for tuning purposes.

The chip plus matching network has a return loss equal to10 dB at 900 MHz, implying that only 10% of the incident power is reflected back towards the antenna.

Figure 11.10 shows the measured S_{21} versus frequency under the following conditions: V_{DD} = 2 V, I = 8 mA, high gain mode. The peak value is 16dB, corresponding to 22 dB voltage gain. In this plot, there is no correction for the insertion loss of the input and output BALUN, which, from previous measurements, was found equal to 1.5 dB. The remaining difference between simulation and measurement results is due to package and board parasitics, not completely taken into account in the design, and an over-estimated inductor quality factor.

When the control voltage equals the cascode gate bias, half of the current is injected in the load and half in the supply: as expected, a 6 dB gain reduction is found.

Figure 11.11 shows the noise figure from post-layout simulations and the measured noise figure at 8 mA current consumption and 2 V supply voltage. The measured values are in good agreement with the predicted curve. The minimum measured noise figure value is 2.0 dB. This is the lowest measured noise figure for a fully differential LNA biased with 8 mA current.

As highlighted earlier, increasing the biasing current can yield even lower values for the noise figure. This was verified experimentally: NF = 1.85 dB at 12 mA and NF = 1.7 dB at 16 mA. Increasing the source resistance – if possible – is another way to lower the noise figure, however due to the higher input voltage, the IIP_3 will go down and the circuit may be more sensitive to instability.

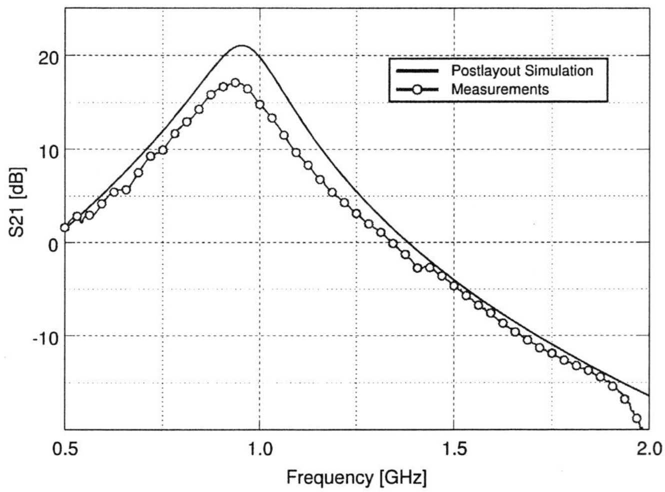

Figure 11.10 S_{21} versus frequency; measurements and post-layout simulations

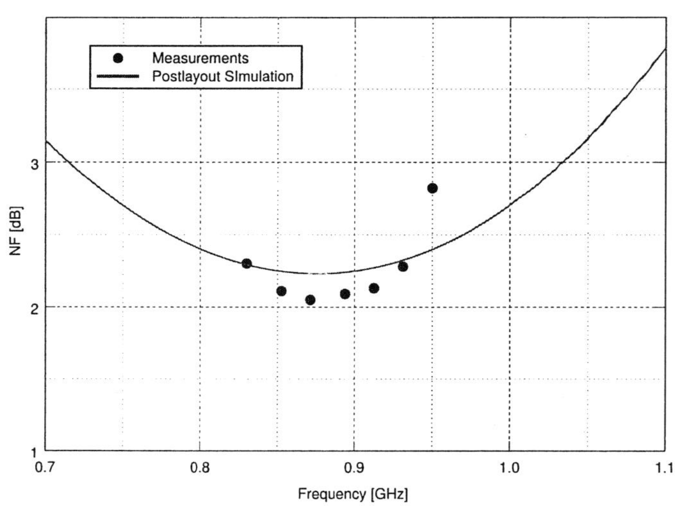

Figure 11.11 Comparison between measured and simulated noise figure.

The intermodulation measurement shows an IIP_3 of -6 dBm (simulated IIP_3= –5 dBm), which is also limited by the output buffers.

11.5 CONCLUSIONS

An introduction to the main LNA design issues has been provided, with particular emphasis on CMOS implementations. The inductively degenerated CMOS LNA proves to be best suited to achieve low noise under matching conditions. A new topology, presenting a PMOS-NMOS in shunt configuration has been presented, which better exploits the improved and converging performance of both devices with technology scaling. A fully-differential 900 MHz prototype, realized in 0.35μm RFCMOS technology, has 2dB NF, 16dB S_{21} with 8 mA current consumption at 2 V.

References

[1] Uhlrich L. Rohde, Jerry C. Whitaker, "Communication Receivers: DSP, Software Radios, and Design", Third Edition, New York: McGraw-Hill, 2001

[2] Thomas H. Lee, "The Design of CMOS Radio-Frequency Integrated Circuits", Cambridge, UK:Cambridge University Press, 1998

[3] David Ballo, "Network Analyzer Basics", Hewlett-Packard Company, 1997 Back to Basics Seminar http://www.tm.agilent.com/ data/downloads/eng/tmo/techinfo/ pdf/comptest_nabasics.pdf

[4] D.K.Shaeffer and T.H.Lee, "A 1.5-V, 1.5-GHz CMOS Low Noise Amplifier", Journal of Solid-State Circuits,vol.32, May 1997.

[5] J.C. Rudell, J.J.Ou, and P.Gray, "An integrated GSM/DECT Receiver:Design Specifications", UCB Electronic Research Laboratory, 1998.

[6] A.A. Abidi, "High-Frequency noise measurements on FETs with small dimensions", IEEE Transactions on Electron Devices, vol. 33, November 1986.

[7] R. Benton et al., "GaAs MMICs for an integrated GPS-front end", in GaAs-IC Symp. Dig. Tech. Papers, 1992.

[8] N.H. Sheng et al., "A 30 GHz bandwidth AlGaAs-GaAs HBT direct coupled feedback amplifier", IEEE Microwave Guided Wave Lett., vol. 1, August 1991.

[9] J.C. Rudell, J.J.Ou, T.B. Cho, G.Chien. F.Brianti, J.A.Weldon, and P.R.Gray, "A 1.9 GHz wide-band IF double conversion CMOS receiver for cordless telephone applications", IEEE Journal of Solid-State Circuits, vol. 32, Decembe, 1997.

[10] I.C. Kizilyalli, M.M. Rambaud, A. Duncan, S.A. Lytle, and M.J. Thoma, "Theresold voltage-minimum gate length trade-off in buried channel PMOS devices for scaled supply voltage CMOS technologies", IEEE Electron Device Letters, vol. 16, October 1995.

[11] A.N. Karanicolas, "A 2.7 V CMOS RF front-end IC for a direct-conversion wireless receiver", ISSCC Digest of Technical Paper, vol. 39, 1996.

12 PRACTICAL HARMONIC OSCILLATOR DESIGN

John F.M. Gerrits

Centre Suisse d'Electronique et de Microtechnique S.A., Neuchâtel, Switzerland

E-mail: john.gerrits@csem.ch

An oscillator is what you get when you try to build an amplifier. This is a popular definition of an oscillator. There is some truth in this statement. Surely, it takes a good amplifier circuit to create an oscillator. It also takes a good resonator to build a harmonic oscillator. On-chip passive components have evolved considerably the last few years. Flip-chip techniques have enabled high Q on-board resonators.

Meanwhile supply voltage of communications equipment is decreasing whereas phase noise requirements are becoming more and more severe. The design of an oscillator for telecommunication applications still constitutes a major challenge for the electronic designer. This chapter proposes a well-structured way to oscillator design by giving both theoretical considerations and practical oscillator implementation examples.

12.1 INTRODUCTION

An oscillator is an active electrical circuit that can generate periodic waveforms out of constants [1]. This short definition is illustrated in Figure 12.1.

DC energy from the power supply is transformed into the time varying oscillator output signal

$$A(t) = \hat{A}\Psi(\omega_0 t + \theta) \qquad (12.1)$$

characterized by the following parameters

- Waveform Ψ
- Amplitude $\hat{A}$
- Frequency ω_0

P. Wambacq et al. (eds.),
Low-Power Design Techniques and CAD Tools for Analog and RF Integrated Circuits, 267–290.

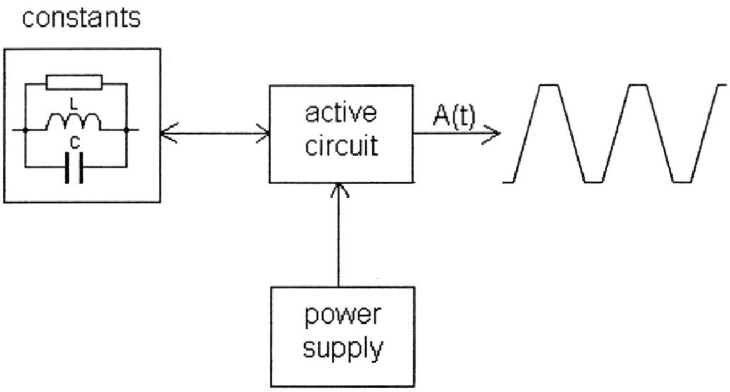

Figure 12.1 An oscillator generates a periodic waveform out of constants

In an ideal oscillator circuit, the oscillation frequency ω_0 depends only on the constants and not on the active part. In a practical oscillator circuit, the active part does have an influence the oscillation frequency. Part of the frequency determining constants may be constituted by the active circuit's parasitic capacitances.

The oscillator can be more precisely characterized by its frequency accuracy and both long and short-term stability. Especially the short-term frequency fluctuations, often referred to as phase noise have become a driving factor for the oscillator circuits used in today's communication equipment. Low noise oscillators use frequency selective devices or resonators to determine the oscillation frequency.

12.2 THE HARMONIC OSCILLATOR

The harmonic oscillator uses a timing reference, a passive circuit whose transfer function H has at least two poles, to provide the constants that determine the oscillator frequency. Figure 12.2 shows the oscillator's mathematical model.

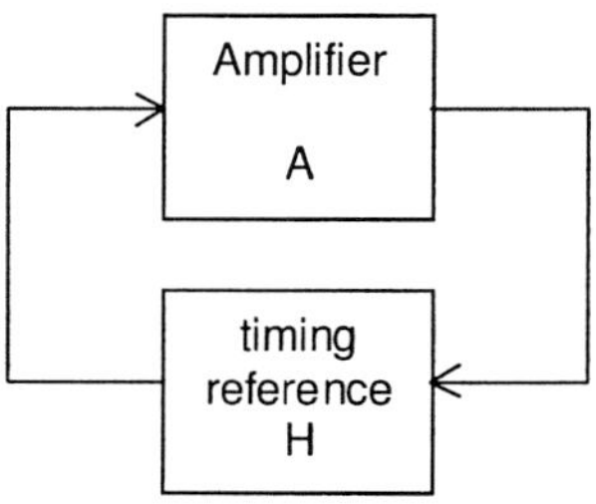

Figure 12.2 Simple model of the harmonic oscillator

The timing reference defines the frequency where the oscillation conditions can be met. The amplifier circuit provides the gain required for start-up and steady state oscillation:

$|AH| > 1$ and arg(AH) = 0 for start-up
$|AH| = 1$ and arg(AH) = 0 in the steady state

Various passive circuits can be used as timing reference in harmonic oscillators [3]. These circuits have poles that can be either real or complex. In this chapter we address oscillators whose timing reference is characterized by two complex conjugated poles and one real zero. This pole-zero pattern is usually encountered in the frequency selective resonators and resonator circuits used as timing reference in low-noise oscillator circuits.

A mechanism is required to reduce the loop gain as the oscillation amplitude increases in order to guarantee a well-defined steady state for the oscillator. Two possibilities are available:

- amplitude regulator (ALC)
- well defined non-linearity in active part

Figure 12.3 illustrates the two approaches.

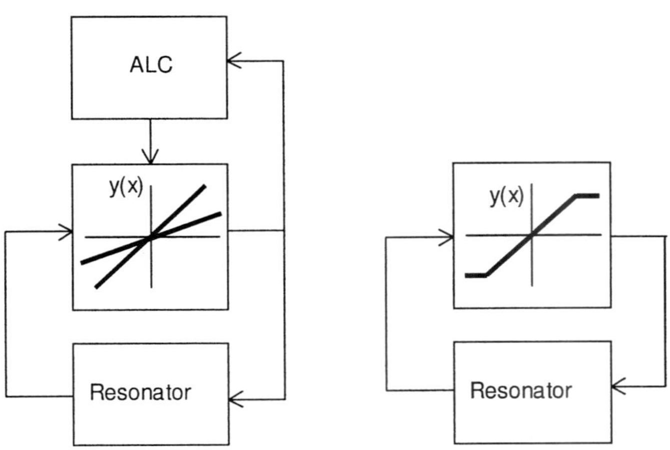

Figure 12.3 Oscillator amplitude control using an ALC loop or a well-defined non-linearity in the amplifier part.

The first solution measures the amplitude of the oscillator output signal and uses a feedback structure to control the amplifier gain. The amplifier always operates in its linear region. Oscillators with ALC circuits yield

- accurate amplitude

- low harmonic distortion

Amplitude regulators or Automatic Level Control (ALC) circuits were used in the very first quartz oscillators for watch applications [4] and have regained popularity recently [5]. Oscillator phase noise originating from the down-conversion of noise at harmonics of the oscillator frequency does not occur in this type of oscillator. The challenge in designing these circuits is to minimize the noise added by the ALC circuit.

The alternative way to go is to implement a well-defined non-linearity in the amplifier part of the oscillator to control the oscillation amplitude. This approach results in simpler circuits whose phase noise performance is slightly degraded with respect to the linear oscillator. Figure 12.4 shows two circuits that can be used to implement a controlled non-linearity [3].

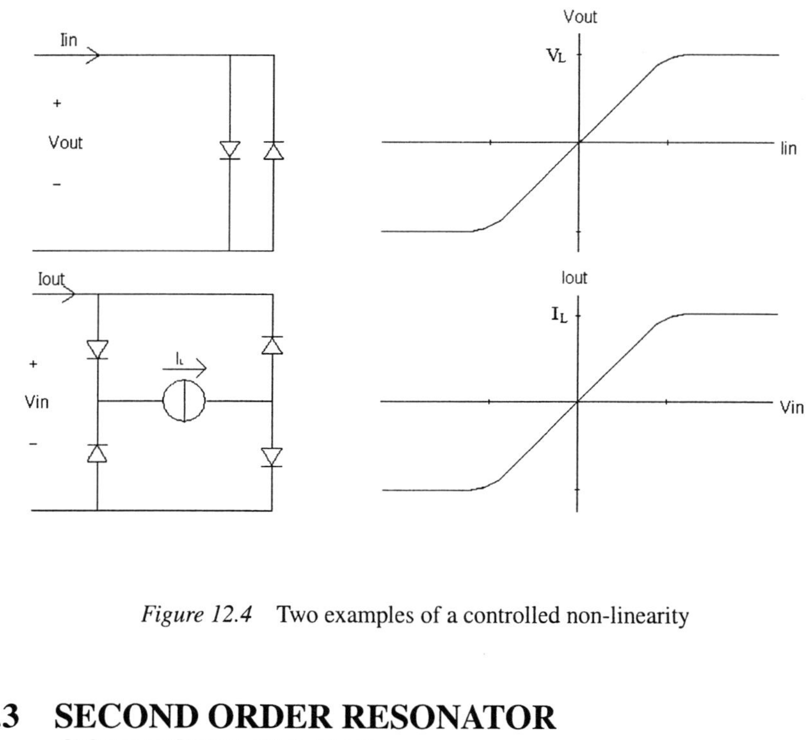

Figure 12.4 Two examples of a controlled non-linearity

12.3 SECOND ORDER RESONATOR CONFIGURATIONS

A resonator is a linear frequency selective system that stores energy in one or several resonance modes. The resonance is characterized by its frequency f_0 and quality factor Q defined as the ratio between stored and dissipated energy per oscillation period T_0.

$$Q = 2\pi \frac{E_{stored}}{E_{dissipated}} = \frac{E_{stored}}{P_{res} T_o} \tag{12.2}$$

The most commonly used resonators in harmonic oscillators have a band-pass transfer function H described by two complex poles and one real zero. With p = $j\omega$, it follows

$$H = H_0 \frac{\frac{\omega_0}{Q} p}{p^2 + \frac{\omega_0}{Q} p + \omega_0^2} = \frac{H_0}{1 + jQ\frac{\omega}{\omega_0} - jQ\frac{\omega_0}{\omega}} \tag{12.3}$$

Some properties of an oscillator signal, like its phase noise, are usually presented as a function of the frequency offset $\Delta\omega$ from the oscillator frequency ω_0. A performance comparison between resonators or oscillators operating at different frequencies also requires normalization with respect to the oscillator frequency. These two features can be easily obtained when the frequency ω is replaced by a variable named detuning ν defined as

$$v = \frac{\omega}{\omega_0} - \frac{\omega_0}{\omega} \approx 2\frac{\omega - \omega_0}{\omega_0} = 2\frac{\Delta\omega}{\omega_0} \tag{12.4}$$

The resonator impedance can now be rewritten as

$$H = \frac{H_0}{1 + jQv} \tag{12.5}$$

This is a convenient form for further mathematical manipulations. The 2nd order resonator is fully characterized by its resonant frequency f_0, quality factor Q and the maximum value of its transfer function H_0. Figure 12.5 shows the magnitude and phase of the 2nd order resonator transfer function. This transfer function can describe either a one-port or a two-port.

Figure 12.6 shows such a resonator circuit made out of a series connection of an inductor L, a capacitor C, and an equivalent series loss resistance R_s. In practical resonators, inductor losses usually dominate and R_s is associated with the inductor.

When this resonator is driven by a voltage source V, the resonator admittance Y equals

$$Y = \frac{I}{V} = \frac{1}{R_s\left(1 + j\omega\frac{L}{R_s} - j\frac{1}{\omega R_s C}\right)} = \frac{1}{R_s}\frac{1}{1 + jQ\frac{\omega}{\omega_0} - jQ\frac{\omega_0}{\omega}} = \frac{1}{R_s}\frac{1}{1 + jQv} \tag{12.6}$$

It can be seen that the inductor and capacitor voltage are a factor Q larger than the driving voltage. The resonant frequency f_0, resonator quality factor Q and resonator power P_{res} are given by

$$f_0 = \frac{1}{2\pi\sqrt{LC}} \tag{12.7}$$

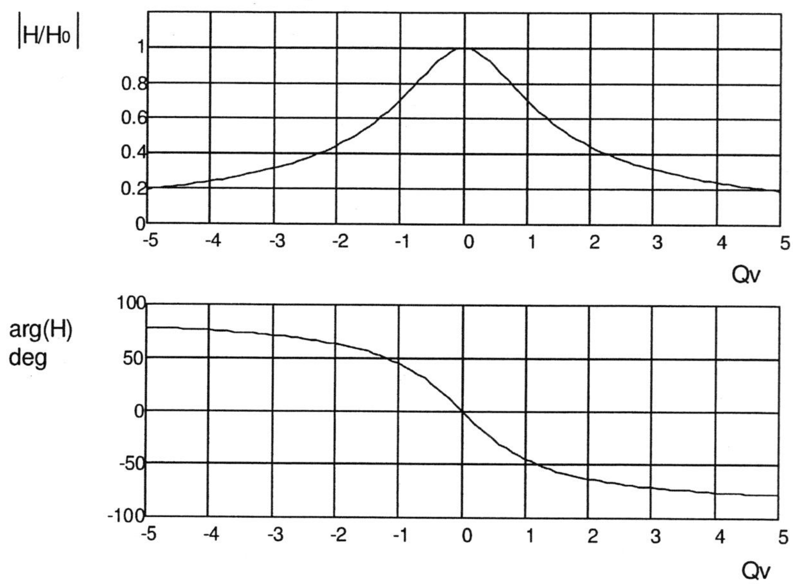

Figure 12.5 Transfer function of a second order resonator

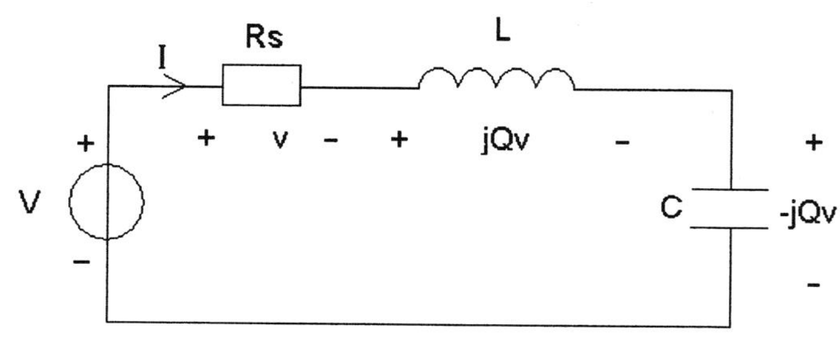

Figure 12.6 A series resonant LC circuit

$$Q = 2\pi \frac{E_{stored}}{E_{dissipated}} = 2\pi \frac{{}^1\!/_2 L\hat{I}^2}{{}^1\!/_2 \hat{I}^2 R_s} f_0 = \frac{\omega_0 L}{R_s} = \frac{1}{R_s}\sqrt{\frac{L}{C}} \tag{12.8}$$

$$P_{res} = \frac{{}^1\!/_2 V^2}{R_s} = {}^1\!/_2 \hat{I}^2 R_s \tag{12.9}$$

When the same passive components are connected in parallel and driven by a current source as shown in Figure 12.7, the impedance Z has a resonant character.

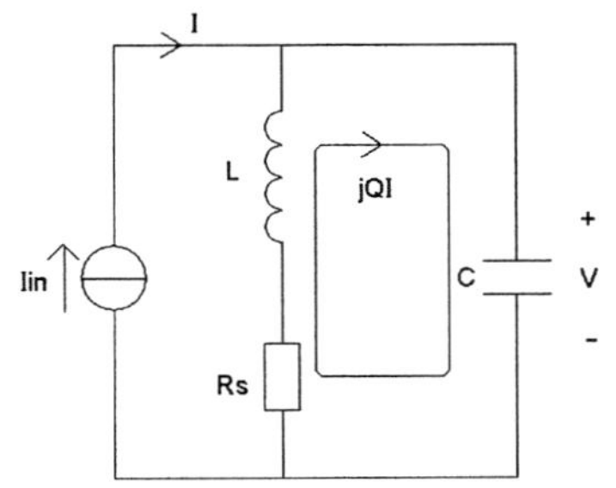

Figure 12.7 A parallel resonant LC circuit

For frequencies $\omega \approx \omega_0$, the series connection of inductance L and series loss resistance R_s can be replaced by an inductance L_p in parallel with an equivalent parallel loss resistance R_p as shown in Figure 12.8.

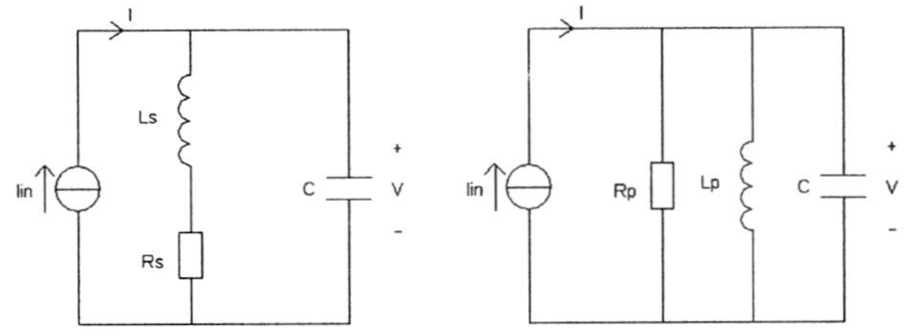

Figure 12.8 Equivalent circuits (near ω_0)

For the parallel resonant circuit, the internal current flowing through L_p and C is a factor Q larger than the external current. The values for the inductance L_p and loss resistance R_p are

$$R_p = \left(1 + Q^2\right) R_s \approx Q^2 R_s \tag{12.10}$$

$$L_p = \frac{\left(1 + Q^2\right)}{Q^2} L_s \approx L_s \tag{12.11}$$

This yields for the impedance of the parallel resonant circuit

$$Z = \frac{V}{I} = \frac{R_p}{1 + j\omega R_p C - j\frac{R_p}{\omega L}} = \frac{R_p}{1 + jQ\frac{\omega}{\omega_0} - jQ\frac{\omega_0}{\omega}} = \frac{R_p}{1 + jQv} \tag{12.12}$$

The resonant frequency, quality factor and resonator power are given by

$$f_0 = \frac{1}{2\pi\sqrt{LC}} \tag{12.13}$$

$$Q_p = 2\pi \frac{E_{stored}}{E_{dissipated}} = 2\pi \frac{\frac{1}{2} L \left(Q\hat{I}\right)^2}{\frac{1}{2}\left(Q\hat{I}\right)^2 R_s} f_0 = \frac{\omega_0 L}{R_s} = \frac{R_p}{\omega_0 L} = R_p \sqrt{\frac{C}{L}} \tag{12.14}$$

$$P_{res} = \frac{\frac{1}{2} V^2}{Q^2 R_s} = \frac{1}{2} \hat{I}^2 Q^2 R_s \tag{12.15}$$

It is a fact that for meeting an oscillator phase noise specification, a minimum resonator power is required. Figure 12.9 shows two basic oscillator circuits using either a series or parallel resonant resonator circuit. Which circuit is to be preferred given this minimum resonator power?

When a parallel resonant circuit is used, this results in a higher voltage swing across the resonator. This may be the preferred solution when the active circuit has enough voltage headroom.

When a series resonant circuit is used, this results in higher current required for driving the resonator. This may appear the only choice when supply voltage is low and large voltage headroom is not available.

12.3.1 Resonator tapping

In a resonator practical oscillator circuit, where the resonator components are imposed by e.g., IC technology, neither of the two extremes shown in Figure 12.9 may yield a circuit with the required specifications. Resonator tapping [3] transforms the resonator impedance into an intermediate value R_T with the restriction

$$R_s < R_T < R_p = Q^2 R_s \tag{12.16}$$

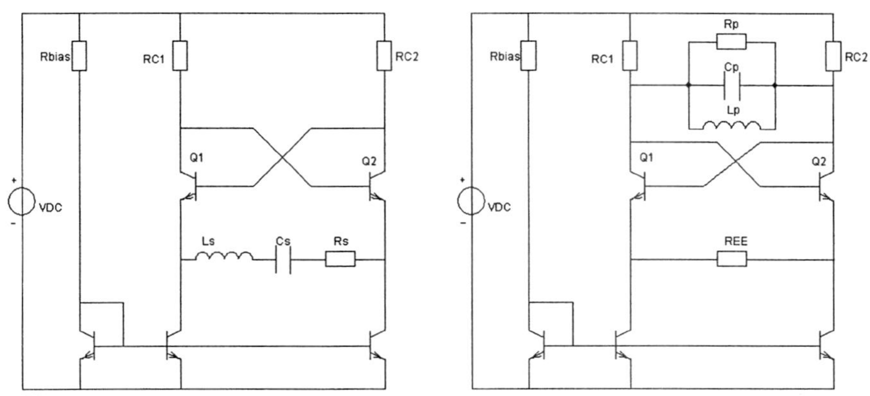

Figure 12.9 Two basic oscillator circuits using one-port resonators: a) series resonant and b) parallel resonant

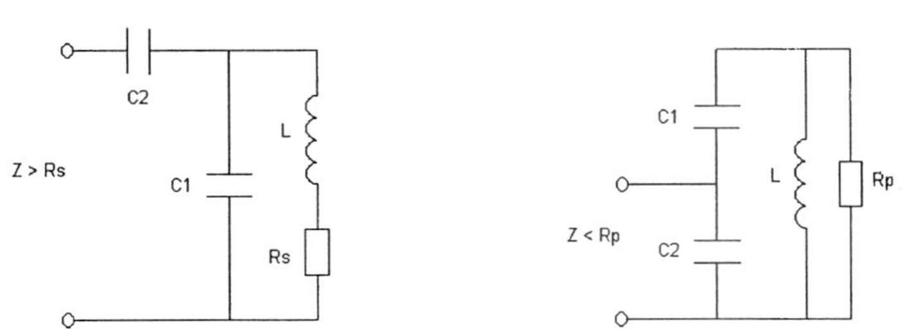

Figure 12.10 Capacitive resonator tapping examples for series and parallel resonant circuits.

Tapping can be interpreted as a sliding transition between the series and parallel resonant circuit. Figure 12.10 shows how to implement capacitive resonator tapping for both series and parallel resonant circuits.

Tapping does not have to be capacitive. Inductive tapping can be implemented by

- two separate uncoupled inductors
- one single inductor with a real physical tap.

The choice between inductive and capacitive tapping is in principle free. In practice, the choice depends on the availability of components, their quality factor (which may be a function of the component value), and the available board space or chip area.

Very often, the advantage is going in the direction of capacitive tapping, or inductive tapping using a physical tap. The latter solution adds no extra components, however, it requires that the tap is accessible.

Tapping of a one-port resonator still yields a one-port resonator. The additional node created by tapping can be used to change the resonator into a two-port. This can be advantageous in some situations as will be shown in the section dealing with practical oscillator design.

12.3.2 Two-port resonators

A more flexible approach is to use a two-port resonator structure where separate input and output terminals exist. The use of a two-port resonator structure makes resonator input and output impedance independent of each other. The first can be chosen for maximum resonator power whereas the latter can be optimized for minimum noise [3].

Figure 12.11 shows an example of a two-port resonator that can be interpreted as a parallel resonant circuit tapped on both the input and output port. This resonator combined with a transconductance amplifier stage is usually referred to as a Collpitts oscillator.

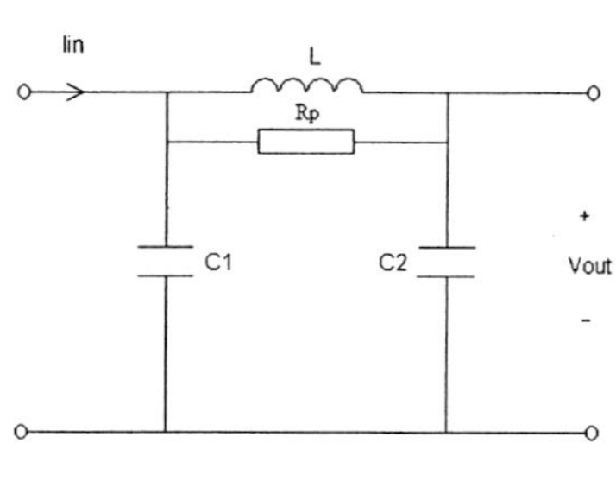

Figure 12.11 Two-port resonator example

Its transfer function at resonance H_0, its resonant frequency f_0 and its quality factor Q are given by

$$H_0 = -R_p \frac{C_1}{C_1 + C_2} \frac{C_2}{C_1 + C_2} \quad (12.17)$$

$$Q_p = \frac{R_p}{\omega_0 L} \quad (12.18)$$

$$f_0 = \frac{1}{2\pi\sqrt{L\frac{C_1 C_2}{C_1 + C_2}}} \quad (12.19)$$

When the inductor is replaced by a quartz crystal, and an inverter circuit is added, a widely used (not necessarily the best) clock oscillator emerges. Figure 12.12 shows the resulting circuit. The oscillator frequency is in between the quartz's resonance and anti-resonance frequency, where its impedance is inductive.

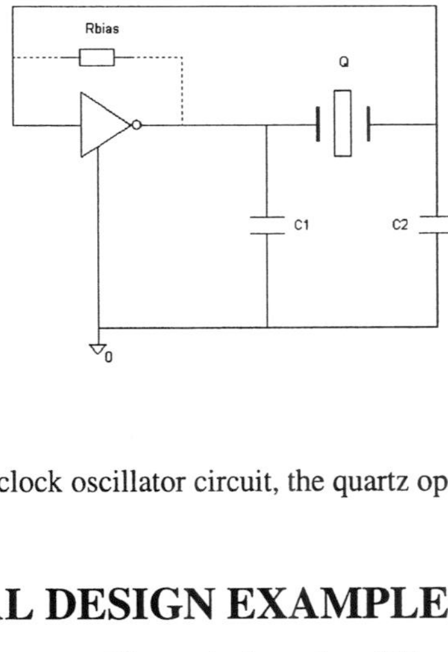

Figure 12.12 A legendary clock oscillator circuit, the quartz operates in its inductive region

12.4 PRACTICAL DESIGN EXAMPLES

This section presents two oscillator designs for different applications realized in different IC technologies. The common factor is the operating frequency of 900 MHz. Table 12.1 shows the most important characteristics of the two oscillator circuits.

First we will calculate the required resonator power for the two applications. Manipulation of Leeson's formula [6], and taking into account the frequency folding as a result of the small-signal loop gain $AH_o > 1$, yields for the resonator power P_{res}

$$P_{res} = 2kT \frac{10^{-\frac{L(\upsilon)}{10}}}{\upsilon^2} \frac{NF.AH_0}{Q^2} \quad (12.20)$$

In this formula $L(\nu)$ is expressed in dBc/Hz. The minimum required resonator power occurs for a hypothetical linear oscillator with a noise-free active part. In that case the thermal noise energy kT of the resonator determines the phase noise.

Table 12.1 Main requirements for the two oscillator circuits

Oscillator characteristic	*Oscillator 1*	*Oscillator 2*
IC Technology	Bipolar	CMOS
Resonator	Off-chip, Q = 25, one-port	On-chip, Q $\approx$ 5, two-port
Phase noise	-100 dBc/Hz	-101 dBc/Hz
	@ Δf = 100 kHz	@ Δf = 25 kHz
Supply current	< 1 mA	< 10 mA
Supply voltage	1.5 V	3 V

Figure 12.13 shows the results for the two practical oscillators. Since oscillator 2 has a more severe phase noise requirement and its resonator Q is lower, it requires a resonator power that is 400 times higher (16 due to L(ν) and 25 due to Q).1.2 mW versus 3 μW.

In a real world oscillator, this minimum value is not sufficient. Especially when the oscillator circuits used are not linear oscillators with an ALC circuit but circuits with a controlled non-linearity to fix resonator power. The small-signal loop gain AH_o of the oscillator is chosen equal to two resulting in reliable start-up and rejection of AM noise [2]. Moreover, the noise factor of the active circuit cannot be neglected and may be as high as 5 (7 dB). This results in a resonator power that may be 10 times higher than the theoretical minimum value.

12.4.1 Oscillator in bipolar technology using off-chip resonator

The major challenge in this oscillator design is constituted by the external tank circuit. It has a high resonator quality factor but the fact that it is off-chip complicates the resonator structure. Figure 12.14 shows the equivalent circuit seen by the on-chip active circuit. It comprises the on-chip bond pads, ESD diodes, bond wires, the package and finally the external resonator. The bond wires constitute high-quality inductors (Q $\approx$100). Together with the on-chip capacitance C_{int}, a second parallel resonance occurs that may be stronger than the desired one defined by the external tank circuit.

Figure 12.15 shows the magnitude and phase of the impedance seen on-chip by the active circuit. The undesired mode at about 4 GHz is stronger than the wanted mode when the series resistance of the on-chip capacitance C_{int} is sufficiently low.

This will result in an unwanted oscillation frequency of 4 GHz. Lowering the bandwidth of the active circuit by adding a low-pass filter as shown in Figure 12.16, solves this problem.

Resistors R_{B1}, R_{B2} and capacitor C_1 lower the loop gain at 4 GHz thus ensuring the correct oscillation frequency of this oscillator circuit. The external tank circuit is constituted by SMD components of the following values:

L_p = 5 nH (j33 Ω at 1 GHz)

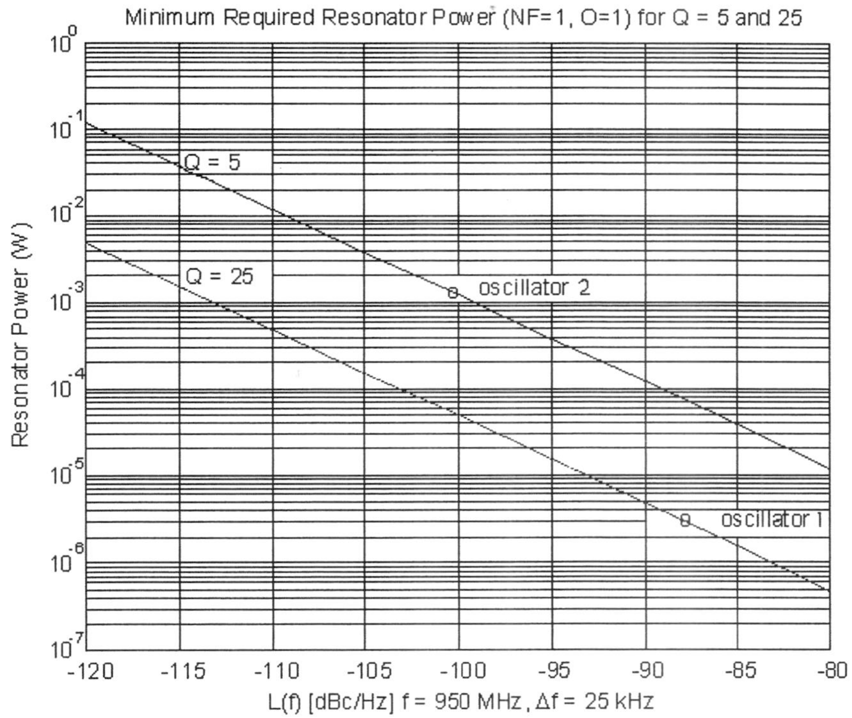

Figure 12.13 Minimum required resonator power.

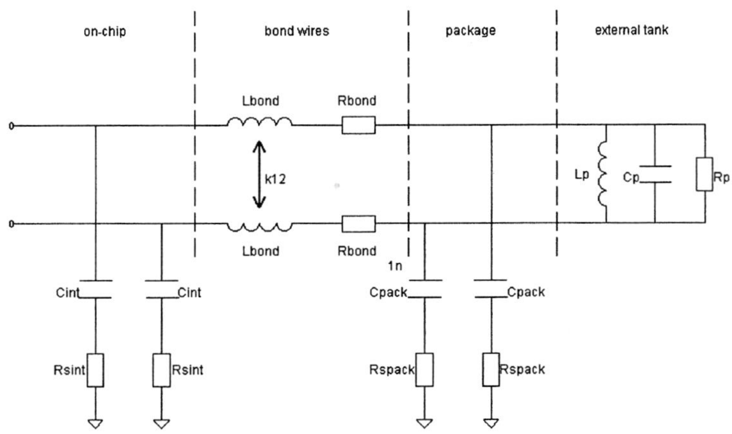

Figure 12.14 Complete resonator structure seen by the on-chip oscillator circuit.

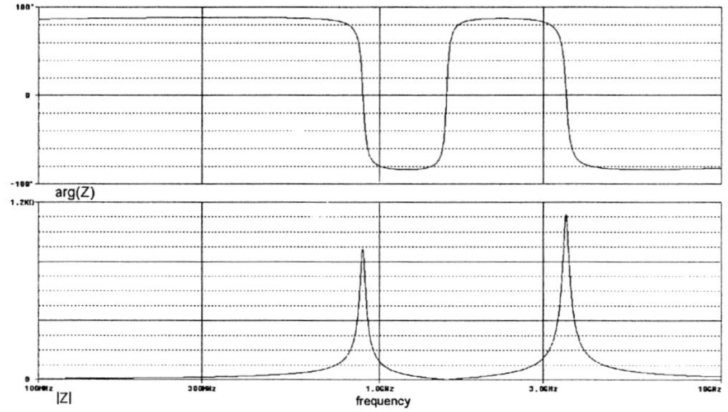

Figure 12.15 Resonator impedance seen on-chip.

$C_p = 5$ pF
$R_p = 800\ \Omega$

The peak resonator current is 300 μA resulting in a resonator power $P_{res} = 40\ \mu$W. Voltage swing equals 240 mV peak. This swing is compatible with the differential pair. No resonator tapping is required.

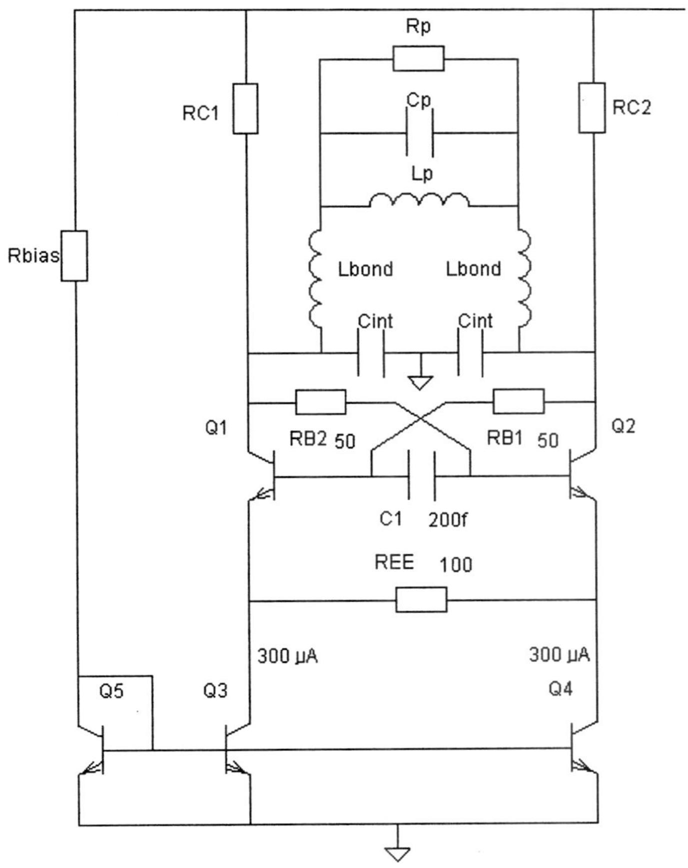

Figure 12.16 Schematic diagram of the bipolar oscillator; components L_p, C_p and R_p constitute the external tank circuit.

The negative impedance Z_N seen by the external tank circuit equals

$$Z_N = -R_N = -\left(R_{EE} + 2\frac{kT}{q}I_E\right) \tag{12.21}$$

Substituting 300 μA for the emitter current I_E of each transistor yields an impedance of -370 Ω. This corresponds to an overdrive $AH_0 = 2.2$.

Phase noise can be calculated using the modified Leeson's formula taking into account the increase of phase noise due to the non-linearity of the active element. Calculation yields -101 dBc/Hz at 100 kHz distance from the carrier.

$$L(v) = 10^{10}\log\left[\frac{2kT}{P_{res}}\frac{NF}{Q^2 v^2}AH_0\right] dBc/Hz \qquad (12.22)$$

The oscillator was realized in a bipolar technology with a transition frequency f_T = 20 GHz. Measurements with various external tank circuits showed no parasitic oscillations in the GHz range. Oscillator phase noise at 100 kHz distance from the carrier was found to be -102 dBc/Hz at an oscillator frequency of 767 MHz and -95 dBc/Hz at an oscillator frequency of 1044 MHz.

12.4.2 Oscillator in CMOS technology using on-chip resonator

The second oscillator is realized in a CMOS technology and uses on-chip Inductors and varactor diodes. This puts some limitations on the available component values. As an example, Figure 12.17 shows the available on-chip inductor values and their respective quality factor. It can be seen that the quality factor is more or less proportional to the square root of he inductance value.

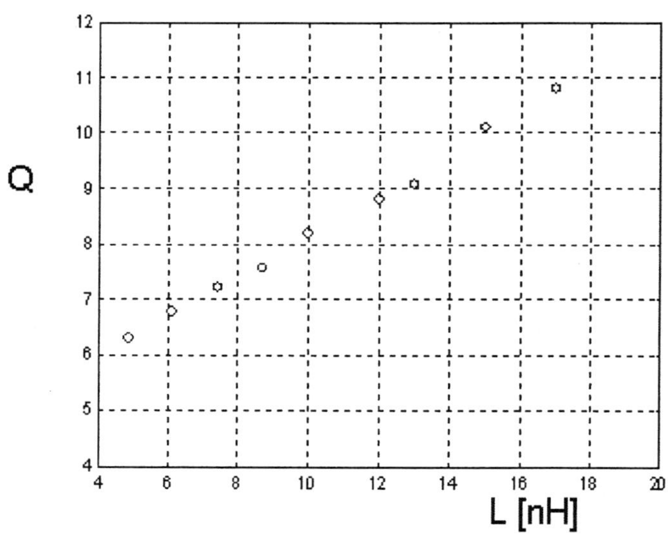

Figure 12.17 Available on-chip inductors for oscillator 2

A resonator power of 10 mW is required for this oscillator. Figure 12.18 shows the resulting resonator voltages and currents for one-port series and parallel resonant circuits.

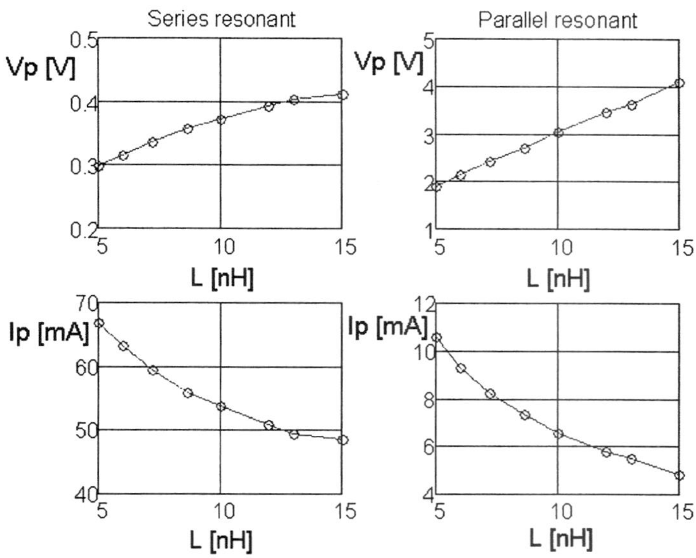

Figure 12.18 Resonator voltage and current for series and parallel resonant circuits

It can be seen that the series resonant circuit requires considerable current whereas the parallel resonant circuit requires a voltage swing at the limit of available voltage headroom- even for the lowest inductor value. Considering the wish for low current consumption a tapped parallel resonant circuit with low inductance value was chosen to reduce the swing at the output of the active circuit.

Figure 12.19 shows the tapping scheme that was used. Not only is the resonator tapped, but it is also used as a two-port since the input of the amplifier is connected to the top of the resonator.

The transfer function for the two-port resonator is the same as that of the one-port.

$$H = \frac{V_{out}}{I_{in}} = \frac{R_p}{1 + jQv} \tag{12.23}$$

As a result the small-signal loop gain AH_o of the oscillator is not lowered by the tapping action, which is advantageous for both current consumption and oscillator noise. However, the input impedance seen at the input of the two-port equals $\frac{1}{2}R_p$. This results in half the voltage swing.

Figure 12.20 shows the respective transfer functions of the two resonator circuits. Figure 12.21 shows the complete oscillator schematic diagram. A differential resonator structure helps to further increase voltage headroom. Capacitors C_{DD1}, and C_{DD2} together with resistor R_{DD2} constitute the on-chip supply decoupling A 45 pF capacitor has an impedance of -j4 Ω at 900 MHz. The series resistance R_{DD2} low-

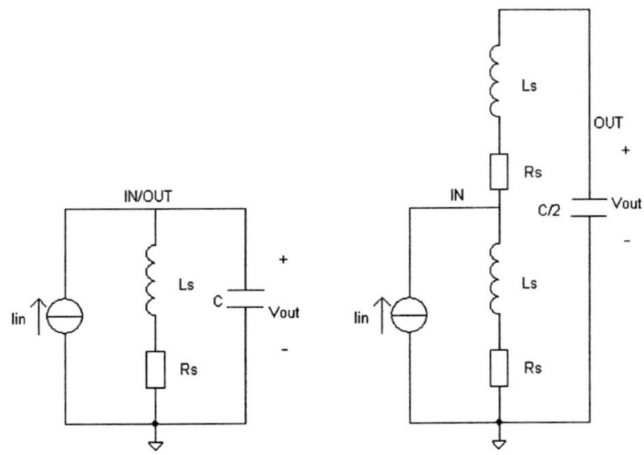

Figure 12.19 A one-port and a two-port circuit with the same transfer function yet different input voltage swing.

ers the quality factor of the decoupling capacitance and helps to suppress resonance phenomena caused by the inductance of the power supply bond wire.

Figure 12.22 shows some of the waveforms occurring in this oscillator circuit.

One may wonder whether the resistive source degeneration by resistors R_{S1} and R_{S2} does not degrade the noise performance of the circuit. Figure 12.23 shows the simulated noise of a differential pair for various transistor sizes. The equivalent noise resistance is the hypothetical resistor value that results in the same current noise.

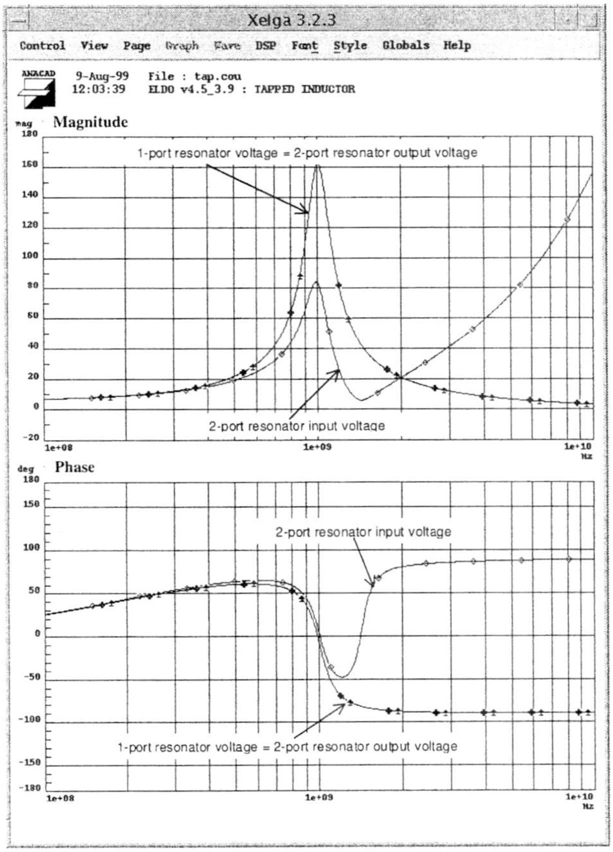

Figure 12.20 Transfer function of one-port and two-port resonator with reduced input voltage swing.

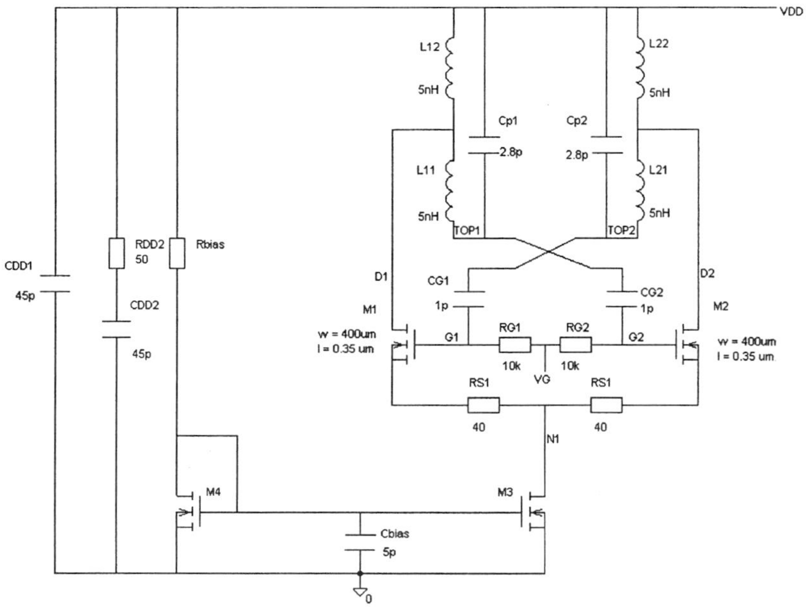

Figure 12.21 Complete oscillator schematic.

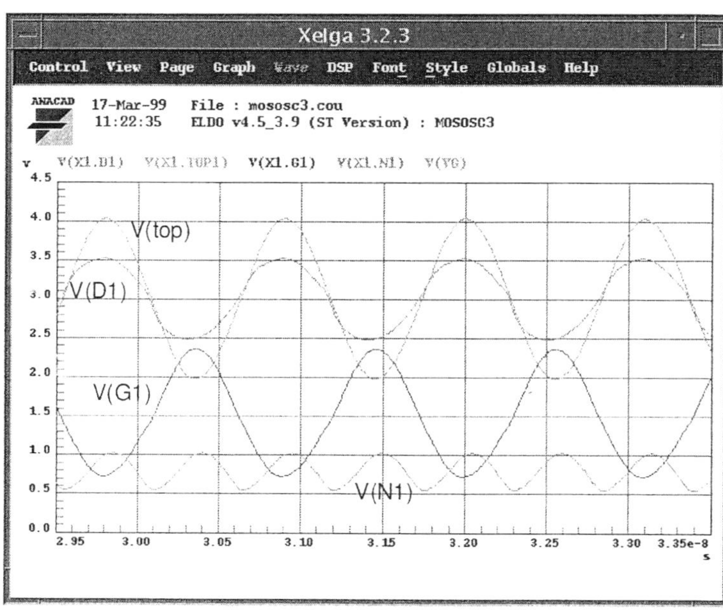

Figure 12.22 Waveforms in the oscillator circuit; node N1 is at the connection of the two source degeneration resistors.

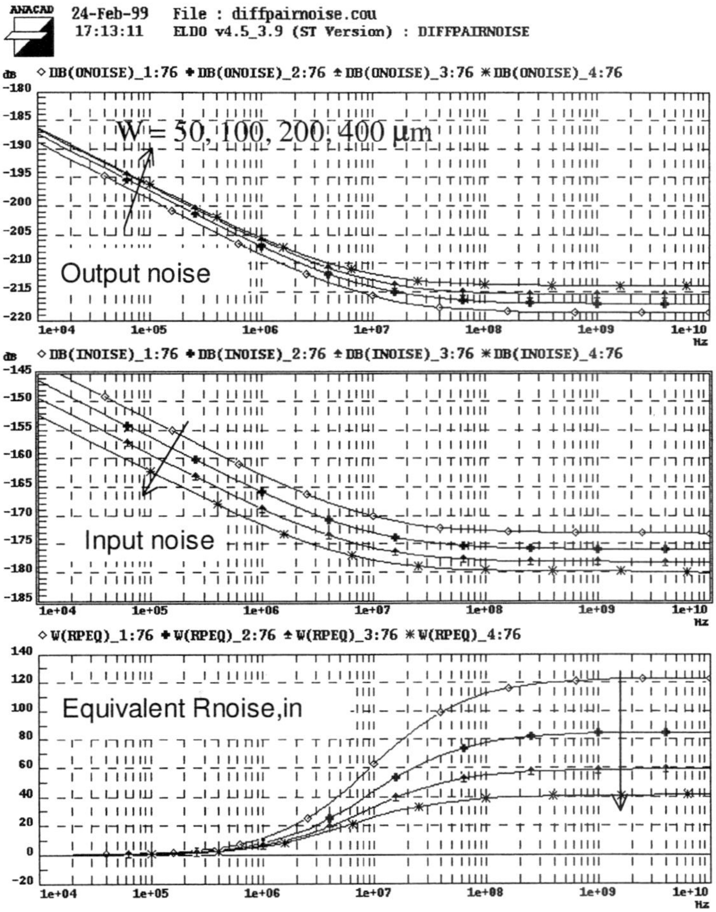

Figure 12.23 Output noise, input noise and equivalent output noise resistance for various transistor widths. The arrow indicates the direction of increasing transistor width.

It can be seen that the output white noise increases with the transistor width, since the transconductance g_m is proportional to the transistor width. The 1/f noise at the input decreases with increasing transistor width. Figure 12.24 shows the effect of resistive degeneration upon the noise.

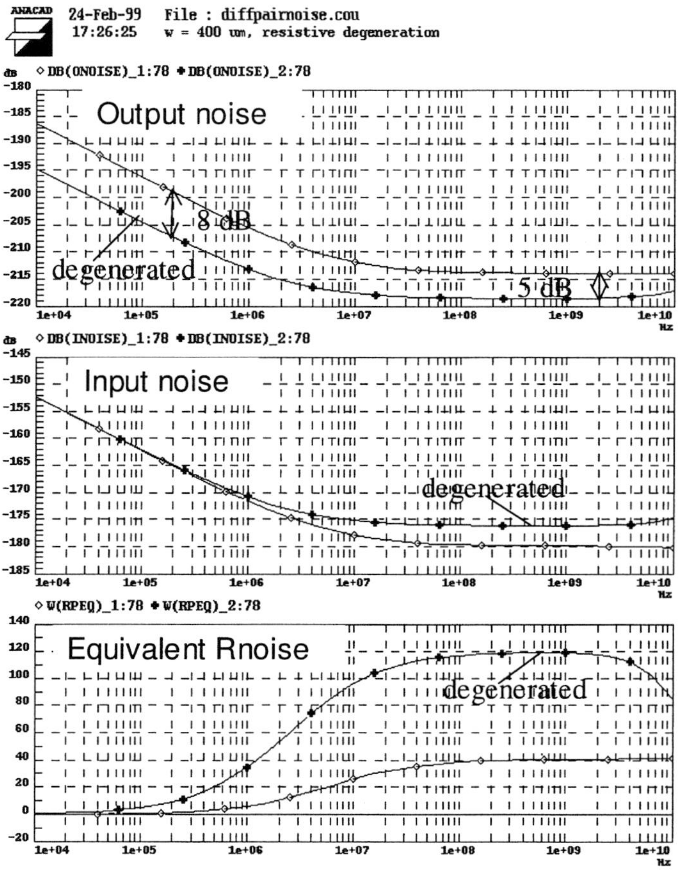

Figure 12.24 Effect of resistive degeneration upon the noise

Comparison of the white output noise level of the differential pair with transistor width W = 50 μm and that of the resistively degenerated pair with W = 400 μm, shows exactly the same output noise value, equivalent to a 120 Ω resistor. Comparison of the 1/f input referred noise however, yields a gain of 8 dB. Therefore, we expect better close-in phase noise performance from the oscillator with the larger transistors and resistive degeneration.

The oscillator noise figure can be deduced from the white part of the output noise

$$NF = \frac{R_p}{R_{noise}} = 3. \tag{12.24}$$

The oscillator was integrated in a 0.35 μm CMOS process. Figure 12.25 shows the measured phase noise. It can be seen that total phase noise equals -93 dBc/Hz at 25 kHz offset from the carrier. The contribution of the white noise equals -100 dBc/Hz which is close to the calculated value of –101 dBc/Hz.

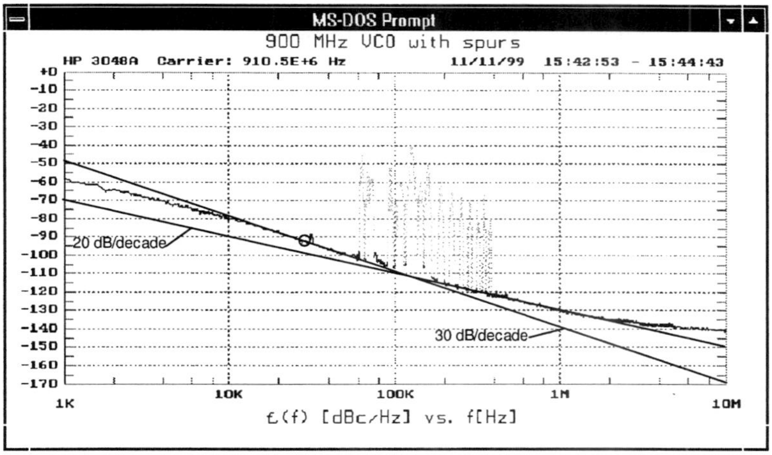

Figure 12.25 Measured phase noise of the CMOS oscillator circuit

12.5 CONCLUSIONS

Two oscillator circuit examples clearly illustrate that it is possible to design oscillator circuits using either external or on-chip inductors. The low quality factor of on-chip inductors results in high resonator power and overall power consumption. A good alternative is to use an inductor on a separate substrate flipped onto the chip or an inductor etched on the PCB of the package (BGA).

References

[1] C.J.M Verhoeven, "First order oscillators" PhD. Thesis, Delft University of Technology, 1989

[2] C.A.M. Boon, "Design of high-performance negative-feedback oscillators", PhD. Thesis, Delft University of Technology, 1989.

[3] Jan R. Westra, Chris J.M. Verhoeven, Arthur H.M. van Roermund, "Oscillators and Oscillator Systems - Classification, Analysis and Synthesis", Boston: Kluwer Academic Publishers, 1999.

[4] E. A. Vittoz, M.G.R. Degrauwe, and S. Bitz, "High-Performance Crystal Oscillator Circuits: Theory and Application", IEEE J. Solid-State Circuits, vol. 23, pp. 774- pp. 783, June 1988.

[5] M.A. Margarit, J.L. Tham; R.G. Meyer, and M.J. Deen, "A Low-Noise, Low-Power VCO with Automatic Amplitude Control for Wireless Applications", IEEE J. Solid-State Circuits, vol.34, pp.761-771, June 1999.

[6] D.B. Leeson, "A simple model of feedback oscillator noise spectrum", Proc. IEEE, vol. 54, pp. 329-330, 1966.

[7] Ernst H. Nordholt and Corlex A.M. Boon "Single-Pin Integrated Crystal Oscillators", IEEE Transactions on circuits and Systems, vol. 37, No.2, February 1990, pp 175 -182

[8] A. Hajimiri, and T.H. Lee, "A General Theory of Phase Noise in Electrical Oscillators", IEEE J. Solid-State Circuits, vol. 33, pp. 179-194, February 1998.

Index